3

Material properties and stress analysis in biomechanics

Material properties and stress analysis in biomechanics

Edited by A. L. Yettram

Manchester University Press

Manchester and New York

Distributed exclusively in the USA and Canada by St. Martin's Press

Published by Manchester University Press
Oxford Road, Manchester M13 9PL, UK
and Room 400, 175 Fifth Avenue,
New York, NY 10010, USA

Distributed exclusively in the USA and Canada
by St. Martin's Press, Inc.,
175 Fifth Avenue, New York, NY 10010, USA

British Library cataloguing in publication data
Material properties and stress analysis
in biomechanics.
1. Biomechanics.
I. Yettram, A. L.
612′. 76

Library of Congress cataloging in publication data
Material properties and stress analysis in biomechanics.
Includes index.
1. Biomechanics. 2. Stresses and strains.
3. Biomedical materials. I. Yettram, A. L.
[DNLM: 1. Biocompatible Materials. 2. Biomechanics.
3. Stress, Mechanical. QT 34 M4256]
QP303.M34 1989 612′.7 88-27380

ISBN 0 7190 25214 *hardback*

Printed in Great Britain
by Biddles Ltd, Guildford and King's Lynn

Contents

	Preface	vii
	List of contributors	ix
1	Cardiac valve prosthesis: materials selection and assessment *P. J. E. Bischler, S.C. Bischler and M. Harrison*	1
2	The mechanical properties of natural and chemically modified bovine pericardium for use with heart valve substitutes *A. Trowbridge, C. Crofts, P. Lawford and M. M. Black*	14
3	Biaxial mechanical properties of passive myocardium *J. D. Humphrey and F. C. P. Yin*	25
4	Fibre reinforcing in articular cartilage *D. W. L. Hukins and R. M. Aspden*	44
5	Temperature dependence of the mechanical properties of human dentine *D. C. Watts*	60
6	Mechanical properties of vertebral cancellous bone *K. Short*	76
7	Correlation between apatite crystal orientation in bone and stress trajectories: a preliminary study *A. L. Yettram*	98
8	Structure, function and mechanical failure of the meniscus *R. M. Aspden and D. W. L. Hukins*	109
9	Contact friction effects in flexural strength tests of alumina for orthopaedic implants *D. C. Fricker*	123
10	Mechanical behaviour of an osteoarticular replacement material *B. F. Shahgaldi, A. A. Amis and F. W. Heatley*	139
11	The structure and mechanical properties of a non-woven vascular prosthesis *T. V. How and D. Annis*	150

12 The role of finite element modelling in biomechanics 163
H. S. Ranu

13 Application of the finite element method to the reconstructed tooth and middle ear 174
K. R. Williams, J. S. Rees and T. H. J. Lesser

14 The loading of cemented femoral hip stems: a finite element investigation *M. J. Fagan and A. J. C. Lee* 187

15 A non-linear finite element model for interface stresses in total hip replacements 203
A. Rohlmann, E. J. Cheal, W. C Hayes and G. Bergmann

16 A loading and stress analysis of the patella *R. J. Minns* 214

17 The distribution of stresses in the human lumbar spine 240
H. S. Ranu

18 Stress–strain characteristics of flat bones in relation to non-penetrating injuries of the head *D. E. M. Taylor, J. S. Whamond, G. J. Cooper, V. A. Evans and R. L. Maynard* 250

Index 272

Preface

In recent years much attention has been paid to applying physical principles and engineering methods to the behaviour of parts of the human body, considering it as both a structure and a machine. However, it soon becomes apparent that what could be taken for granted in conventional engineering no longer necessarily applies. Considering the most basic property involved in the application of stress analysis – the size and shape of a particular structure – this clearly differs from individual to individual and is never easy to obtain. For hard-tissue elements such as bones and teeth, radiography and, in particular, computer tomography can provide some *in vivo* information, although this may well not be of the quality or quantity one might wish. In the case of soft tissues (internal organs, muscles, etc.) it is even more difficult to obtain realistic geometric data. Furthermore the structural components of the body are living organisms which respond to stress and strain biologically as well as mechanically.

Forces which act on anatomical structures are not just applied externally, they also arise from internal muscle action. Indeed muscles are structural elements which are both active and passive at the same time. The passive material properties of tissues are not simply linearly elastic although much useful stress analysis work can be carried out on that basis. In reality they are non-linear, viscoelastic, non-homogeneous and often highly anisotropic.

Modern methods of medical imaging enable the hidden structures of the body to be mapped better than ever before, and highly developed techniques of experimentation can be used on excised cadaveric tissue *in vitro* to obtain some idea of mechanical properties such as compliance. Also, in the field of general engineering, methods of analysis have been devised based on large digital computers, which enable solutions to be obtained to complex situations without too much dependence on simplifying assumptions. The finite element method, which is widely used in biological engineering, is typical, although it may often require large resources of computer time and memory. Its main advantage is that it can deal with objects of irregular shape and consisting of non-linear materials which may or may not be isotropic. It has been applied to mandibles, scapulae, long bones, teeth, heart, lungs, skin – to practically every part of the human anatomy. Of course the more sophisticated and general the method of analysis, the more necessary it becomes to have data of the best quality as input.

In spite of all the inherent difficulties, many researchers worldwide are applying their efforts to examining the behaviour of biological structures. It is hoped that this type of work may produce output of real benefit to clinicians in determining differences between normal tissue and those in various diseased states, and hence assist in the work of diagnosis. There have already been important contributions to the design of artificial joints and other internal prostheses.

A conference on Material Properties and Stress Analysis in Biomechanics was held at Brunel University, Uxbridge, Middlesex in the spring of 1985. It was organised by the Stress Analysis Group of the Institute of Physics (in collaboration with the Materials and Testing Group of the Institute and also the Biological Engineering Society). Presenters at that conference have been invited to contribute to this book in order to describe some of their recent work in the field of biomechanics and thus to bring it to a wider audience.

A. L. Yettram

Contributors

Dr A. A. Amis, Department of Mechanical Engineering, Imperial College of Science and Technology, Exhibition Road, London SW7 2BX

Dr D. Annis, Institute of Medical and Dental Bioengineering, University of Liverpool, PO Box 147, Liverpool L69 3BX

Dr R. M. Aspden, Department of Medical Biophysics, University of Manchester, Manchester M13 9PT

Dr G. Bergmann, Department of Orthopaedic Surgery, Free University Berlin, Oskar-Helene-Heim, Biomechanics Laboratory, Clayallee 229, D-1000 Berlin 33, Federal Republic of Germany

Dr P. J. E. Bischler, 'The Bothy', Shute Street, Kings Stanley, Gloucester GL10 3JS

Dr S. C. Bischler, 'The Bothy', Shute Street, Kings Stanley, Gloucester GLIO 3JS

Professor M. M. Black, Department of Medical Physics and Clinical Engineering, University of Sheffield, The Royal Hallamshire Hospital, Sheffield S10 2JF

Dr E. J. Cheal, Harvard Medical School, Beth Israel Hospital, Orthopaedic Biomechanics Laboratory, 330 Brookline Avenue, Boston, MA 02215, USA

Dr G. J. Cooper, Medical Division, Chemical Defence Establishment, Porton Down, Salisbury, Wiltshire

Dr C. Crofts, Department of Medical Physics and Clinical Engineering, University of Sheffield, The Royal Hallamshire Hospital, Sheffield S10 2JF

Dr V. A. Evans, Medical Division, Chemical Defence Establishment, Porton Down, Salisbury, Wiltshire

Dr M. J. Fagan, Department of Engineering Design and Manufacture, University of Hull, North Humberside HU6 7RX

Dr D. C. Fricker, Howmedica International Inc., Raheen Industrial Estate, Limerick, Ireland

Dr M. Harrison, UKAEA, Risley, Warrington, Cheshire WA3 6AT

Dr W. C. Hayes, Harvard Medical School, Beth Israel Hospital, Orthopaedic Biomechanics Laboratory, 330 Brookline Avenue, Boston, MA 02215, USA

Mr F. W. Heatley, Orthopaedic Academic Unit, Rayne Institute, St. Thomas's Hospital, London

Dr T. V. How, Institute of Medical and Dental Engineering, University of Liverpool, PO Box 147, Liverpool L69 3BX

Dr D. W. L. Hukins, Department of Medical Biophysics, University of Manchester, Manchester M13 9PT

Dr J. D. Humphrey, Department of Mechanical Engineering, University of Maryland, Baltimore, MD 21228, USA

Dr P. Lawford, Department of Medical Physics and Clinical Engineering, University of Sheffield, The Royal Hallamshire Hospital, Sheffield S10 2JF

Dr A. J. C. Lee, Department of Engineering Science, University of Exeter, Exeter EX4 4QF

Mr T. H. J. Lesser, ENT Department, University Hospital of Wales, Heath Park, Cardiff CF4 4XN

Dr R. L. Maynard, Medical Division, Chemical Defence Establishment, Porton Down, Salisbury, Wiltshire

Dr R. J. Minns, Regional Medical Physics Department, Durham Unit, Dryburn Hospital, Durham DH1 5TW

Professor H. S. Ranu, St. Luke's/Roosevelt Hospital Center, Columbia University College of Physicians and Surgeons, Amsterdam Avenue at 114th Street, New York, NY 10025, USA

Mr J. S. Rees, Department of Prosthetic Dentistry, University of Bristol Dental School, Lower Maudlin Street, Bristol BS1 2LY

Dr A. Rohlmann, Department of Orthopaedic Surgery, Free University Berlin, Oskar-Helene-Heim, Biomechanics Laboratory, Clayallee 229, D-1000 Berlin 33, Federal Republic of Germany

Ms B. F. Shahgaldi, Department of Mechanical Engineering, Imperial College of Science and Technology, Exhibition Road, London SW7 2BX

Dr K. Short, PDA Engineering International Ltd, Winton House, Winton Square Basingstoke, Hampshire RG21 1EN

Professor D. E. M. Taylor, Department of Applied Physiology and Surgical Sciences, Institute of Basic Medical Sciences, Royal College of Surgeons of England, 35/43 Lincoln's Inn Fields, London WC2A 3PN

Dr A. Trowbridge, Department of Medical Physics and Clinical Engineering, University of Sheffield, The Royal Hallamshire Hospital, Sheffield S10 2JF

Dr D. C. Watts, Biomaterials Science Unit, University Dental Hospital of Manchester, Higher Cambridge Street, Manchester M15 6FH

Dr J. S. Whamond, Department of Applied Physiology and Surgical Sciences, Institute of Basic Medical Sciences, Royal College of Surgeons of England, 35/43 Lincoln's Inn Fields, London WC2A 3PN

Dr K. R. Williams, Department of Basic Dental Science, Dental School, University of Wales College of Medicine, Heath Park, Cardiff CF4 4XY

Dr A. L. Yettram, Department of Mechanical Engineering, Brunel University, Kingston Lane, Uxbridge, Middlesex UB8 3PH

Dr F. C. P. Yin, Departments of Medicine and Physiology, Johns Hopkins Medical Institutions, Baltimore, MD 21205, USA

1

Cardiac valve prostheses: materials selection and assessment

P. J. E. Bischler, S. C. Bischler and M. Harrison

1.1 Introduction

The clinical acceptability of any material used as a replacement for a natural part of the body is dependent on it successfully meeting certain requirements. Briefly, the ideal implant material should be chemically and biologically inert so as to be non-biodegradable and non-toxic. It should also possess adequate physical and mechanical properties to ensure that the implant will function competently for the required period of time. Prosthesis design is also important as this will determine the efficiency of operation and hence the quality of life of the recipient.

In the following sections the main criteria for materials' selection and the design of cardiac valve prostheses are outlined. The application of these criteria are illustrated with reference to the most popular of mechanical valves, the Björk–Shiley 60° tilting disc prosthesis. Using this cardiac valve as an example, it is hoped to show how the use of post-failure analysis can help to identify design or manufacturing problems. The investigation also outlines the importance in biomedical applications of good materials engineering and design.

1.2 Factors affecting acceptance of prosthetic cardiac valve materials

The reaction of host tissue to the insertion of a foreign object within the circulatory system is to try to assimilate it into the body by encapsulating it in living tissue. This reaction is especially significant in the cardiovascular system since, as well as the adverse effects such as stiffening of leaflet heart valves and the obstruction of valves, it can be beneficial in developing blood compatible surfaces. The acceptance of a prosthesis, and the creation of intimal linings on its surface, is affected by factors which depend on variable characteristics of the implant itself. The most important factors in this respect are the constituent materials and physical construction of the prosthesis. Most prosthetic materials are not truly inert. If the implant induces adverse stimuli, chemical or physical, rejection will result while the stimulus remains.

1.3 Criteria for materials selection

Compliance with certain criteria must be considered in the selection of materials for prosthetic cardiac valves.

1.3.1 *Biocompatiblilty*

The prosthetic materials should not present any haemolytic or embolic threat, being made from materials that are compatible with blood and tissue. The effects of chemical stimuli, wear and fatigue behaviour must all be considered in relation to biocompatibility. For example a chemical stimulus may be produced by:

(1) the evolution of ions during the corrosion of metal implants;

(2) the leaching from polymers of monomers, catalyst residues, impurities, filler materials, plasticisers, antioxidants or other additives, such as mould release agents, machining oils, etc.

Virtually any chemical at the interface between the implanted prosthetic valve and the blood/heart tissue may cause some degree of stimulus. The success and acceptance of an implant material can also be affected by non-toxic chemical interactions between the material and physiological substances. These can lead to changes in the surface properties and mechanical functioning of the implant. An example of this is prostheses containing components manufactured from silicone rubber: these absorbed lipids from the blood which led to serious valve misfunction (McHenry *et al.*, 1970; Joyce *et al.*, 1978).

Corrosion influences compatability because it concerns the interaction between the host environment and the implanted material. Thrombosis, for example, is recognised as an interfacial reaction between blood and material. Prosthetic materials used in cardiovascular applications, such as artificial heart valves, must also satisfy the additional basic criterion of blood compatibility. The materials should not only be non-thrombogenic but must not adversely affect cells, platelets, blood proteins, enzymes or electrolytes. Important variables that affect the blood compatability of prostheses include the distribution of surface charges on the material and the degree of hydrophilicity. These effects of materials on blood have been widely reviewed over the years (Baier & Dutton, 1969; Biggs, 1969; Williams, 1971; Grode *et al.*, 1972; Gilchrist, 1976).

1.3.2 *Mechanical integrity*
The mechanical design of a prosthesis and the nature of the materials required depend on the design stresses and the required lifetime of the implant. Thrombogenesis is the most critical aspect of blood–material interaction but haemolysis is also of some significance with respect to the design of the prosthesis, making mechanical factors and the materials used in construction a most important consideration (see 1.4 Design criteria).

1.3.3 *Sterility and durability*
The prosthetic materials must be easily sterilised and not subject to deterioration if stored for long periods. The durability of the constituent materials should be of consistently high quality, exceeding the normal life expectancy of the patient.

1.4 **Design criteria**

The principal areas of concern in the design assessment of cardiac valve prostheses include the area of attachment, efficiency of mechanical function and efficiency of hydraulic performance.

The technique of attachment to a possibly diseased and variable tissue bed should be without leak, relatively quick, and an accurate procedure. The prosthetic valve should not, by its shape, adversely affect the normal function of any surrounding tissue. Mechanical and geometrical factors may contribute to the rejection of an implant. Tissue response can result from cellular irritation through movement of the prosthesis with respect to adjacent tissue. Mechanical trauma is an important factor in the tissue reaction and can result, for example, from the continuous extension and contraction of the heart muscles in pumping blood. The size and shape of the prosthesis relative to the surrounding tissue is another important factor: serious and occasionally fatal cardiac arrhythmia has resulted from some caged-ball prostheses implanted in the mitral position whose bulk could not be accomodated in the left ventricular cavity. This led to obstruction of the cavity or impingement on the endocardial surfaces. These factors led to the development of low-profile prosthetic valves.

The more rigid mechanical cardiac valve prostheses have a higher incidence of thromboembolic complications than leaflet prosthetic valves. To aid the attachment and acceptance of prosthetic cardiac valves, woven textiles such as Teflon and Dacron are utilised in the sewing ring that allows suturing of the prosthesis to the heart. Factors such as formation time and thickness of intimal tissue growth are largely dependent upon the physical nature of the fabric, i.e. its depth of weave, rather than on the chemistry of the material. With less-porous materials capillary formation within the tissue deposit is prevented and growth becomes self-limiting, thus gross fibrous encapsulation does not occur, which could itself become an irritating factor. With a more open-mesh weave (or porous material) capillary growth continues unchecked until the vessel becomes occluded. The restoration of tissue continuity and cellular interactions at the implant interface are important factors affecting acceptance of a prosthetic material.

The prosthetic valve should open inaudibly with minimum resistance, causing minimal turbulence and pressure gradient. It should also close inaudibly with an effective seal, causing negligible trauma to the blood's constituents.

1.5 The importance of implant retrieval and post failure analysis

An effective way of improving implant design and performance is to use data from implants which have been in service. These can generally be divided into two categories; those of which are examined after failure and those which are removed in full working order for other reasons.

In the case of an implant failure, valuable information can be obtained by the materials engineer from a close examination of the failed parts. Recent advances in fracture analysis using techniques such as scanning electron microscopy and electron microprobe analysis, allow the smallest of components to be examined thoroughly. It is not always necessary to have all the fractured components to determine the failure mode. With post-failure analysis techniques it may be possible to indicate the cause of failure which in turn may lead to modifications in implant design, materials or manufacturing procedure. Conversely, such a study may show that failure was brought about by the device being damaged or incorrectly fitted.

Occasionally it becomes necessary to remove an implant which has not failed, for example when problems have arisen due to suture failure. Studies of such retrieved devices have provided much information on biocompatiblity and real-time wear rates. Implant manufacturers have recognised the opportunities this has given them to gain further knowledge of performance and have used them to evaluate implant integrity.

1.6 Application of materials selection and design criteria to the Björk–Shiley tilting-disc prosthesis

1.6.1 *Description of the valve*

The Björk–Shiley tilting-disc cardiac valve consists of a cobalt–chromium (Stellite) low-profile 'cage', housing a suspended pyrolytic carbon (Pyrolite) disc occluder. The housing is covered with a carbon coated polyethene terephthalate (Dacron) sewing ring (Figure 1.1).

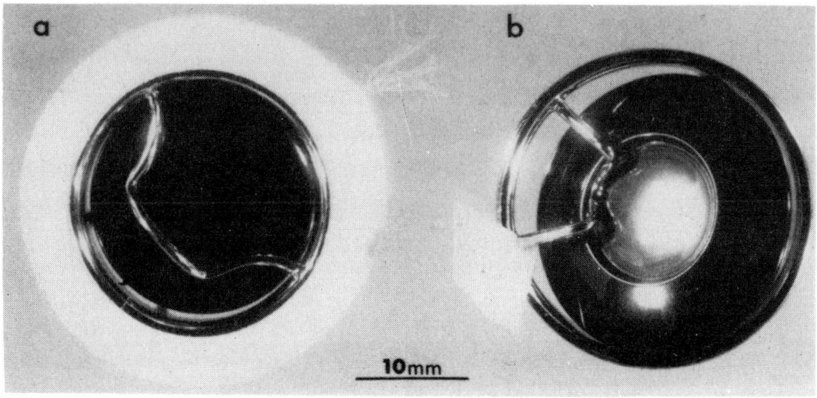

Fig. 1.1 The Björk–Shiley cardiac valve: (a) shows the valve complete with pyrolytic carbon disc and sewing ring from the inlet side (note 'major' disc retaining strut); (b) illustrates the valve with its sewing ring removed from the outlet side (note that one of the minor struts has fractured.)

The 'free' floating disc tilts, in the open position, to an angle dependent on the design of the major (inflow) and minor (outflow) disc retaining struts. This was fixed at 60° (see the following section on development). The valve has been designed to allow the disc to rotate one turn in every 100 to 200 cycles, (Björk & Ovenfors, 1971), contact wear from the retaining struts is then evenly distributed and disc life is extended (Björk, 1971). The closed valve configuration allows the disc occluder to fit into the orifice with no overlap, a feature designed to reduce mechanical trauma to blood cells.

1.6.2 *Development of the Björk–Shiley prosthetic cardiac valves*

The original caged-ball design of prostheses had two major drawbacks: large profile height, and excessive occluder-induced turbulence in the flow through and distal to the valve. The first successful implant of a caged-ball prosthesis (survival greater than 3 months), was the replacement of a mitral valve (Starr, 1960). This was followed by several models of valve prosthesis whose design owed much to the involvement of non-medical researchers from many scientific backgrounds. The fruit of this successful collaboration between clinicians and scientists was rapid progress towards optimising prosthesis design. Attempts to reduce the height of the valve by using a disc occluder in a flatter cage saw the introduction of the Kay–Shiley caged-disc valve in 1965. This design, however, suffered from disc wear and systemic embolism. One major disadvantage of mechanical valves is the need for continuous anticoagulation therapy to minimise the risk of thrombosis and thromboembolic complications.

During 1969 the most significant development in mechanical valve design was the introduction of the Björk–Shiley tilting-disc valve. This is currently the most popular design of valve although central orifice tissue valves are gaining in usage. The advantage of the tilting-disc occluder over the caged ball and the caged disc is that it acts like an aerofoil in the blood flow when in the open position, inducing predominantly central laminar flow and thus substantially less disturbance. Various improvements to the Björk–Shiley tilting-disc valve have been made, concentrating on the disc's geometry, component material and on the disc retaining system itself. These are typical of the developments in the field of prosthetic cardiac valve design in the last 15 years.

The delrin discs, used since the introduction of the tilting-disc valve in 1969, were superceded in 1971 by a straight pyrolytic carbon disc. This change was effected to reduce both the handling and preparation problems endured with the delrin discs. Delrin is a polyacetal resin formed by condensation polymerisation of formaldehyde to give the polyether: polyoxymethylene. Delrin possesses very good all round mechanical properties especially with respect to fatigue, impact and creep resistance. It is not, however, as resistant to chemical attack as some other polymers. Water absorption is relatively high and care is needed with sterilisation since autoclaving can cause a temporary enlargement.

Carbon is generally more expensive to fabricate than metals or polymers but, despite this, it is used for prosthetic cardiac valves because of its highly desirable compatability characteristics (Jenkins, 1980). It also exhibits good resistance to fatigue, wear and degradation. Carbon is virtually inert, with no ions or low molecular weight organic molecules being leached *in vivo*.

The most wear-resistant carbons are obtainable by co-depositing silicon, from the gas silane (SiH_4), with high-density isotropic carbon in a fluidised bed at temperatures below 1500°C. A radio-opaque tantalum disc embedded in the pyrolytic carbon disc allows for non-invasive observation of its performance. The disc is able to rotate, thus preventing any excessive wear to it, but the increased wear of the struts by the harder carbon occluders also had to be allowed for in the overall design.

A carbon-coated dacron sewing ring was also introduced to reduce tissue overgrowth (as previously discussed) and to lower the risk of thrombosis at the cloth/metal junction. These rings were fabricated approximately 1 mm larger in diameter since autoclaving causes about 3% shrinkage. They replaced the previously used Teflon rings whose carbon-coating was not feasible at that time.

Stellite, a cobalt–chromium alloy, is utilised for the cages of the Björk–Shiley prosthetic valves. It has very good thrombo-resistance and tissue acceptibility. This is largely due to the very low corrosion rates of cobalt–chromium alloys rather than to the low toxicity of their reaction products. Co^{2+} and Cr^{3+} ions are highly cytotoxic and even a very small degree of leaching out of these ions is sufficient to cause rejection of the implant by some patients who demonstrate a low tolerance to these particular metals. The mechanical properties of cobalt–chromium alloys are greatly influenced by processing techniques.

1.6.3 *Performance of the Björk–Shiley prostheses to date*

Instances of breakage of the larger disc-supporting strut were first reported in 1973. Investigation of such failures identified the probable cause as a combination of factors related to the welded areas. From 1976 a convexo-concave design of disc was devised, to lower the rate of thrombosis and thromboembolic complications. In response to strut failures, the larger inflow strut was made an integral part of the valve ring in this modified design of prosthesis in 1977. The minor strut, however, was still attached by welding. All valves prior to this had both the inflow and outflow struts welded into the valve ring.

The first reported strut fractures with the post-1977 convexo-concave design occurred mostly in the small aortic valve so it was assumed that the opening forces were a major contributory factor and the opening angle was subsequently increased to 70°. Up to October 1984, of 78 309 implanted Björk–Shiley 60° convexo-concave valves of sizes 21 to 33, 114 fractures of the welded outflow strut had occurred (Björk, 1985). Failure began by the fracture of a single outlet strut followed by the failure of the second strut within a matter of days of the first, usually with fatal results. Shiley investigated the cause of the strut fractures, initially concentrating on the stress loading in the struts but these were found to be very low and were not considered significant. Production processes were studied by the manufacturer with improvements undertaken where possible, concentrating on the strut welds.

A significant proportion of the failures mentioned above came from implants manufactured between February 1981 and March 1982 which were subsequently withdrawn by the manufacturers. Two failed and one apparently unfailed valve manufactured between these dates were recovered at Oxford for examination. Very little has previously been reported about the mechanism predisposing to valve failure, though in one report metallurgical analysis (Brubakk *et al.*, 1981) suggested a brittle fracture of the weld joining the outlet strut to the valve ring. A detailed metallurgical analysis was conducted to determine the mechanism of

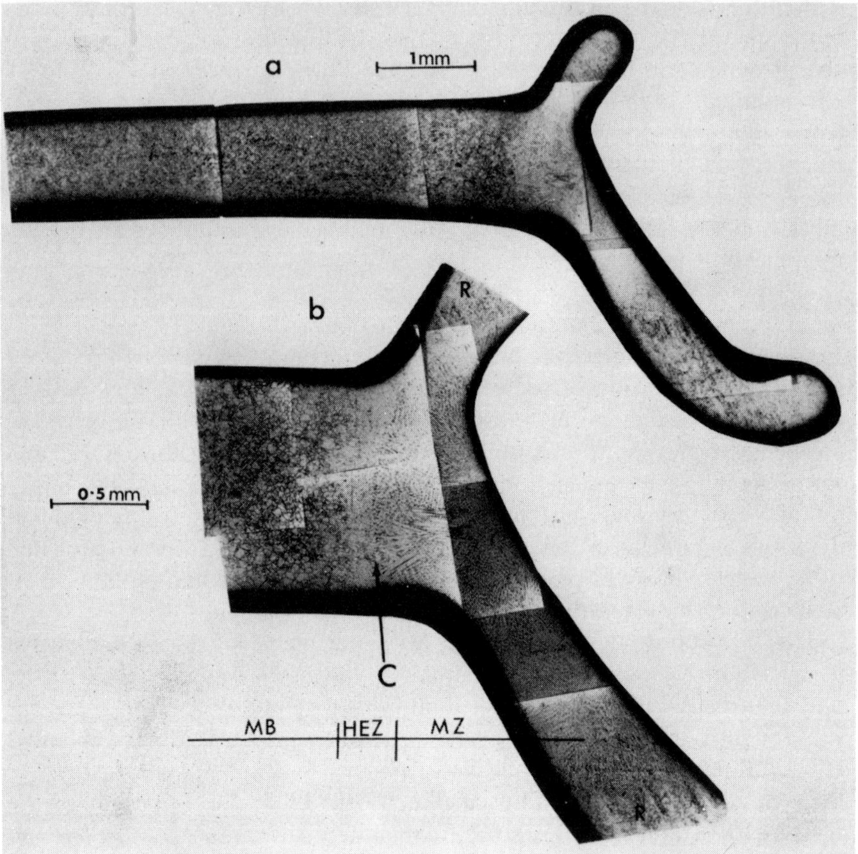

Fig. 1.2 (a) Metallographic section through the minor strut, weld and annulus of an unused Björk–Shiley prosthesis. An enlarged section of the weld area is shown in (b): MB = main body of short strut; HEZ = heat-affected zone; MZ = molten zone; C = carbide deposits; and R = parts of the ring.

failure of these valves. The findings, which are outlined below, illustrate how valuable data may be obtained from implants that are either retrieved in working order or have failed.

1.7 Failure analysis of fractured Björk–Shiley prostheses

A number of valves, both failed and unfailed, have been examined to further elucidate the causes of failure. Figure 1.2 is a longitudinal metallurgical section through the root of an unused valve outlet strut and its point of attachment to the valve annulus. A number of features may be observed:

(1) The bulk of the strut (MB) is composed of fine ($d \simeq 25$ μm) equi-axed recrystallised grains, which derive from the manufacturing process of the wire stock from which is formed.

(2) The annulus (R) is machined from solution-annealed bar stock and exhibits a larger grain size than the strut ($d \simeq 50$ μm) and evidence of an aligned structure of segregated carbide.

(3) The weld melt zone (MZ) consists of a coarse-grained ($d \simeq 250$ μm) dendritic structure with a dispersion of coarse carbide particles throughout the section.

(4) At the boundary between the molten zone and the main boundary of the strut is a narrow heat-affected zone (HEZ) delineated by a band of coarse carbides at the melt interface (arrowed C). The grain structure within this zone is poorly defined due to the re-solution of the fine carbide which usually outlines the grain boundaries.

(5) There is a constriction in the diameter of the strut at its point of attachment to the valve annulus.

Two failed valves have been examined. Fractographic analysis was undertaken by scanning electron microscopy, with metallographic sections being taken to confirm the location of the fracture as coincident with the HEZ/weld metal interface.

Fracture surfaces were composed of a facetted structure with microstructural features consistent with the character of the molten zone. In one case the attachment point of one strut leg was heavily 'smeared' indicating extensive wear, thus inferring that a significant time elapsed between failures of the individual strut legs (Figure 1.3a). At higher magnifications, remains of the facetted structure, observed all over the other point of attachment, could be seen where smearing of the surface had not taken place, i.e. within deep valleys of the original fracture surface (Figure 1.3b). This behaviour was also apparent in a third valve which was removed prematurely due to a break down in the suturing. One leg of the strut was found to be fractured and displaced laterally due to the relief of residual welding stress on fracture (Figure 1.1b). In common with the two failed valves examined, a number of transgranular cracks were found at the attachment point of the strut (Figure 1.3c), together with a dispersion of cavities on the external surface in the melt zone region (Figure 1.3d).

The valve is subjected to a large number of operational cycles in service and thus is subject to a high-cycle mode of fatigue loading. The failure of

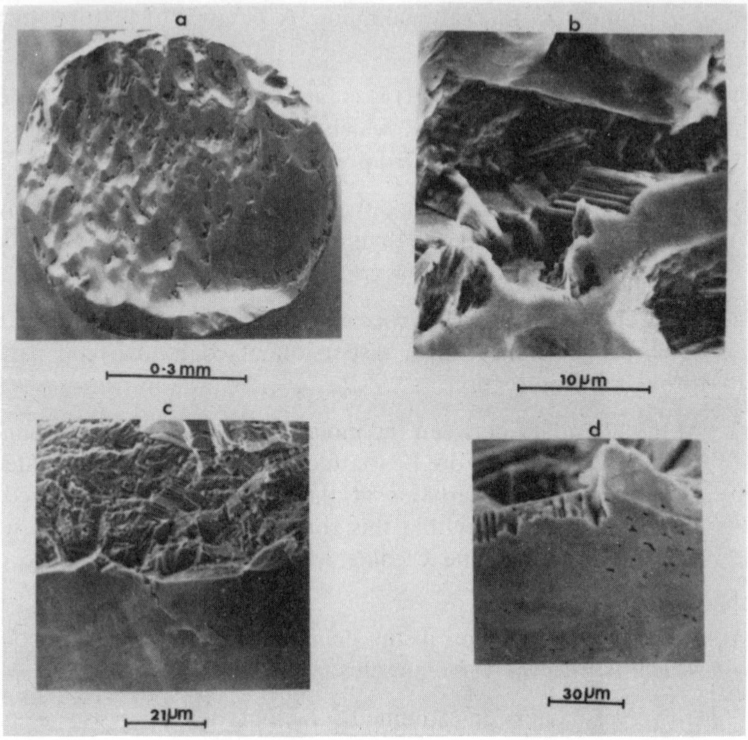

Fig. 1.3 Scanning electron micrographs illustrating: (a) smearing on the first failure site due to extensive wear after fracture; (b) facets seen within unsmeared areas of (a); (c) transgranular cracks in the weld at the final fracture site; and (d) a dispersion of cavities near the point of failure.

the strut at its point of attachment is a classical case of fatigue failure, initiated by the surface cavities.

The specific mechanism of failure is difficult to ascertain due to the number of potentially deleterious features within the weld region. Any one of the following factors may have initiated, or contributed to the failure.

An autogenous tungsten inert-gas (TIG) welding process was used to attach the strut to the ring. The lack of filler material in this process results in a reduced cross-sectional area in the weld zone. This is due to the redistribution of the melt to accommodate the strut/annulus fit-up clearance and is driven by the surface tension of the melt pool causing a concave constriction. The principal imposed stress tends to cause bending of the strut which results in a maximum shear stress at the point of attachment; due to the reduced cross-section there is a general increase in net section stress at this point.

The welding process results in a cast microstructure which has not been heat treated. Therefore the strength of the material in this region is significantly reduced for the following reasons:

(1) The coarse carbide distribution provides only limited dispersion hardening which is the principal strengthening mechanism in wrought Co–Cr alloys.

(2) The large grain size also results in a general reduction in strength as the constraint to plastic deformation is reduced. Due to the hexagonal structure of the alloy, physical and mechanical properties are likely to vary with crystallographic orientation. The concept of engineering materials normally relies on their physical and mechanical properties being macroscopically homogeneous and isotropic. This is often the case with small-grained materials. However in this case it is likely that elastic moduli, for example, may vary significantly through the cross-section resulting in large strain discontinuities.

(3) The coarse carbides present in the weld structure are about 10 μm in diameter and may act as stress-raising sites thus giving rise to a local stress intensification. This is superimposed on the general increase in stress level from the reduction in cross-sectional area at the point of attachment.

The welding process also creates a heat-affected zone, which in this case is marked by the re-solution of the grain-boundary carbides. However this zone is narrow and there is a sharp transition in materials properties from the weld metal to the unaffected main strut region which will further exacerbate the possible strain discontinuity referred to above.

The final process in the manufacture of the prosthesis is a polishing stage. It is evident that the cavities on the surface of the melt zone are associated with the coarse carbide distribution. It is probable that they have arisen due to carbide pull-out during polishing. Due to the nature of the service loading, the principal loading stress on the strut is a bending stress. The stress distribution, which varies through the cross-section, is a maximum in the outer 'fibres' of the material. The cavities therefore take on a renewed significance acting as local stress-raisers in a high-stress field with an easy crack trajectory afforded by the weak boundaries of the large as-cast grains in the melt zone. The branching surface cracks observed at the root of the strut in both fractured and retrieved valves support this opinion.

Although the specific failure mechanism cannot be unambiguously determined, it is evident that the principal cause of failure is in the weld region. Failures have previously been reported in the large strut when this was also welded but these have been eliminated by machining the major strut as one with the annulus. A new version of the Björk–Shiley tilting-disc prosthesis was brought out in 1984 which was machined from a single piece of metal and thus contained no welds at all. To date the authors are not aware of any instances of failure in the 'monostrut' prostheses.

The design/failure analysis of the Björk–Shiley valve raises a number of important points which must be applied to ensure the long-term integrity of prostheses. It is most important to have an appreciation of the change in materials properties which may be initiated by a manufacturing process. It is also important to verify that the basic assumption of engineering materials properties being linear, isotropic and homgeneous is still valid after manufacture.

The value of adequate stress analysis of the loaded structure should also be recognised, particularly in the case of potential fatigue loadings, but where high peak loads may be experienced, it is also necessary to adopt a defect-tolerant approach to design. This must recognise that even in relatively low load-bearing situations a combination of geometrical and microstructural stress-raisers may increase local 'crack tip' stresses by an order of magnitude. This latter point is particularly important in the case cited here where it appeared to be the polishing action which initiated arguably the most serious defect; a stress-raiser at the point of peak stress.

1.8 **Conclusions**

The design performance of a cardiac valve prosthesis must be assessed against four principal criteria: haemodynamic performance; tissue compatibility; low toxicity; and mechanical integrity. Good design must seek to satisfy all criteria.

The current design of the Björk–Shiley 60° tilting-disc prosthesis is seen to satisfy all of these criteria.

The post-failure analysis of prosthetic cardiac valves outlined herein, demonstrates how factors affecting mechanical integrity assist in the identification of design and manufacturing problems.

Engineering materials are generally considered to be isotropic and homogeneous. The present work stresses the importance of verifying this assumption.

Acknowledgements

The authors wish to thank Professor Sir Peter Hirsch for the laboratory facilities used to undertake the failure analyses, Dr John Martin for helpful comments and Dr S. H. Sacks for providing clinical information.

References

Baier, R. E. & Dutton, R. C. (1969). Initial events in interaction of blood on a foreign surface. *Journal of Biomedical Research*, **3**, 191–206.

Biggs, R. (1969). Materials and the study of blood coagulation. *Biomedical Engineering*, **4**, 157–9.

Björk, V. O. (1971). Disc wear in the Björk–Shiley tilting disc valve. *Scandinavian Journal of Thoracic and Cardiovascular Surgery*, **5**, 87–91.

Björk, V. O. (1985). Metallurgic and design development in response to mechanical dysfunction of Björk–Shiley heart valves. *Scandinavian Journal of Thoracic and Cardiovascular Surgery*, **19**, 1–12.

Björk, V. O. & Ovenfors, C. O. (1971). The disc rotation in patients operated upon with the Björk–Shiley tilting disc valves with radio-opaque markers. *Scandinavian Journal of Thoracic and Cardiovascular Surgery*, **5**, 83–6.

Brubakk, O., Simonsen, S., Kalman, L. & Fredriksen, A. (1981). Strut fracture in the new Björk–Shiley mitral valve prosthesis. *Journal of Thoracic and Cardiovascular Surgery*, **29**, 108–9.

Gilchrist, T. (1976). Blood compatible materials. In *Biocompatibility of Implant Materials*, ed. D. Williams. Part V, pp. 21, 174. Sector Publishing Ltd, London.

Grode, G. A., Falb, R. D. & Crowley, J. P. (1972). Biocompatible materials for use in the vascular system. *Journal of Biomedical Materials Research*, **6**, Symposium 3, 77–84.

Jenkins, G. M. (1980). Biomedical applications of carbons and graphites. *Clinical Physics*, Paper 095, review article.

Joyce, L. D., Emery, R. W. & Nicoloff, D. M. (1978). Ball variance and fracture of mitral valve prostheses causing recurrent thromboemboli. *Journal of Thoarcic and Cardiovascular Surgery*, **75**, 309.

McHenry, M. M., Smeloff, E. A. & Fong, W. Y. (1970). Critical obstruction of prosthetic heart valves due to lipid absorption by Silastic. *Journal of Thoracic and Cardiovascular Surgery*, **59**, 413–25.

Williams, D. F. (1971). The properties and medical uses of materials. Part 3:2, The effect of materials on blood. *Biomedical Engineering*, **6**, 205–8.

2

The mechanical properties of natural and chemically modified bovine pericardium for use with heart valve substitutes

A. Trowbridge, C. Crofts, P. Lawford and M. M. Black

2.1 **Introduction**

The pericardium is a sac surrounding the heart. It is composed of an outer layer, the fibrous pericardium, and an inner layer, the serous pericardium. Bovine fibrous pericardium is a cone-shaped bag, which is connected to the diaphragm by a loose fibrous diaphragmatic attachment and to the sternum by the sternopericardial ligaments. The serous pericardium is a closed sac which lines the inner surface of the fibrous pericardium. It consists of a parietal and a visceral layer. The parietal layer is contiguous with the fibrous pericardium and is separated from the visceral layer, which covers the muscular wall of the heart, by the pericardial cavity. Most of the natural functions of the pericardium are mechanical. They include the prevention of excessive dilatation of the heart chambers and limit the magnitude of hypertrophy of the heart under conditions of strenuous activity.

The main constituents of pericardial tissue are collagen and elastin embedded in a viscoelastic gel of ground substance and water. Collagen is a helical crystalline protein. It consists of polypeptide chains which are organised ultimately into fibrils of approximately 100 nm in diameter. In contrast, elastin is a globular protein. An amorphous gel of mucopolysaccharides, glycoproteins and soluble proteins in water lies between the mesh of fibrous proteins. Natural bovine pericardium contains approximately 76% by weight of water – the majority of which is chemically unbound. Hence, pericardium can be considered as a laminated, pliant, biological composite.

Following chemical modification with glutaraldehyde, bovine pericardium is used in the construction of the leaflets of heart valve substitutes.

This type of bioprosthesis has been in clinical use for over a decade (Angell *et al.*, 1979). In addition, chemically modified pericardium is often employed in cardiovascular patches (Wessex Medical, 1985). More recently its use in reconstructive hand surgery has been advocated (Smith *et al.*, 1986). This chapter discusses the mechanical properties of both natural and chemically modified bovine pericardium.

2.2 **Chemical modification by glutaraldehyde fixation**

For clinical use, it is desirable to minimise the antigenicity of the tissue and limit host reactivity, as well as providing biochemical stability, through prevention of autolysis and subsequent cellular degradation. Chemically treating pericardium with glutaraldehyde produces a material with these properties.

Glutaraldehyde is a simple dialdehyde. It modifies soft tissue chemically by introducing cross-links between the collagen molecules. The reaction involves the primary amines of the lysine, hydroxylysine or N-terminal amino acid residues present in the collagen polypeptide chains. These cross-links, as well as introducing stability into the material, are also thought to have greatly enhanced resistance to proteolytic degradation, therefore reducing the antigenicity of the tissue. In addition, although this treatment does not produce a biologically inert material, the magnitude of host reactivity to glutaraldehyde-fixed pericardium appears to be less than for other types of tissue. Glutaraldehyde fixation may also remove some of the ground substance from soft tissue (Ferrans *et al.*, 1978; Lee *et al.*, 1984).

A hydrothermal shrinkage test is one way of assessing the increase in cross-link formation. The hydrothermal shrinkage temperature, T_S, is related to the molecular stability of the triple helix of the tropocollagen molecule. Synthesis of new linkages leads to a more stable structure and results in values of T_S which are higher than in the natural tissue. Fixation with buffered glutaraldehyde raises the T_S of natural bovine pericardium from 60 to 84.5°C.

Glutaraldehyde fixation causes tissue shrinkage in the linear dimension of the order of 10% (Crofts, 1986). Pericardial tissue fixed under tension demonstrates an increase in material stiffness (Reece *et al.*, 1982; Crofts, 1986). Hence, to obtain chemically modified pericardium without locked-in stresses, the tissue needs to be fixed free of tension with due consideration given to the degree of shrinkage. This condition can be achieved by draping the pericardial sacs loosely on embroidery frames (150 mm in diameter) and immersing them in a large volume of 0.2% solution of glutaraldehyde (BDH Chemicals Limited) buffered to pH 7.4 in 0.15 $mol{\cdot}L^{-1}$ phosphate buffer (Sorensen) for 7 days.

2.3 The deformation response to uniaxial load

All experimental results reported in this chapter have been obtained from pericardial sacs harvested from 16–20-week-old calves. After transportation from the abattoir in ice-cold isotonic saline (0.9% NaCl) each sac was inspected to verify the presence of both the diaphragmatic attachment and sternopericardial ligaments. All visible fat was stripped by hand from the tissue and then each sac was mounted on an embroidery frame. Since natural pericardium is a biological composite, it is likely to demonstrate both animal-to-animal and anatomical variation in its mechanical properties. Glutaraldehyde fixation introduces a further source of variation into the chemically modified material. To allow statistical assessment of all of these sources of variation, the experimental design must include a method of identifying the same sites in each sac studied, as well as using a sufficiently large sample of pericardial sacs. To standardise the anatomical variation, a circular plastic template of diameter 150 mm was constructed. Five strips of length 80 mm and width 12 mm were cut out, in a random way, around the centre of the plastic disc and numbered from 1 to 5 as shown in Figure 2.1. By using the diaphragmatic attachment and the

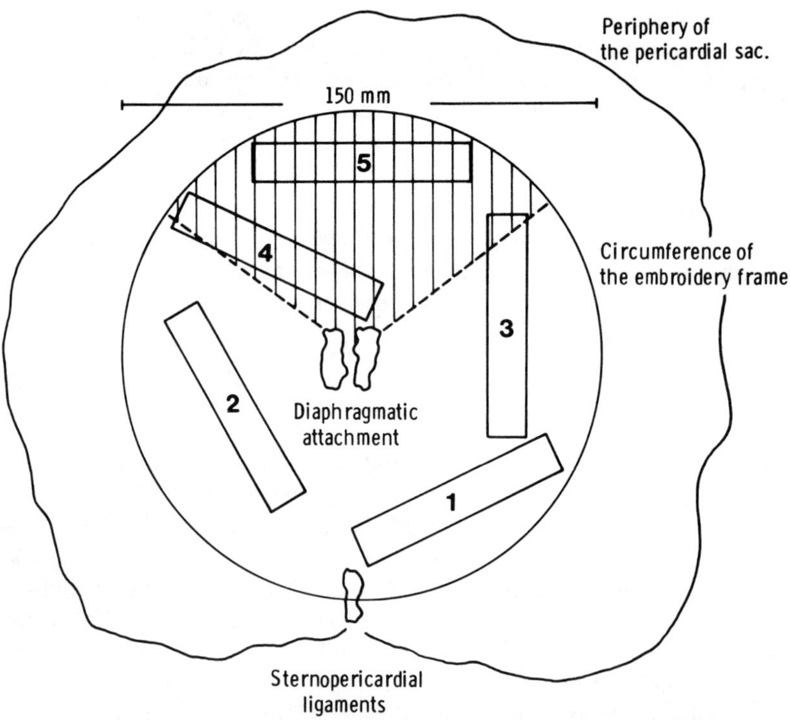

Fig. 2.1 A schematic representation of the five sites harvested from the pericardial sac. The hatched area indicates the region of highest extensibility.

sternopericardial ligaments as a reference frame, the same sites could be identified in every sac.

For non-destructive testing all specimens were loaded to a maximum stress of $0.6\,N{\cdot}mm^{-2}$. This was judged to be the tension associated with a physiological pressure of 100 mmHg (Trowbridge *et al.*, 1986). By measuring the width and thickness of the specimen before loading, the undeformed cross-sectional area could be computed and then the tissue was loaded to the force associated with the Piola–Kirchoff stress of $0.6\,N{\cdot}mm^{-2}$. Figure 2.2 shows the family of response curves to uniaxial

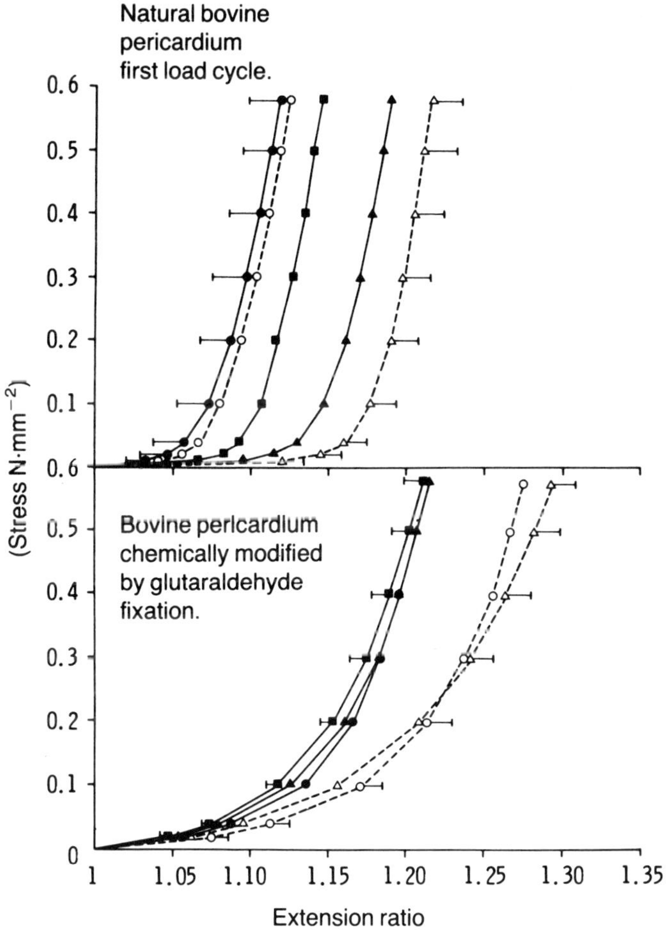

Fig. 2.2 The family of response curves to uniaxial load for natural (upper half; means ± SEM, $n = 9$) and chemically modified (lower half; means ± SEM, $n = 10$) bovine pericardium: (●) position 1; (■) position 2; (▲) position 3; (○) position 4; (△) position 5.

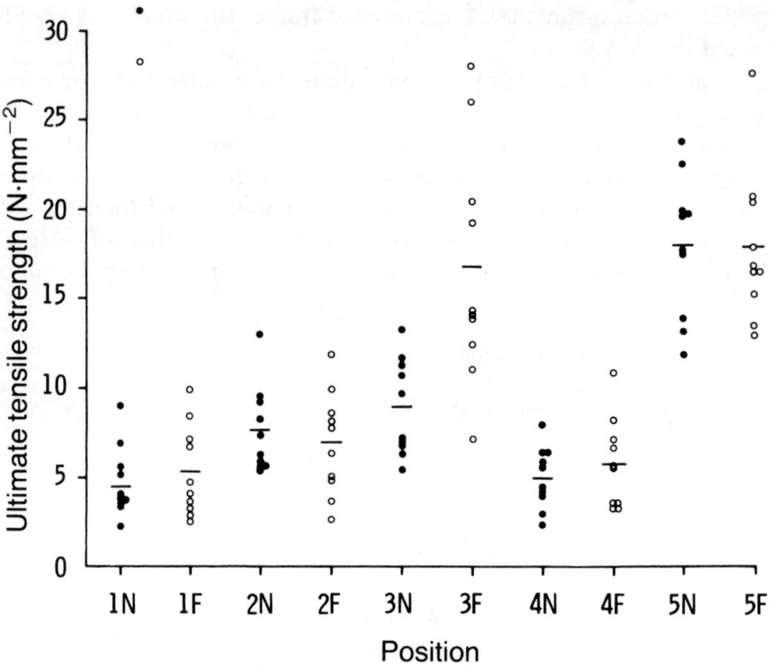

Fig. 2.3 The tensile strength of natural (●) and chemically modified (○) bovine pericardium.

load for the natural and chemically modified tissue. The natural tissue demonstrates a marked extension for very small stresses, followed by an abrupt stiffening with little change in extension at the higher loads. The chemically modified material has similar properties, but the stiffening with load is less abrupt. Analysis of variance shows that there is no significant variation in extensibility from sac to sac in either the natural or the glutaraldehyde fixed tissue. However, chemical modification produces a more extensible material than that found in the natural state. Furthermore, tissue harvested from site 5 (see Figure 2.1) is significantly more extensible than the other sites in the sac. This anatomical variation in extensibility has been attributed to differences in the elastin ultrastructure at these sites (Trowbridge *et al.*, 1986).

2.4 **Tensile strength**

An identical method of site selection has been used to investigate the tensile strength and strain at rupture of both natural and chemically modified bovine pericardium (Crofts & Trowbridge, 1987). Figures 2.3 and

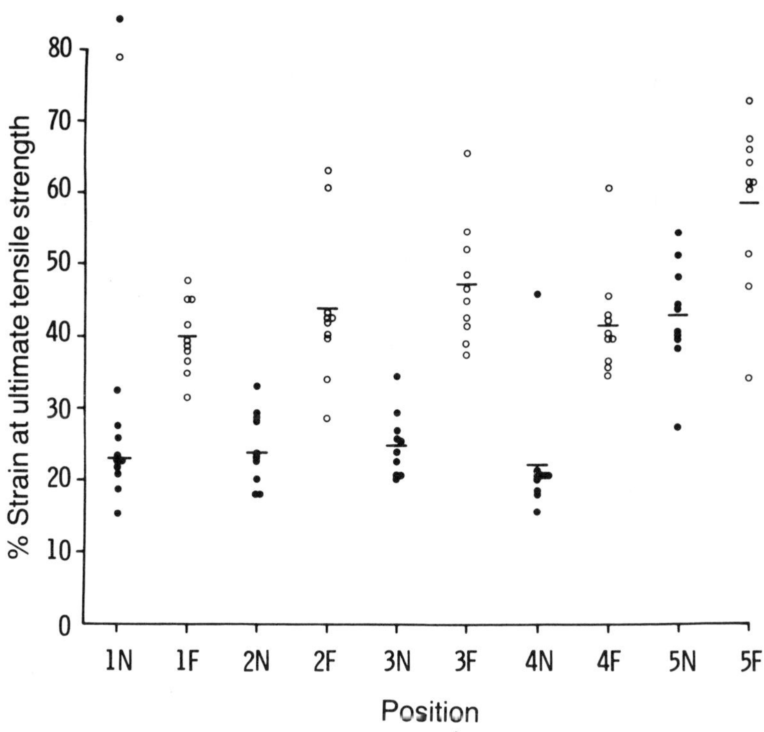

Fig. 2.4 The percentage strain at rupture of natural (●) and chemically modified (○) bovine pericardium.

2.4 show the tensile strength and strain at rupture of samples harvested from sites 1–5. Chemical modification does not alter the strength of the natural tissue, but the strain at rupture is significantly increased. The tissue fails by laminate debonding through shear and subsequent fibre slippage.

2.5 **Viscoelasticity**

Both the natural and chemically modified tissue demonstrate a decay in stress, which varies logarithmically with time, when impulsively loaded specimens are maintained at a constant length. Figure 2.5 shows that relaxation of stress is greater in the natural tissue than after chemical modification.

Linear viscoelastic materials usually display a hysteresis loop during loading and unloading which depends on the strain rate. Bovine pericardium is characterised by a hysteresis loop during cyclic load, indicating a measure of dissipation in mechanical energy (Trowbridge & Crofts, 1987). However, there is no evidence of strain rate dependence (see Figure 2.6).

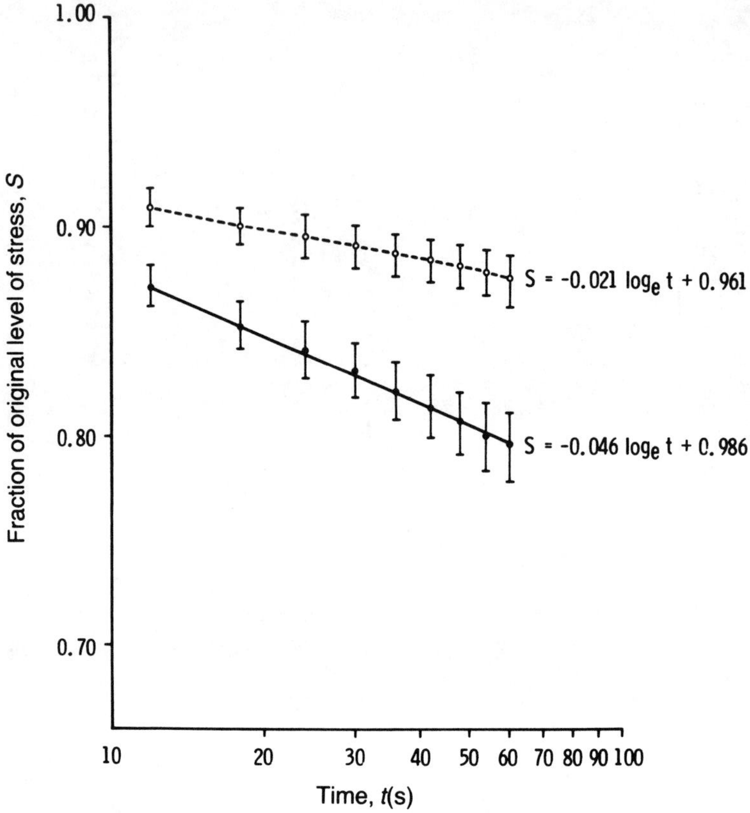

Fig. 2.5 The relaxation of stress in natural (●) and chemically modified (○) bovine pericardium (means ± SEM, $n = 15$).

Lee and Boughner (1981) have suggested that the viscous ground matrix is thixotropic. Hence, extension rate independence of stress during uniaxial loading may arise from shear thinning of the embedding gel as the collagen fibres are aligned in the direction of load.

2.6 **Plasticity**

No evidence of permanent plastic deformations have been observed under uniaxial loads up to a maximum stress of $0.6\,\mathrm{N{\cdot}mm^{-2}}$ (Trowbridge & Crofts, 1986a). However, the pericardium does possess a 'cyclic creep' property. The undeformed length of the test sample gradually increases as the number of load cycles is increased. The chemically modified tissue had lengthened by about 11% after 36 load cycles (Trowbridge & Crofts, 1986a). A similar phenomenon has been observed after 2048 cycles (Trowbridge & Crofts, 1986b). When the load is removed the tissue quickly

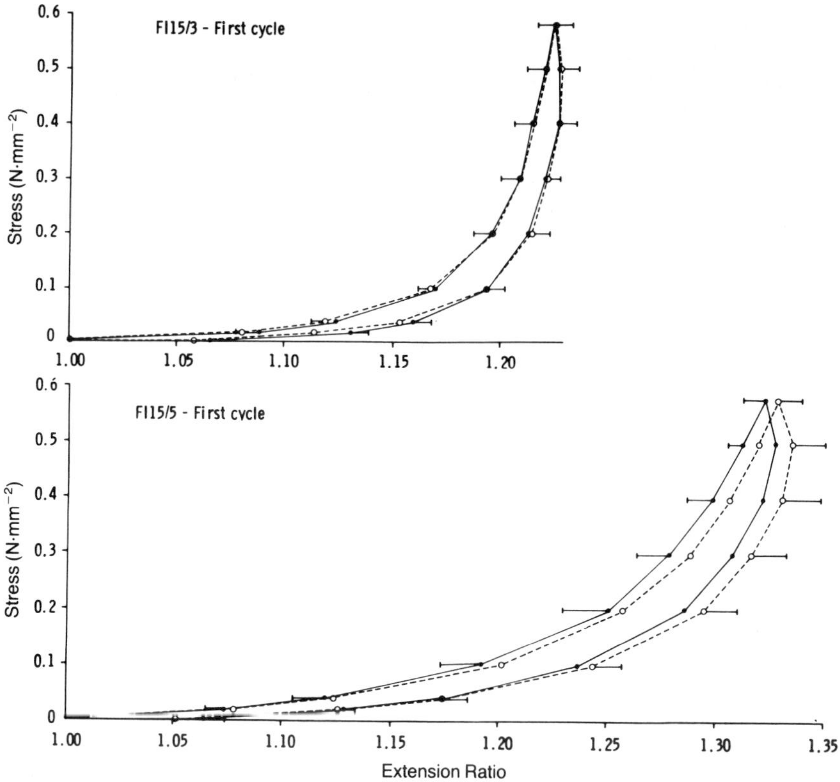

Fig. 2.6 The first cycle mean loading curves at two different extension rates for chemically modified bovine pericardium: (●) 2.1 mm·min^{-1}; (○) 120 mm·min^{-1}.

returns to its original undeformed length (see Figure 2.7). Subsequent mechanical tests indicated that all knowledge of the earlier deformations had been erased from the memory of the tissue (Trowbridge & Crofts, 1986a). Hence, bovine pericardium can be accredited with a rapidly fading memory for strain history.

2.7 **Temperature**

The majority of tests that have been reported in this article were performed at room temperature (approximately 20°C). Leaflets of heart valve substitutes are subjected to loads at a body temperature of 37°C. Figure 2.8 shows that the response to uniaxial load of chemically modified tissue is independent of temperature over the range 20–37°C. This observation may be associated with the increased stability induced by glutaraldehyde fixation (Crofts & Trowbridge, 1988).

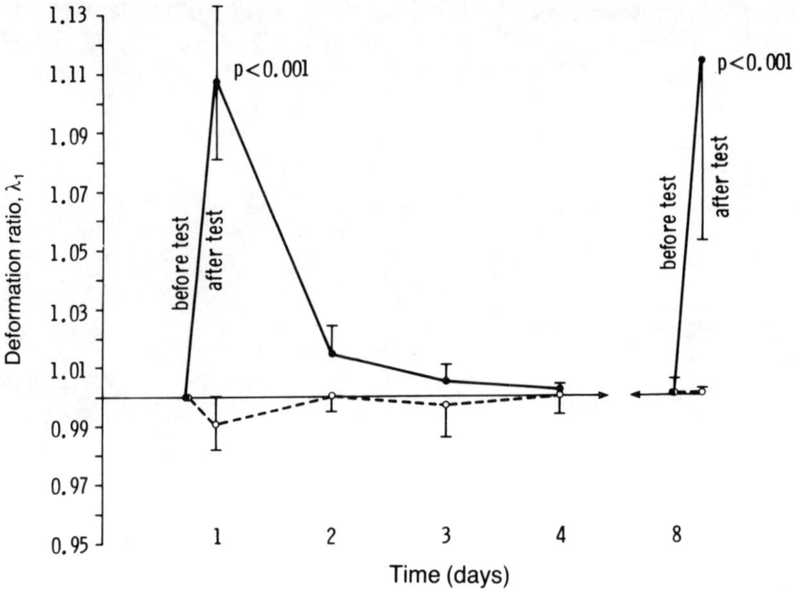

Fig. 2.7 The variation in the length of chemically modified bovine pericardium with time after mechanical testing: (●) test; (○) control samples; mean ± SEM, $n = 5$).

2.8 Conclusions

Bovine pericardium is a laminated pliant biological composite. Glutaraldehyde fixation increases the number of cross-links between collagen molecules. This manoeuvre results in a greater material extensibility and strain at rupture than that found in the natural tissue, but it does not improve the strength of the material. Pericardium displays viscoelastic properties which are extension rate and temperature independent. At loads associated with the physiological closing pressures applied to heart valve substitutes, there is no evidence of permanent plastic deformation. However, the tissue does increase its undeformed length as it is mechanically worked. Only after removal of the cyclic load does the material return to its original undeformed state. All of these features must be considered in the design of those bioprostheses constructed from chemically modified bovine pericardium.

Acknowledgements

This work has been made possible by a training award and project grant from the Medical Research Council. We are grateful to Wessex Medical Limited for providing a constant supply of pericardium.

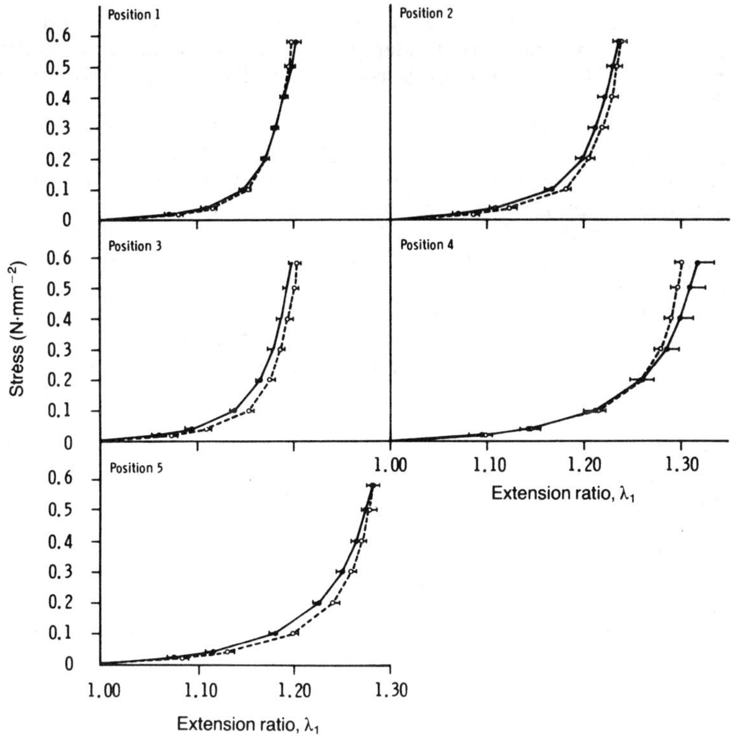

Fig. 2.8 The first cycle mean loading curves (means ± SEM, n = 5) at two different temperatures for chemically modified bovine pericardium: (●) room temperature; (○) body temperature.

References

Angell, W. W., Angell, J. D. & Kosek. J. C. (1979). Clinical and experimental comparisons establishing the glutaraldehyde-treated xenograft as the standard for tissue heart valve replacement. In: *Tissue Heart Valves*, ed. M. I. Ionescu, p. 89. Butterworth and Co. (Publishers) Limited, London.

Crofts, C. E. (1986). Pericardial heterografts. Tissue mechanical properties and their implications for valve design. PhD Thesis, University of Sheffield.

Crofts, C. E. & Trowbridge, E. A. (1987) The tensile strength of natural and chemically modified bovine pericardium. *Journal of Biomedical Materials Research*, **22**, 89–98.

Crofts, C. E. & Trowbridge, E. A. (1988) The response to uniaxial load of chemically modified bovine pericardium at different temperatures. *Journal of Materials Science*, **23**, 1510–3.

Ferrans, V. J., Spray, T. L., Billingham, M. E. & Roberts, W. C. (1978). Structural changes in glutaraldehyde-treated porcine heterografts used as substitute cardiac valves: Transmission and scanning electron microscopic observations in 13 patients. *American Journal of Cardiology*, **41**, 1159–84.

Lee, J. M. & Boughner, D. R. (1981). Tissue mechanics of canine pericardium in different test environments. Evidence for time-dependent accommodation, absence of plasticity and new roles for collagen and elastin. *Circulation Research*, **49**, 533–44.

Lee, J. M., Boughner, D. R. & Courtman, D. W. (1984). The glutaraldehyde-stabilised porcine aortic valve xenograft. II Effect of fixation with or without pressure on the tensile viscoelastic properties of the leaflet material. *Journal of Biomedical Materials Research*, **18**, 79–87.

Reece, I. J., van Noort, R., Martin, T. R. P. & Black, M. M. (1982). The physical properties of bovine pericardium. A study of the effects of stretching during chemical treatment in glutaraldehyde. *Annals of Thoracic Surgery*, **33**, 480–6.

Smith, D. J., Jones, C. S., Hull, M., Robson, M. C. & Kleinert, H. E. (1986). Bioprosthesis in hand surgery. *Journal of Surgical Research*, **41**, 378–87.

Trowbridge, E. A. & Crofts, C. E. (1986a). Evidence that deformations which occur during mechanical conditioning of bovine pericardium are not permanent. *Biomaterials*, **7**, 49–54.

Trowbridge, E. A. & Crofts, C. E. (1986b). The standardisation of gauge length. Its influence on the relative extensibility of natural and chemically modified pericardium. *Journal of Biomechanics*, **19**, 1023–33.

Trowbridge, E. A. & Crofts, C. E. (1987). The extension rate independence of the hysteresis in glutaraldehyde-fixed bovine pericardium. *Biomaterials*, **8**, 201–6.

Trowbridge, E. A., Roberts, K. M., Crofts, C. E. & Lawford, P. V. (1986). Pericardial heterografts. Toward quality control of the mechanical properties of glutaraldehyde-fixed leaflets. *Journal of Thoracic and Cardiovascular Surgery*, **92**, 21–8.

Wessex Medical Limited (1985). Information pamphlet. Peachey House, Bepton Road, Midhurst, West Sussex, CU29 9LU.

3

Biaxial mechanical properties of passive myocardium

J. D. Humphrey and F. C. P. Yin

3.1 Introduction

Mechanical stresses in the heart can affect coronary blood flow, myocardial oxygen consumption, the development of hypertrophy, the stability of cardiac aneurysms and, of course, overall cardiac performance. Mechanical stresses cannot be measured directly in the intact heart, however, but must be inferred from biomechanical analyses which are based on descriptions of the material properties (Huisman *et al.*, 1980). Thus, fundamental to the analysis of heart mechanics and physiology is a detailed understanding of the mechanical properties of cardiac tissues, including delineation of the active and passive components of the behaviour. This is probably best accomplished by first quantifying the passive properties, since the non-contractile state is simpler and is an important determinant (Frank Starling mechanism) of the active behaviour (Glantz & Parmley, 1978; Janicki & Weber, 1980). Consequently, in this chapter the mechanical properties of passive myocardium and, in particular, observations based on biaxial experiments are briefly reviewed.

3.2 Myocardial ultrastructure

Myocardium is composed of, among other constituents, a complex arrangement of muscle, collagen, elastic fibres, amorphous substances and microthreads (Caulfield & Borg, 1979; Robinson, 1983). The gross mechanical properties depend, of course, upon both the internal composition and the architecture of the tissue. Streeter (1979) has shown that the muscle fibres are locally parallel in certain regions within the heart wall, and that their orientations change continuously with radial location (Figure 3.1). The changing muscle fibre directions in the equatorial region of the left ventricle can, for example, be described concisely by a unit vector, $\boldsymbol{N}$,

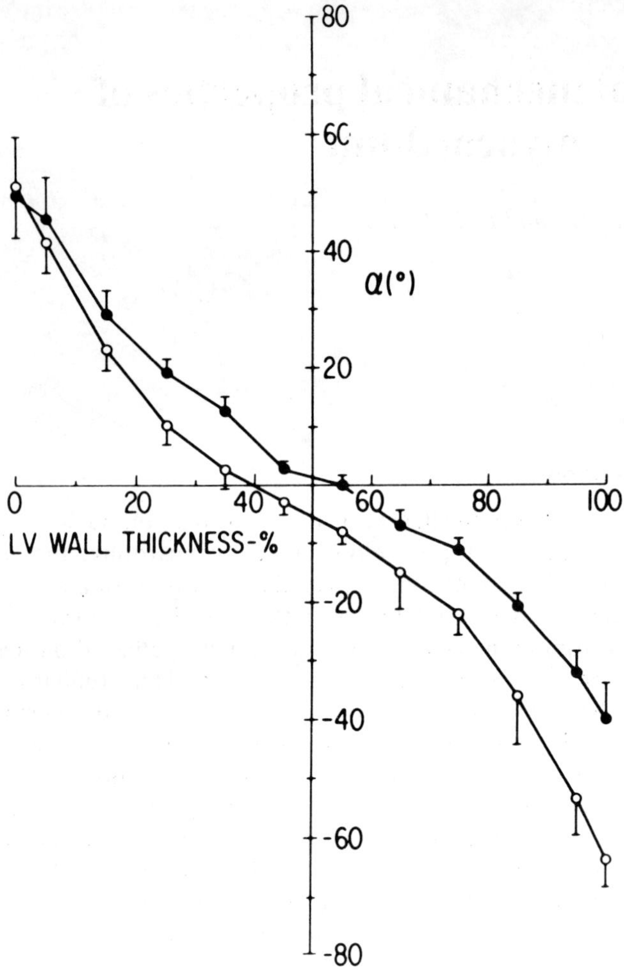

Fig. 3.1 Transmural distribution of muscle fibre angles (α) through the heart wall as reported by Streeter (1979) for the equatorial region in a canine heart. Note that the angle α in the figure is the same as Φ in the text. (With permission from the American Heart Association.)

which is written in terms of cylindrical coordinates referred to an initial configuration. That is,

$$N = \cos \Phi(R) \boldsymbol{E}_{\Theta} + \sin \Phi(R) \boldsymbol{E}_Z \tag{3.1}$$

where Φ is an initial muscle fibre angle measured from the circumferential direction, $\boldsymbol{E}_A$ are orthonormal base vectors, and (R, Θ, Z) are cylindrical coordinates in the initial configuration. $\Phi(R)$ has, for example, been assumed to be of the following form (Chadwick, 1982; Tozeren, 1983),

$$\Phi(R) = \Phi_o\{[2R - (R_o - R_i)]/[R_o - R_i]\}^3 \qquad (3.2)$$

where R_i and R_o are endocardial and epicardial radii, respectively, and Φ_o is the value of Φ at the epicardium (typically -60 to $-90°$). Eqn (3.2) yields a transmural distribution of muscle fibre angles similar to that in Figure 3.1.

Scanning electron micrographs of myocardium confirm the existence of locally parallel muscle fibres, but further reveal the presence of a dense plexus of collagen, elastin and microthreads (Figure 3.2). The non-muscular constituents appear to be fairly randomly distributed in certain configurations (see Caulfield & Borg, 1979, Figures 2–10) even though two major groups of collagen follow the muscle fibre architecture: (i) a collagen weave, or sheath, encloses the muscle fibres and runs along their axes; and (ii) collagen struts provide myocyte-to-myocyte and myocyte-to-capillary connections. Collectively, the organisation, properties and interactions of the collagen, elastin, microthreads, amorphous substances, and passive muscle fibres yield the overall passive mechanical properties of myocardium.

3.3 **Background**

Equations which characterise the behaviour of a material in response to applied loads are called constitutive relations since they describe the gross behaviour resulting from the internal constitution of the material (Malvern, 1969). There are, in general, three steps in the determination of a constitutive relation: prescription of a general form of the relation; identification of a specific functional form; and determination of the particular material parameters. Whereas the first step is based primarily on basic principles or postulates of mechanics (e.g. material frame indifference) and on considerations of material symmetries and kinematic constraints, the remaining steps are based largely on experimental data interpreted and analysed within an associated theoretical framework. Except for the simplest of materials, the interaction between theory and experiment must be iterative in the determination of a constitutive relation (Fung, 1981).

Most of our knowledge of the mechanical properties of cardiac tissue has come from uniaxial experiments on papillary muscles and trabeculae carnae (Sonnenblick, 1964; Pinto & Fung, 1973; Kitabatake & Suga, 1978; Yin *et al.*, 1980). Uniaxial information is useful in determining general characteristics of the behaviour of cardiac tissue in both health and disease, and an excellent review of many uniaxial studies can be found in Mirsky (1976).

The heart experiences multiaxial deformations (Waldman *et al.*, 1985),

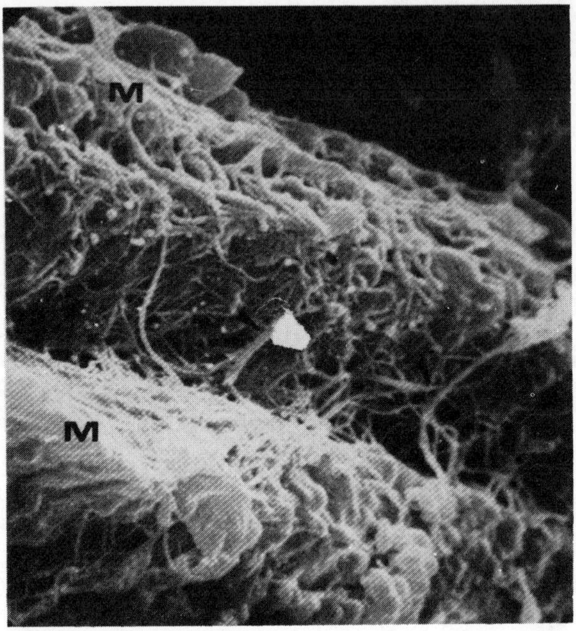

Fig. 3.2 Scanning electron micrograph of myocardium. Collagen fibres (arrowheads) are shown interconnecting hamster myocytes (M), and running along the myocyte axes thereby enclosing them in a weave. Magnification is × 7800. (Taken with permission from J. B. Caulfield & T. K. Borg, *Laboratory Investigation*, **40**, 364–72, 1979.)

however, and uniaxial data cannot be generalised to describe multiaxial behaviour (Fung, 1973; Bergel & Hunter, 1979; Grood *et al.*, 1979; Moriarty, 1980; Yin, 1981; Humphrey & Yin, 1987b). For example, fluid-filled muscle fibre-reinforced constitutive models (Feit, 1979; Chadwick, 1982; Tozeren, 1983; Jan, 1985) are based largely on data from papillary muscles, and incorrectly predict that passive myocardium can support loads only in the direction of the muscle fibres. Similarly, Pao *et al.* (1980) deduced uniaxial constitutive relations for passive myocardium on the basis of data from uniaxial tests on full-thickness strips of ventricular wall. They also incorrectly assumed that myocardium is composed solely of 'slippery-skinned rods' of muscle fibres, and thereby neglected the non-muscular constituents. That myocardium can bear loads transverse to the directions of the muscle fibres is supported intuitively by scanning electron micrographs of its ultrastructure (Figure 3.2), and is demonstrated by multiaxial data which are discussed subsequently.

Various constitutive relations have also been proposed for passive myocardium based on pressure–volume data from intact ventricles

(Mirsky, 1973; Demiray, 1976, 1981; Abe & Nakamura, 1982). Pressure–volume data reveal only an average, global behaviour, and are not sufficient for characterising completely a constitutive relation for cardiac tissue (Moriarty, 1980).

Multiaxial data are required, therefore, to rigorously determine constitutive relations for passive myocardium. Well-controlled and easily interpreted triaxial experiments on cardiac tissue are difficult to define and perform, and have only recently been attempted (Halperin *et al.*, 1987). Fortunately, however, biaxial data can be used to determine certain multidimensional stress–strain relations since myocardium, like many biological tissues, can be assumed to be nearly incompressible (Fung, 1981). Consequently, existing biaxial experimental data on passive myocardium, and constitutive relations that have been proposed based upon these data will be reviewed.

3.4 Biaxial experimental data

3.4.1 *Data collection*

Biaxial data have been reported for canine myocardium from various regions in 49 hearts (Demer & Yin, 1983), and on subepicardial specimens from one region in five hearts (Yin *et al.*, 1987). Briefly, in both studies thin (1–2 mm) rectangular slabs of potassium-arrested myocardium were excised tangential to the epicardial surface of canine left ventricles using a commercial meat slicer. In this way, the muscle fibre orientations (Figure 3.1) remained in the plane of the specimens which were typically 30 × 30 × 1 mm in dimension. That is, if (X_1, X_2, X_3) denotes material particles within a specimen in an undeformed configuration, then the muscle fibre orientations can be described by the unit vector $\boldsymbol{N}$, where

$$\boldsymbol{N} = \cos \Psi(X_3)\boldsymbol{E}_1 + \sin \Psi(X_3)\boldsymbol{E}_2 \tag{3.3}$$

where Ψ is measured from the X_1 direction, and the 3-direction is out-of-plane.

A specimen was then mounted in a biaxial test apparatus by placing it in an oxygenated physiological solution, and coupling it to the apparatus through four arrays of 000 sutures (Figure 3.3). The apparatus consists of two pairs of orthogonally positioned carriages which are driven independently by two linear motors. Various stretching or loading protocols can be performed through the motions of the carriages which are prescribed via a function generator and a controller unit. Force transducers are mounted on one carriage of each pair, and the specimen deformation is monitored optically through two video dimension analysers (Yin *et al.*, 1972). Video dimension analysers yield a voltage output which is proportional to distances (in X_1 and X_2) between parallel markers affixed to the surface of

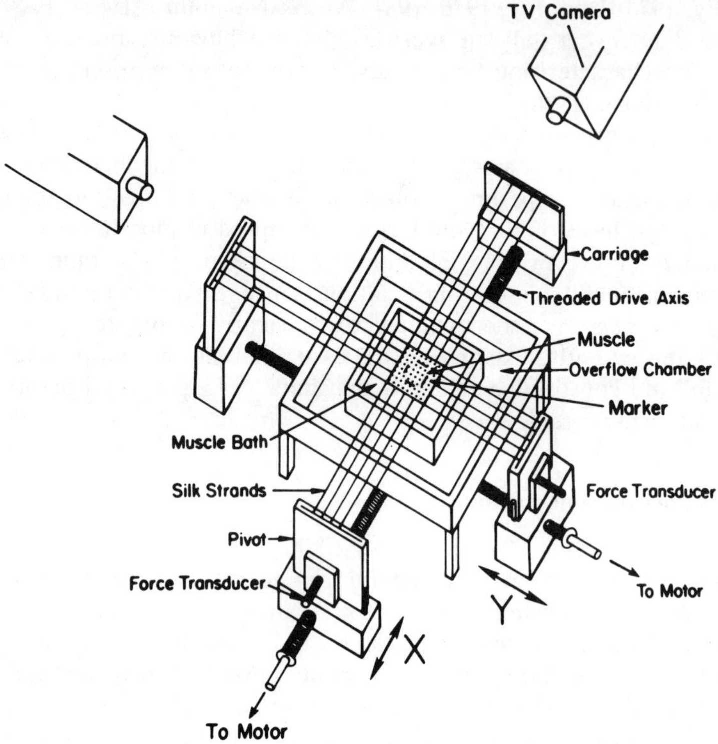

Fig. 3.3 Schematic of the improved biaxial apparatus used by Yin *et al.* (Taken with permission from F. C. P. Yin, R. K. Strumpf, P. H. Chew & S. L. Zeger, *Journal of Biomechanics*, **20**, 577–99, 1987.)

the specimen (Figure 3.3). The deformations are measured in the central region to avoid edge effects. The experimental protocols are defined based on feedback from either the measured forces or the extensions in the orthogonal stretching directions.

3.4.2 *Data reduction*

The deformation in the central region of a biaxially tested myocardial specimen can be assumed to be a general homogeneous deformation including biaxial extension and in-plane shear. Thus

$$x_1 = \lambda_1 X_1 + \varkappa_1 X_2, \quad x_2 = \lambda_2 X_2 + \varkappa_2 X_1, \quad x_3 = \lambda_3 X_3 \tag{3.4}$$

wherein material particles originally at (X_1, X_2, X_3) are mapped into (x_1, x_2, x_3) after deformation. The λ_i are stretch ratios and $\varkappa_i$ are measures of in-plane shear. The in-plane stretch ratios are measured easily with video dimension analysers, whereas both the stretch ratios and the shears can be determined by tracking multiple markers affixed to the surface of the specimen (Humphrey *et al.*, 1987).

The physical components of the deformation gradient, $\boldsymbol{F}$, are determined from $F_{iA} = \partial x_i/\partial X_A$, where lower-case and upper-case indices refer to the deformed and undeformed configurations, respectively. The value of λ_3 can be determined from the in-plane quantities by enforcing the incompressibility requirement that the det $\boldsymbol{F} = 1$. Finally, the Cauchy–Green deformation tensors, $\boldsymbol{B} = \boldsymbol{F} \cdot \boldsymbol{F}^T$ and $\boldsymbol{C} = \boldsymbol{F}^T \cdot \boldsymbol{F}$, the Green strain tensor $\boldsymbol{E} = (\boldsymbol{C} - \boldsymbol{I})/2$, and the displacement gradient ($\boldsymbol{G} = \boldsymbol{F} - \boldsymbol{I}$) are all easily determined from the deformation gradient, $\boldsymbol{F}$.

The normal physical components of the Cauchy, or true, stress are inferred in the central region of the specimen by first calculating the first-Piola Kirchhoff stresses, P_{11} and P_{22}, as the normal forces per undeformed area in the central region. The Cauchy stress, $\boldsymbol{t}$, is then calculated as $\boldsymbol{t} = \boldsymbol{F} \cdot \boldsymbol{P}$ since the det $\boldsymbol{F} = 1$. Thus, biaxial stress–strain data are easily generated from the experimentally measured deformation ($\lambda_i, \varkappa_i$) and applied normal forces.

3.4.3 *Representative results*

Yin *et al.* (1987) studied subepicardial specimens from five canine hearts. The specimens were mounted within the apparatus such that the predominant muscle fibre direction coincided with a stretching axis (i.e. $\Psi = 0$ in eqn (3.3)), and the stretching protocols were performed such that ratios of the in-plane extensional Green strains remained constant at particular values (see Figure 2 in Yin *et al.*, 1987). Finally, $\varkappa_i$ were assumed to be negligible as judged by eye. Typical stress–stretch data under nearly equibiaxial stretching are shown in Figure 3.4. In particular, note the large deformations, the non-linear relationship between Cauchy stress and stretch, the stiffer behaviour in the predominant muscle fibre direction, and the myocardium's ability to resist loads in a direction transverse to the predominate muscle fibre direction. Additionally, Yin *et al.* reported that the potassium-arrested slabs of myocardium exhibited hysteresis under cyclic loading (not shown in Figure 3.4). It can be concluded, therefore, that passive myocardium behaves similarly to many soft biological tissues (Fung, 1981). Thus, the mechanical behaviour of potassium-arrested myocardium may be described via a pseudostrain-energy function, W, wherein material parameters are determined separately for the tissue in loading and unloading.

3.5 **Multiaxial constitutive relations**

Although many different constitutive relations have been proposed as descriptors of passive cardiac behaviour, only three basic approaches have

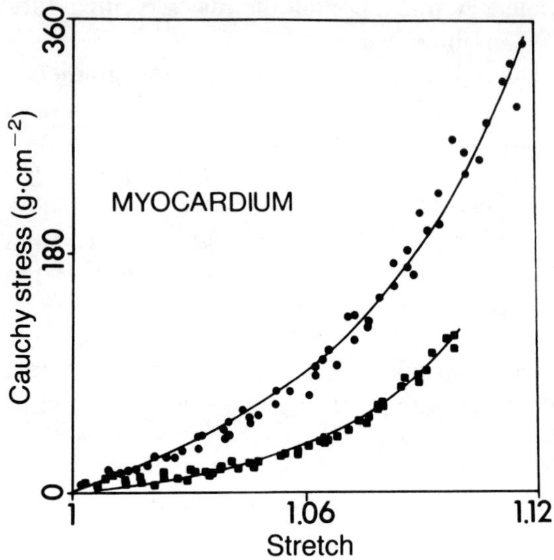

Fig. 3.4 Biaxial Cauchy stress *versus* stretch ratios for a nearly equi-biaxial experiment on a thin slab of myocardium: (●) data from the muscle fibre direction; (■) data in the orthogonal, or cross-fibre, direction. The theoretical fit (solid lines) is based on eqn (3.13). (Taken with permission from J. D. Humphrey & F. C. P. Yin, *Biophysical Journal*, **52**, 563–70, 1987.)

been employed to formulate the relations: purely phenomenological, structurally-motivated phenomenological, and microstructural. Various relations based on each of these three formulations are discussed below.

3.5.1 *Phenomenological relations*

Most myocardial constitutive relations have been formulated phenomenologically, whereby experimental data are used to identify general characteristics of the tissue behaviour (e.g. anisotropy), and then various reasonable forms of pseudostrain-energy functions are postulated to describe the data. The relations which provide the best fit to the data in a least-squares sense are then chosen as descriptors of the behaviour.

For example, Yin *et al.* (1987) observed from the aforementioned biaxial data that myocardium behaved similarly to many soft tissues (Fung, 1981). Consequently, they employed a general pseudostrain-energy function, W, proposed by Fung (1973) for soft tissues,

$$W = c[e^Q - 1], \qquad \text{with } Q = 0.5\, a_{MNKL} E_{MN} E_{KL} \tag{3.5}$$

The as and c are material parameters, E_{MN} are components of the Green strain, and summation is implied over like indices. Since Q is of the same

Table 3.1 Values of the parameters in equations (3.6) and (3.7) for one specimen.

	Equation (3.6a)				Equation (3.6b)			Equation (3.7)		
E_{11}/E_{22}	a_1	a_2	a_{12}	c	a_1	a_2	c	a_1	a_2	c
5.4	87.7	0.06	37.4	4.51	96.8	199.4	2.2	49.3	39.7	0.66
2.4	62.9	6.8	31.3	5.99	75.6	85.0	3.0	40.9	27.4	0.87
1.1	65.8	41.6	2.7	7.01	68.2	44.4	3.6	33.9	21.2	1.10
0.5	24.6	15.2	14.0	12.4	51.4	19.5	6.9	36.1	21.8	1.0
0.3	0.1	17.8	19.6	10.9	0.9	0.2	598.6	34.7	19.9	1.19

E_{11}/E_{22} is the ratio of the muscle fibre to cross-fibre direction extensional Green strains; $n = 1.429$ for eqn (3.7). The units of c are $g{\cdot}cm^{-2}$ ($1\ g{\cdot}cm^{-2} = 98.07\ N{\cdot}m^{-2}$) whereas a_{ij} are dimensionless. (Taken with permission from F. C. P. Yin, R. K., Strumpf, P. H. Chew & S. L. Zeger, *Journal of Biomechanics*, **20**, 577–99, 1987.)

form as Hooke's law, material symmetry arguments developed for linearly elastic materials can be employed for these special exponential materials.

Specifically, Yin and colleagues examined the fit to biaxial data of two slightly different forms of eqn (3.5),

$$W = 0.5c \exp\left[a_1 E_{11}^2 + a_2 E_{22}^2 + 2a_{12} E_{11} E_{22}\right] \tag{3.6a}$$

$$W = 0.5c \exp\left[a_1 E_{11}^2 + a_2 E_{22}^2\right] \tag{3.6b}$$

where the parameters (c, a_1, a_2, a_{12}) were determined from data in a non-linear least-squares sense (Yin *et al.*, 1986). Whereas the data were fitted equally well by both functions for a given experimental protocol, there was an apparent ambiguity in the interpretation of the directional properties of the tissue. As seen in Table 3.1, one might conclude from eqn (3.6a) that the tissue is stiffer in the fibre direction (axis 1) since a_1 was larger than a_2 for all the different loadings except the case where the 2-direction strain was much larger than the 1-direction strain. Using eqn (3.6b), however, the exact converse would hold. While it is possible that, depending on how the specimen was loaded, the tissue constituents may have realigned and thereby produced different responses for the different loadings, it is difficult to reconcile the directionally opposite interpretations as to the direction of greater stiffness. A more detailed study of the stiffness issue would require examination of experimental $\Delta t/\Delta V$ (Humphrey & Yin, 1986) where $V^2 = \boldsymbol{B}$, but this is not done here.

These ambiguities led Yin and his colleagues to propose the following form of W, in terms of non-integral powers of strain,

$$W = c \exp\left[a_1 E_{11}^n + a_2 E_{22}^n\right] \tag{3.7}$$

wherein the values of n were determined as follows. Consider the ratio of the two in-plane normal components of the second Piola–Kirchhoff stress ($\boldsymbol{S} = \partial W/\partial \boldsymbol{E}$) based on eqn (3.7)

$$S_{11}/S_{22} = (a_1/a_2)(E_{11}/E_{22})^{n-1} \tag{3.8}$$

Thus, n can be determined directly from data wherein a ratio of S_{11} to S_{22} remains constant for a prescribed ratio of E_{11} to E_{22}, by plotting the logarithms of the stress ratios versus the strain ratios. That is, $n - 1$ is the slope of such a log–log plot and (a_1/a_2) is the associated intercept. The particular values of c, a_1 and a_2 are determined in a least-squares sense. The material parameters based on eqn (3.7) are also given in Table 3.1. The reader is referred, however, to Yin *et al.* (1986) for a complete description of the non-linear regression technique as well as a new non-parametric statistical approach for the evaluation of the significance of the parameters.

Although eqn (3.7) seemed to produce a more consistent description of the tissue for the different stretching protocols, there is still no assurance that the interpretations based on it truly reflect the behaviour of the slabs of myocardium. Finally, eqn (3.7) is only two-dimensional, is difficult to employ in certain models of the heart, and cannot describe data wherein E_{11} or E_{22} is identically equal to zero.

Other phenomenological constitutive relations have been proposed for describing passive myocardial behaviour, although only eqns (3.6a), (3.6b) and (3.7) have been based directly on biaxial data. For example, Bogen *et al.* (1984) suggested the following form of W,

$$W = (\mu/k)(\Lambda_1^k + \Lambda_2^k + \Lambda_3^k - 1) \tag{3.9}$$

where μ and k are material constants evaluated from either uniaxial papillary muscle or pressure–volume data, and Λ_i are principal stretch ratios. Bogen (1987) has recently modified eqn (3.9), which exhibits isotropic material symmetry, in an attempt to describe better the available biaxial data which clearly indicates anisotropic behaviour.

Alternatively, McCulloch (1986) suggested a form of W in terms of two invariant measures of the deformation, consistent with an assumption of transverse-isotropy with respect to the local muscle fibre directions. Due to a lack of biaxial data available to him, however, McCulloch determined material parameters on the basis of uniaxial papillary muscle data and pressure–volume results. The disadvantages of inferring multiaxial relations from uniaxial or pressure–volume data have already been discussed, and McCulloch noted the need for more complete biaxial data.

The advantages of the phenomenological approach are obvious: one does not require a detailed knowledge of the structure of the tissue and the expressions for the constitutive relations are mathematically simple. Nevertheless, this approach is not entirely satisfactory. Phenomenological parameters do not have direct physical counterparts, and load-dependent behaviour is often difficult to describe adequately (Yin *et al.*, 1986; Humphrey & Yin, 1987a).

3.5.2 *Structurally-motivated phenomenological relations*

Humphrey & Yin (1987a) recently proposed a new constitutive formulation wherein limited ultrastructural observations can be used in conjunction with limited biaxial data to prescribe a general form of a constitutive relation. Using this approach, Humphrey and Yin (1987b) then formulated a constitutive relation for passive myocardium. Briefly, based on scanning electron micrographs (Figure 3.2), it was assumed that myocardium could be idealised as being composed of both an isotropic matrix and locally parallel non-interacting muscle fibres. The matrix collectively represented the randomly oriented non-muscular constituents. Moreover, it was assumed that both the muscle fibres and the matrix follow a gross deformation, that the overall material symmetry was transversely-isotropic with respect to the local muscle fibre direction, and that a pseudostrain-energy function W could be written as $W(I_1, \alpha)$. I_1 is the first principal invariant of the Cauchy–Green deformation tensors ($I_1 = \mathrm{tr}\boldsymbol{C} = \mathrm{tr}\boldsymbol{B}$) and α is a stretch ratio associated with the muscle fibre directions ($\alpha^2 = \boldsymbol{N} \cdot \boldsymbol{C} \cdot \boldsymbol{N}$).

Since sufficient data were not available to determine rigorously the specific functional form of W directly from data, it was further assumed that the matrix and muscle fibres were mechanically decoupled, and that the total strain-energy function could be idealised into two portions: one representing the energy stored in the collection of randomly oriented structural constituents (i.e. matrix), and the other attributable to the energy stored in the muscle fibres:

$$W = W_{\mathrm{m}}(I_1) + W_{\mathrm{f}}(\alpha) \tag{3.10}$$

It should be emphasised that this decomposition of the pseudostrain-energy was strictly for convenience, and that the approach is phenomenological, albeit structurally-motivated. Thus, material parameters are not to be construed nor interpreted as the physical properties of a specific tissue constituent. For example, of the collagen fibres comprising the weave that surrounds the muscle fibres (Figure 3.2), those which are more or less randomly oriented may contribute to the 'matrix energy', W_{m}, whereas those aligned along the muscle fibres may contribute to the 'fibre energy', W_{f}.

Nevertheless, a constitutive relation for eqn (3.10) can be written as,

$$\boldsymbol{t} = -p\boldsymbol{I} + 2W_1\boldsymbol{B} + (W_\alpha/\alpha)\boldsymbol{F} \cdot \boldsymbol{N} \otimes \boldsymbol{N} \cdot \boldsymbol{F}^{\mathrm{T}} \tag{3.11}$$

where p is a Lagrange multiplier enforcing incompressibility, $W_1 = \partial W/\partial I_1$, $W_\alpha = \partial W/\partial\alpha$, and $\otimes$ denotes tensor product. For the biaxial data of Yin *et al.* (1987), $\boldsymbol{N}$ is given by eqn (3.3) and $\boldsymbol{F}$ is easily determined

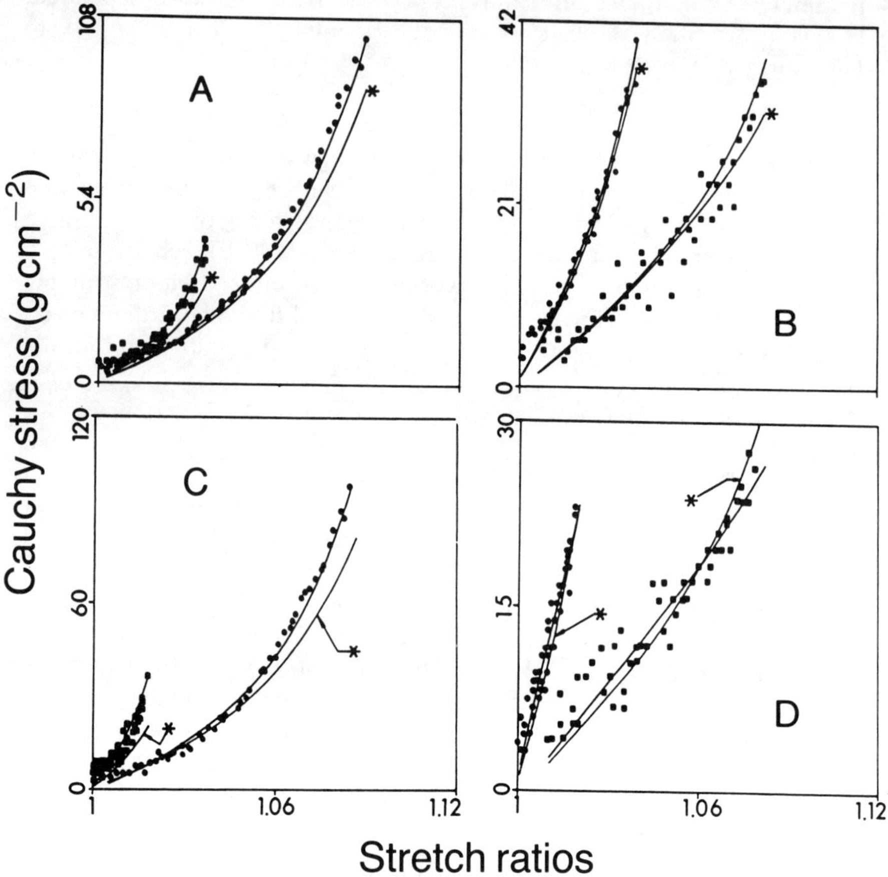

Fig. 3.5 Non-equibiaxial Cauchy stress *versus* stretch ratios for one specimen. Theoretical predictions (*-curves) based on eqn (3.13), with material parameters determined from equibiaxial data for the same specimen, are compared to the fit-to-data with eqn. (3.13) and parameters determined separately for each data set (unstarred curves). The predictions based on the equibiaxially-determined constants are seen to be reasonable. The panels contain data from constant Green strain ratio (E_{11}/E_{22}) tests wherein the ratios are: A (2.42), B (0.45), C (5.38) and D (0.25). Symbols are as in Figure (3.4). (Taken with permission from J. D. Humphrey & F. C. P. Yin, *ASME Journal of Biomechanical Engineering*, **109**, 298–304, 1987.)

from eqn (3.4). The Lagrange multiplier can be determined from the plane stress boundary condition for biaxially loaded thin slabs ($t_{33} = 0$), and eqn (3.11) yields two equations in terms of two unknowns (W_1, W_α) and experimentally measureable quantities ($\alpha, \lambda_i, t_{ij}$). Consequently, the following relations were obtained for the partial derivatives of W with respect to the invariants of the deformation,

$$W_1 = \{t_{22}/[2(\lambda_2^2 - \lambda_3^2)]\} \tag{3.12a}$$

$$W_\alpha = \{t_{11} - t_{22}[(\lambda_1^2 - \lambda_3^2)/(\lambda_2^2 - \lambda_3^2)]\} \tag{3.12b}$$

Thus, by plotting W_1 versus I_1 and W_α versus α, a specific functional form of W was determined directly from experimental data to be,

$$W = c[e^{b(I_1-3)} - 1] + A[e^{a(\alpha-1)^2} - 1] \tag{3.13}$$

where c, b, A and a are material parameters to be determined from data in a non-linear least squares sense.

Since the non-muscular constituents may reorient significantly during certain stretching protocols, the assumption of isotropy of the matrix had to be evaluated. To do this, the parameters were determined from equi-biaxial data wherein the constituents do not reorient (Humphrey & Yin, 1986), and the predictive capability of eqn (3.13) containing a unique set of parameters was evaluated. The data in Figure (3.5) are from four non-equi-biaxial (i.e. $E_{11}/E_{22} \neq 1$) stretching protocols for a specimen from which equi-biaxial parameters were determined. The starred curves (*) in the figures show the fit to data based on the equi-biaxial parameters, whereas the unstarred curves show the fit to data based on parameters determined from each protocol separately. Thus, based on these results, it appears that eqn (3.13) is a reasonable descriptor of the behaviour of potassium-arrested myocardium, at least within the ranges of strain studied by Yin *et al.* (1987). That is, passive myocardium appears to be nearly transversely-isotropic with respect to the local muscle fibre directions as assumed by others (Panda & Natarajan, 1977; Moskowitz, 1985; McCulloch, 1986).

Advantages of the structurally-motivated phenomenological constitutive formulation include the ease of construction of the general form of the relation, the incorporation of ultrastructural information without the need for extreme detail, the retention of mathematical simplicity inherent to the purely phenomenological approach, the determination of a functional form of W directly from experimental data, and a predictive capability wherein parameters obtained from fitting nearly equi-biaxial data can be used to predict reasonably well the data from non-equi-biaxial protocols on the same specimen. This predictive capability has not been achieved with other approaches thus far. Disadvantages, however, include the lack of a one-to-one correspondence between the parameters and the moduli of the individual constituents, and the assumption that the constituents are mechanically decoupled.

3.5.3 *Microstructural relations*

In the microstructural approach (Lanir, 1983), one attempts to prescribe a constitutive relation based directly on the architecture and properties of the constituents comprising a tissue. Horowitz *et al.* (1988) recently employed this approach to describe the passive behaviour of myocardium. Because of the complex myocardial ultrastructure (Figure 3.2), however, a number of simplifying assumptions were invoked to make this approach tractable. They began by assuming that the major structural components within the tissue are muscle and collagen fibres, that both types of fibres have a linear stress–strain behaviour, that the fibres bear loads only when straight, and that the fibres possess varying degrees of waviness. The overall non-linear stress–strain response of the tissue is accounted for, therefore, through the gradual recruitment of straightened muscle and collagen fibres.

Since the available ultrastructural information is largely qualitative, there are no data on the exact spatial distributions of the two fibre networks, nor the degrees of waviness of each. Thus, as a first approximation, Horowitz and colleagues assumed a Gaussian distribution of waviness for each type of fibre. Moreover, for Yin's biaxial experiments, the muscle fibres were assumed to run primarily along the 1-axis, but also to vary linearly through the thickness of the slabs from $\pm 7.5°$ to the 1-axis. The collagen fibres, on the other hand, were assumed to be distributed circumferentially and symmetrically about the axis of the muscle fibres and also at varying inclination angles to the muscle fibre axis. For simplicity, these angles of inclination were assumed to follow a Gaussian distribution.

The total myocardial strain-energy was assumed to be composed of the energies stored in the muscle and the collagen fibres such that,

$$W = \sum_k s_k \int R_k(\theta, \phi)\left[\int^{\gamma'_{11}} D_k(x) f_k(\gamma'_{11})\mathrm{d}x\right]\mathrm{d}\theta\mathrm{d}\phi \qquad (3.14)$$

where $k = 1, 2$ denotes muscle and collagen fibre, respectively; s_k are the volume fractions, $D_k(x)$ are the waviness distribution functions, $R_k(n)$ are the density distribution functions describing the spatial arrangement of the fibres, f_k is the stress in a fibre; and γ'_{11} is the extensional strain in a fibre. The volumetric fractions for the muscle and collagen were further assumed to be homogeneous throughout the volume of the tissue.

The material parameters to be determined from data are a linear elastic coefficient and a mean and variance for the waviness distribution of the muscle fibres and, for the collagen, a linear elastic coefficient, a waviness mean and variance, and a mean and variance for the inclination angle along the axis of the muscle fibres. In order to expedite convergence of the non-linear estimation, the parameters were determined from a combined data set consisting of Yin's biaxial data from all stretching protocols for a given specimen. It was found that the contribution of the muscle fibres relative to the collagen was insignificant, being at least three orders of magnitude smaller. Thus, only five of the eight parameters were determined. The particular values were 2.89 GN·m^{-2}, 0.26, 0.08, 0.78, 0.28 for the collagen modulus, waviness mean and variance, and spatial distribution mean and variance, respectively (mean and variance in radians).

Thus, Horowitz *et al.* suggest that the collagen fibres are considerably undulated in the passive state (a waviness of 1 is a straight fibre), and that the mean inclination angle is about 45°. It is noteworthy that the predicted elastic coefficient (2.89 GN·m^{-2}) is of the same order as that found for pure collagen. Nevertheless, the predictions of the model need to be verified by comparison to histological data. Given the dense collagen network, the difficulty in measuring waviness and angles of the many fibres, and the limited region that is viewed during electron microscopy, this task could prove exceedingly difficult.

The main advantage to the microstructural approach is that the estimated parameters can have a direct physical counterpart and could, in theory, be verified from experimental data. The obvious limitation, however, is the detail in which these histological data need to be known, and the many assumptions which are currently invoked due to the lack of data. At present, there are no methods to quantify the waviness and inclination angles in a particular specimen. Whether or not new techniques will be developed in the future for performing such measurements is unknown.

3.6 Conclusions and future needs

Although much has been learned in the last 25 years in the area of cardiac biomechanics, there remains much that is unknown. Thus, there are many areas of needed research – a few of which are noted here.

According to Truesdell & Noll (1965), '. . . physical theory must be based on experience, but experiment comes after, rather than before, theory. Without theoretical concepts one would know neither what experiment to perform, nor be able to interpret them.' Unfortunately, the majority of the experimental data on the mechanical properties of myocardium have been obtained without the direction of clear theoretical guidelines relevant to the performance and interpretation of the experiments. For example, it is well known that experimentally measured stresses in biaxial experiments on slabs of myocardium (e.g. Demer & Yin, 1983) are average values through the thickness of the specimens by definition. Humphrey & Yin (1989) were the first to consider, however, under which experimental conditions the measured values of stress may represent well the actual stress distribution in a specimen. For example, they showed that in-plane shearing strains and through-the-thickness splays of muscle fibres can grossly affect the stress distribution in a biaxially tested specimen, and that these and other experimental factors must be carefully considered and monitored. There is a need for continued development of theoretical guidelines for myocardial experimentation which will lead to better data, better interpretations, better constitutive relations, and perhaps a more efficient use of experimental animals.

Regional variations in passive myocardial properties are suspected on the basis of theoretical arguments (Janz & Grimm, 1973; Mirsky & Rankin, 1979; Moriarty, 1980; Yin, 1981) and limited data (Demer & Yin, 1983). Moreover, it has been shown theoretically that inhomogenous myocardial properties can significantly influence the regional distribution of stresses in the heart (Moriarty, 1980). That is, relative to predictions based on homogeneous properties, the stresses in the heart walls can become either more uniform or more heterogeneous, depending on the assumed variation in properties. There is a need, however, to determine explicitly whether regional differences in cardiac properties actually exist, and if they do, to quantify the differences (Mirsky, 1979; Yin, 1985). Moreover, there is a need to quantify the properties of the epicardium which differ from those of myocardium (Humphrey *et al.*, 1987). Without these data, one can only speculate on the distribution of stresses in the heart as well as their effects on regional and overall cardiac physiology and pathophysiology.

Myocardium consists of both intra- and extracellular fluids which, in addition to the solid constituents, contribute to the overall mechanical behaviour of the tissue. In fact, it has been suggested that such fluids contribute to the viscoelastic character of the tissue (Tsaturyan *et al.*, 1984), and affect the so-called intramyocardial pressure and thereby coronary flow (Arts & Reneman, 1985; Jan, 1985; Huyghe, 1986). It is becoming increasingly evident that the fluid constituency of myocardium must be better understood and included in future considerations of certain aspects of cardiac mechanics.

Before a complete analysis of stresses in the passive intact heart can be performed, both a complete description of the mechanical properties of tissues involved and an understanding of the residual stresses in the heart (Fung, 1984) must be available. Thus, it should be emphasised that the issue of residual stresses is an important one, deserving attention.

Although this chapter is focused on the mechanical properties of passive cardiac tissue, it should be emphasised that there is a grave need for research on active multiaxial properties as well. In fact, active properties of cardiac tissue are less well understood than passive properties, and yet they are essential to an understanding of the ejecting heart. Most of the data thus far is based on papillary muscle studies (see Pinto, 1987) or pressure–volume data. These data are necessary, but not sufficient, for a complete understanding of active cardiac behaviour, and there is a need for both theoretical guidelines for, and multiaxial experiments on, the active properties of cardiac tissue.

In conclusion, one should be encouraged by the gains that have been made in biomechanical studies of the heart, but one must remain committed to the challenging research that lies ahead.

References

Abe, H. & Nakamura, T. (1982). Finite deformation model for the mechanical behavior of left ventricular wall muscles. *Mathematical Modelling*, **3**, 143–52.

Arts, T. & Reneman, R. S. (1985). Interaction between intramyocardial pressure (IMP) and myocardial circulation. *ASME Journal of Biomechanical Engineering*, **107**, 51–6.

Bergel, D. & Hunter, P. J. (1979) The mechanics of the heart. In: *Quantitative Cardiovascular Studies, Clinical and Research Applications of Engineering Principles*, eds. N. I. C. Hwang, D. R. Gross & D. J. Patel, pp. 151–213. University Park Press, Baltimore, Maryland.

Bogen, D. K. (1987). Strain-energy descriptions of biological swelling. I Single fluid compartment models. *ASME Journal of Biomechanical Engineering*, **109**, 252–6.

Bogen, D. K. & McMahon, T. A. (1984). Do cardiac aneurysms blow out? *Biophysical Journal*, **27**, 301–16.

Caulfield, J. B. & Borg, T. K. (1979). The collagen network of the heart. *Laboratory Investigation*, **40**, 364–72.

Chadwick, R. S. (1982). Mechanics of the left ventricle. *Biophysical Journal*, **39**, 279–88.

Demer, L. L. & Yin, F. C. P. (1983). Passive biaxial mechanical properties of isolated canine myocardium. *Journal of Physiology (London)*, **339**, 615–30.

Demiray, H. (1976) Stresses in ventricular wall. *ASME Journal of Applied Mechanics*, **43**, 194–7.

Demiray, H. (1981) Large deformation analysis of some biological soft tissues. *ASME Journal of Biomechanical Engineering* **103,** 73–8.

Feit, T. S. (1979). Diastolic pressure-volume relations and distribution of pressure and fiber extension across the wall of a model left ventricle. *Biophysical Journal*, **28**, 143–66.

Fung, Y. C. (1973). Biorheology of soft tissues. *Biorheology*, **10**, 139–55.

Fung, Y. C. (1981). *Biomechanics: Mechanical Properties of Living Tissues*, pp. 242–5. Springer-Verlag, New York.

Fung, Y. C. (1984). *Biodynamics: Circulation*, pp. 34, 70. Springer-Verlag, New York.

Glantz, S. A. & Parmley, W. W. (1978). Factors which affect the diastolic pressure–volume curve. *Circulation Research*, **42**, 171–80.

Grood, E. S., Phillips, C. A. & Mates, R. E. (1979). Contractile filament stress in the left ventricle and its relationship to wall stress. *ASME Journal of Biomechanical Engineering*, **101**, 225–31.

Halperin, H. R., Chew, P. H., Weisfeldt, M. L., Sagawa, K., Humphrey, J. D. & Yin, F. C. P. (1987). Transverse stiffness: A method for estimation of myocardial wall stress. *Circulation Research*, **61**, 695–703.

Horowitz, A., Lanir, Y., Yin, F. C. P., Perl, M., Sheinman, I. & Strumpf, R. K. (1988). Three-dimensional structural constitutive law for the passive myocardium. *ASME Journal of Biomechanical Engineering*, **110**, 200–7.

Huisman, R. M., Elzinga, G., Westerhof, N. & Sipkema, P. (1980). Measurement of left ventricular wall stress. *Cardiovascular Research*, **14**, 142–53.

Humphrey, J. D. & Yin F. C. P. (1986). Fiber-induced material behavior in composites. *Mechanics Research Communications*, **13**, 277–83.

Humphrey, J. D. & Yin, F. C. P. (1987a). A new constitutive formulation for characterizing the mechanical behavior of soft tissues, *Biophysical Journal*, **52**, 563–70.

Humphrey, J. D. & Yin, F. C. P. (1987b). On constitutive relations and finite deformations of passive cardiac tissue. I A pseudostrain-energy function. *ASME Journal of Biomechanical Engineering*, **109**, 298–304.

Humphrey, J. D. & Yin, F. C. P. (1989). Biomechanical experiments on excised myocardium: Theoretical considerations. *Journal of Biomechanics*. (In press)

Humphrey, J. D., Chew, P. H. & Yin, F. C. P. (1987). Epicardium: More like pericardium than myocardium. *Circulation*, **76**, 536.

Humphrey J. D., Vawter, D. L. & Vito, R. P. (1987). Quantification of strains in biaxially tested soft tissues. *Journal of Biomechanics*, **20**, 59–65.

Huyghe, J. M. R. J. (1986). Non-linear finite element models of the beating left ventricle and the intramyocardial coronary circulation. PhD Thesis, Technische Universiteit Eindhoven, Netherlands.

Jan, K-M. (1985). Distribution of myocardial stress and its influence on coronary blood flow. *Journal of Biomechanics*, **18**, 815–20.

Janicki, J. S. & Weber, K. T. (1980). Factors influencing the diastolic pressure–volume relation of the cardiac ventricles. *Federation Proceedings*, **39**, 133–40.

Janz, R. F. & Grimm, A. F. (1973). Deformation of the diastolic left ventricle. I Non-linear elastic effects. *Biophysical Journal*, **13**, 689–704.

Kitabatake, A. & Suga, H. (1978). Diastolic stress–strain relation of non-excised blood-perfused canine papillary muscle. *American Journal of Physiology*, **234**, H416–20.

Lanir, Y. (1983). Constitutive equations for fibrous connective tissues. *Journal of Biomechanics*, **16**, 1–12.

Malvern, L. E. (1969). *Introduction to the Mechanics of a Continuous Medium*. Prentice-Hall, New Jersey.

McCulloch, A. D. (1986). Deformation and stress in the passive heart. PhD Thesis, University of Auckland, New Zealand.

Mirsky, I. (1973). Ventricular and arterial wall stresses based on large deformation analyses. *Biophysical Journal*, **13**, 1141–57.

Mirsky, I. (1976). Assessment of passive elastic stiffness of cardiac muscle: mathematical concepts, physiologic and clinic considerations, directions of future research. *Progress in Cardiovascular Diseases*, **18**, 277–308.

Mirsky, I. (1979). Elastic properties of the myocardium: A quantitative approach with physiological and clinical applications. In: *Handbook of Physiology*, ed. R. M. Berne, American Physiological Society, Bethesda, MD.

Mirsky, I. & Rankin, J. S. (1979). The effects of geometry, elasticity, and normal pressure on the diastolic pressure – volume and stiffness – stress relations: How important is the pericardium. *Circulation Research*, **44,** 602–11.

Moriarty, T. (1980). Law of Laplace, its limitations as a relation for diastolic pressure, volume or wall stress of the left ventricle. *Circulation Research*, **46**, 321–31.

Moskowitz, S. E. (1985). Constitutive stress–strain relations for the myocardium in diastole. *Journal of Biomechanics*, **18**, 177–87.

Panda, S. C. & Natarajan, R. (1977). Finite-element method of stress analysis in the human left ventricular layered wall structure. *Medical and Biological Engineering and Computing*, **15**, 67–71.

Pao, Y. C., Nagendra, G. K., Padiyan, R. & Ritman, E. L. (1980). Derivation of myocardial fiber stiffness equation based on the theory of laminated composites. *ASME Journal of Biomechanical Engineering*, **102**, 252–7.

Pinto, J. G. (1987). A constitutive description of contracting papillary muscle and its implications to the dynamics of the intact heart. *ASME Journal of Biomechanical Engineering*, **109**, 181–91.

Pinto, J. G. & Fung Y. C. (1973). Mechanical properties of the heart muscle in the passive state. *Journal of Biomechanics*, **6**, 597–616.

Robinson, T. F. (1983). The physiological relationship between connective tissue and contractile filaments in heart muscle. *The Einstein Quarterly*, **1**, 121–7.

Sonnenblick, E. H. (1964). Series elastic and contractile elements in heart muscle: Change in muscle length. *American Journal of Physiology*, **207**, 1330–8.

Streeter, D. D. (1979). Gross morphology and fiber geometry of the heart. In: *Handbook of Physiology, Section 2: The Cardiovascular System*, Vol. 1, *The Heart*, eds, Berne, R. M. *et al.*, American Physiological Society, Bethesda, MD.

Tozeren, A. (1983). Static analysis of the left ventricle. *ASME Journal of Biomechanical Engineering*, **105**, 39–46.

Truesdell, C. & Noll, W. (1965). The non-linear field theories of mechanics. In: *Handbuch der Physik*, vol. III/3, ed. S. Flugge. Springer-Verlag, Berlin.

Tsaturyan, A. K., Izacov, V. J., Zhelamsky, S. V. & Bykov, B. L. (1984). Extracellular fluid filtration as the reason for the viscoelastic behavior of passive myocardium. *Journal of Biomechanics*, **17**, 749–55.

Waldman, L. K., Fung, Y. C. & Covell, J. W. (1985). Transmural myocardial deformation in canine left ventricle: Normal *in vivo* three-dimensional finite strains. *Circulation Research*, **57**, 152–63.

Yin, F. C. P. (1981). Ventricular wall stress. *Circulation Research*, **49**, 829–42.

Yin, F. C. P. (1985). Application of the finite-element method to ventricular mechanics. *CRC Critical Reviews in Biomedical Engineerng*, **12**, 311–42.

Yin, F. C. P., Chew, P. H. & Zeger, S. L. (1986). An approach to quantification of biaxial tissue stress–strain data. *Journal of Biomechanics*, **19**, 27–37.

Yin, F. C. P., Strumpf, R. K., Chew, P. H. & Zeger, S. L. (1987). Quantification of the mechanical properties of noncontracting canine myocardium under simultaneous biaxial loading. *Journal of Biomechanics*, **20**, 577–99.

Yin, F. C. P., Tompkins, W. R., Peterson, K. L. & Intaglietta, M. (1972). A video dimension analyser. *IEEE Transactions in Biomedical Engineering*, **19**, 376–81.

Yin, F. C. P., Spurgeon, H. A., Weisfeldt, M. L. & Lakatta, E. G. (1980). Mechanical properties of myocardium from the hypertrophic rat hearts. *Circulation Research*, **46**, 292–300.

4

Fibre reinforcing in articular cartilage

D. W. L. Hukins and R. M. Aspden

4.1 **Introduction**

Articular cartilage forms a resilient coating at the ends of bones where they form synovial (diarthrodial) joints. Its importance for the normal function of the joint is illustrated by the problems which occur when it is eroded by osteoarthritis (osteoarthrosis). Osteoarthritis ultimately leads to exposure of the underlying subchondral bone which is accompanied by the formation of bony outgrowths (osteophytes) and bone cysts, the appearance of mineral in the joint fluid, and inflammation of the lining of the joint; these degenerative changes are associated with loss of joint function and pain (Ball, 1987). The normal cartilage also has the further function of acting together with the joint fluid (synovial fluid) to form a low friction articulating surface (Swanson, 1979).

The purpose of this chapter is to consider how collagen fibrils reinforce articular cartilage to maintain mechanical stability during normal function of the joint. Cartilage cells (chondrocytes) are surrounded by an abundant extracellular matrix which is responsible for the mechanical properties of the tissue (Meachim & Stockwell, 1979). About 70% of the extracellular matrix is water; the remainder is mostly collagen (20–30%) and proteoglycans (Muir, 1979). The water and proteoglycan together form a highly hydrated 'ground substance' which has a low shear strength and so is reinforced by collagen fibrils (Hukins & Aspden, 1985). Collagen fibrils have low torsional and flexural stiffness and they readily buckle under compression – as demonstrated by the properties of tendon (Hukins, 1982). But they are stiff and strong in tension so that the restoring stress in strained fibrils is able to balance an applied stress (tendons operate at a stress of about 10 $MN{\cdot}m^{-2}$; Hooley & Cohen, 1979). Therefore, collagen fibrils can provide reinforcement if they are oriented so that they are placed in tension by whatever stresses are applied to a tissue (Hukins, 1984).

4.2 **Ground substance**

Ground substance exerts an internal pressure which balances compressive loads applied to the cartilage coating of the joint surfaces (Maroudas, 1979, 1980; Maroudas & Urban, 1980; Myers *et al.*, 1984). A proteoglycan consists of a protein core with polyanions (glycosaminoglycans) attached to it; the proteoglycans aggregate into a polyanionic network within the cartilage (Bayliss, 1984). This polyanionic network attracts water by the 'Donnan effect' which is described in the remainder of this paragraph. Negative charges associated with the polyanions are balanced by mobile cations (e.g. Ca^{2+}) which are free to diffuse into the tissue. However, the negatively charged groups of the polyanions cannot diffuse out. In contrast, mobile anions (e.g. PO_4^{3-}) may diffuse into the tissue, from the surroundings, and their charges have to be balanced by further mobile cations. As a result, the ground substance of cartilage has a higher concentration of mobile ions than its surroundings and so has a higher osmotic pressure. Because of this high osmotic pressure, water passes into the matrix from its surroundings by osmosis. The result is that the matrix exerts an internal swelling pressure which can support an externally applied load.

However, ground substance on its own is weak in shear and would fail in compression if it were not reinforced by collagen. The shear strength of ground substance has been estimated to be around 0.1 $MN{\cdot}m^{-2}$ (Hooley & Cohen, 1979) as compared with a typical value for the compressive stress applied to articular cartilage of the order of 1 $MN{\cdot}m^{-2}$ (Armstrong *et al.*, 1979). Furthermore, the shear modulus of the ground substance appears to be only about 10 $N{\cdot}m^{-2}$, which is very considerably less than the value of about 10 $MN{\cdot}m^{-2}$ for intact articular cartilage (Myers *et al.*, 1984).

4.3 **Degree of reinforcement by collagen**

Collagen fibrils in cartilage have an unknown length with a typical diameter of about 30 nm (Meachim & Stockwell, 1979); their structure and mechanical properties appear well suited to providing tensile reinforcement to the ground substance in much the same way as glass or carbon fibres reinforce synthetic composite materials (Hukins, 1982). The behaviour of synthetic fibre-reinforced composite materials is well documented (see, for example, Krenchel, 1964; Kelly, 1973; Agarwal & Broutman, 1980; Caddell, 1980) and the principles involved can be applied to articular cartilage (Hukins *et al.*, 1984) as well as other connective tissues (Jeronimidis & Vincent, 1984; Hukins & Aspden, 1985).

Suppose that the collagen fibrils and ground substance form a composite material such that an applied stress leads to a strained material with equal strains in both components (see Caddell, 1980). The stiffness of the composite extracellular matrix, E_t, is then related to the stiffness of the collagen, E_c, and of the ground substance, E_g, by,

$$E_t = \beta E_c + (1 - \beta)E_g \tag{4.1}$$

where β is the volume fraction of collagen. Since collagen has a density of 1.34×10^3 kg·m^{-3} (Woodhead-Galloway *et al.*, 1978) and it contributes at least 20% of the mass of cartilage (see above), β has a minimum value of about 0.15. Hence, given the low shear modulus of ground substance (see above), eqn (4.1) approximates to,

$$E_t \simeq \beta E_c \qquad E_g \ll E_c \tag{4.2}$$

which shows how the collagen content influences the stiffness of cartilage. This approximate result holds for most synthetic fibre-reinforced composite materials, which led Krenchel (1964) to use β as a measure of 'degree of reinforcement'.

4.4 **Collagen fibril orientations**

Collagen, as noted, can only provide reinforcement if its fibrils are oriented so that they are axially strained by a stress which is applied to the tissue. Thus the collagen content, as measured by β, must be inadequate to describe how cartilage is reinforced and has to be supplemented by structural information, i.e. the orientations of its fibrils.

The orientations of the collagen fibrils may be defined by the angles ϕ and ψ of the spherical polar coordinate system of Figure 4.1. Let $n(\phi, \psi)$ define the distribution of the number density of fibres passing through a spherical surface of radius r centred at the origin, O, of the coordinate system. An element of this spherical surface, defined by elemental angles $\delta\phi$ and $\delta\psi$ has an area of,

$$r\,\delta\phi.\ r \sin\phi\,\delta\psi$$

so that the number of fibres passing through it is,

$$n(\phi, \psi)\, r^2 \sin\phi\,\delta\phi\,\delta\psi$$

Let N be the total number of fibres passing through the surface of the sphere, so that,

$$N = \int_{\psi=0}^{2\pi} \int_{\phi=0}^{\pi/2} n(\phi, \psi)\, r^2 \sin\phi\, d\phi\, d\psi \tag{4.3}$$

Integration with respect to ϕ is over an angular range of only $\pi/2$ radians because the range $\pi/2$ to π describes the other ends of the same fibres (see Figure 4.1). Eqn (4.3) can be rearranged to give,

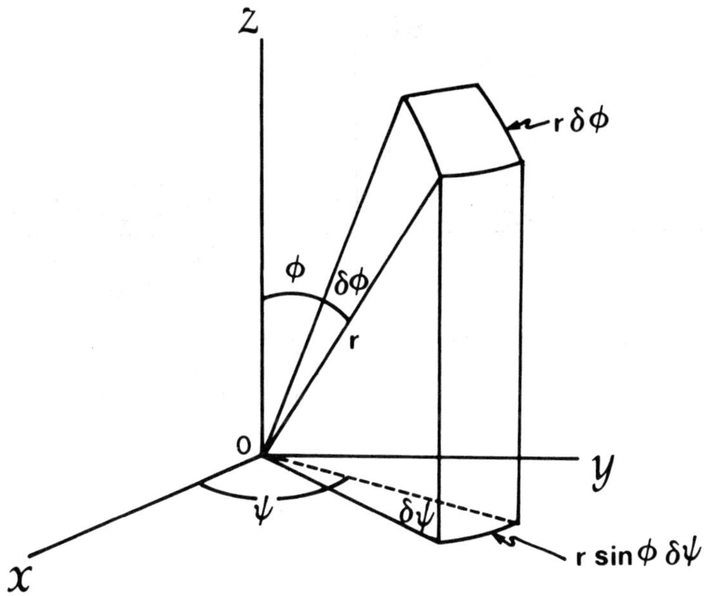

Fig. 4.1 Spherical polar coordinate system used to define the orientations of reinforcing fibres.

$$\int_0^{2\pi}\int_0^{\pi/2} \frac{n(\phi,\psi)}{N}\, r^2 \sin\phi \; d\phi \; d\psi = 1 \tag{4.4}$$

$$\left.\begin{aligned} \Rightarrow \int_0^{2\pi}\int_0^{\pi/2} g(\phi,\psi) \sin\phi \; d\phi \; d\psi &= 1 \\ g(\phi,\psi) &= r^2\, n(\phi,\psi)/N \end{aligned}\right\} \tag{4.5}$$

where $g(\phi, \psi) \sin \phi \, \delta\phi \, \delta\psi$ has the status of a probability of finding a fibre oriented between (ϕ, ψ) and $(\phi + \delta\phi, \psi + \delta\psi)$. The function $g(\phi, \psi)$ is sometimes referred to as an 'orientation distribution function' (Deas, 1952; Aspden & Hukins, 1979; Hukins, 1984). Note that it is dimensionless, as defined in eqn (4.5), since $n(\phi, \psi)$ has the units of reciprocal area.

A maximum in $g(\phi, \psi)$ then represents a direction of preferred orientation for the fibres. In the present application the z-axis is taken to be the axis of preferred orientation. It is then reasonable to assume, in the case of articular cartilage, that all values of ψ are equally probable so that $g(\phi, \psi)$ has cylindrical symmetry about the z-axis. A cylindrically symmetric orientation distribution function is not then a function of ψ and may be represented by $g(\phi)$, with the result that eqn (4.5) can be integrated with respect to ψ to give,

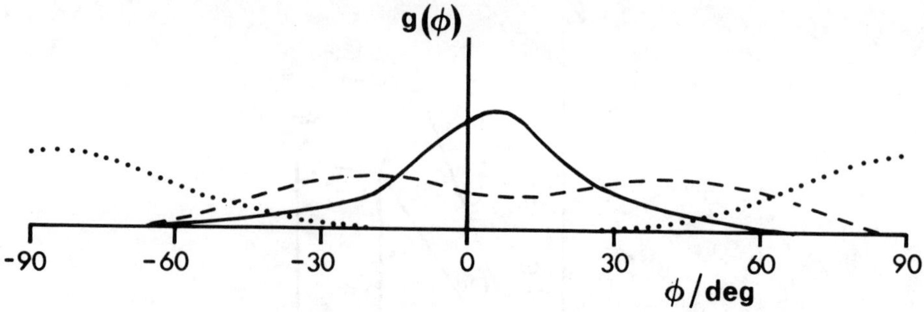

Fig. 4.2 Orientation distribution function, $g(\phi)$, for collagen fibrils in the surface (—) intermediate (--), and deep (. . .) zones of articular cartilage coating a human patella; the zero value of ϕ is defined by the normal to the articular surface.

$$2\pi \int_0^{\pi/2} g(\phi) \sin \phi \, d\phi = 1 \tag{4.6}$$

Eqn (4.6) then represents a normalisation condition for a cylindrically symmetric orientation distribution function.

In principle there are several techniques for determining $g(\phi)$ for the collagen fibrils in tissues (Hukins, 1984). However, X-ray diffraction provides the simplest method for measuring $g(\phi)$ from a fully hydrated tissue sample without making any assumptions about its form; a single diffraction pattern allows $g(\phi)$ to be determined for the tissuc site from which it was recorded (Aspden & Hukins, 1979; Hukins, 1984). We have used this technique to determine $g(\phi)$ as a function of depth from the surface of articular cartilage (Aspden & Hukins, 1981a,b; Yarker *et al.*, 1984). In addition, by making reasonable assumptions about its form, the width of $g(\phi)$ can be calculated from measurements of birefringence of thin sections of cartilage soaked in glycerol (Yarker *et al.*, 1983).

Figure 4.2 shows $g(\phi)$ at three different depths from the surface of human patellar cartilage (Aspden & Hukins, 1981a). Each function plotted in Figure 4.2 was obtained by analysis of a single X-ray diffraction pattern, recorded with the X-ray beam perpendicular to the direction of preferred orientation of the collagen fibrils; the analysis assumed cylindrical symmetry of the orientation distribution function about this preferred direction (Aspden & Hukins, 1979). The functions plotted in Figure 4.2 correspond to a 'surface zone' where there is a peak in $g(\phi)$ (preferred orientation) almost parallel to the articular surface, an 'intermediate zone' in which $g(\phi)$ has a very broad bimodal distribution, and a 'deep zone' in which the preferred orientation is almost perpendicular to its direction at the surface. Very similar results have been obtained from the articular cartilage coating

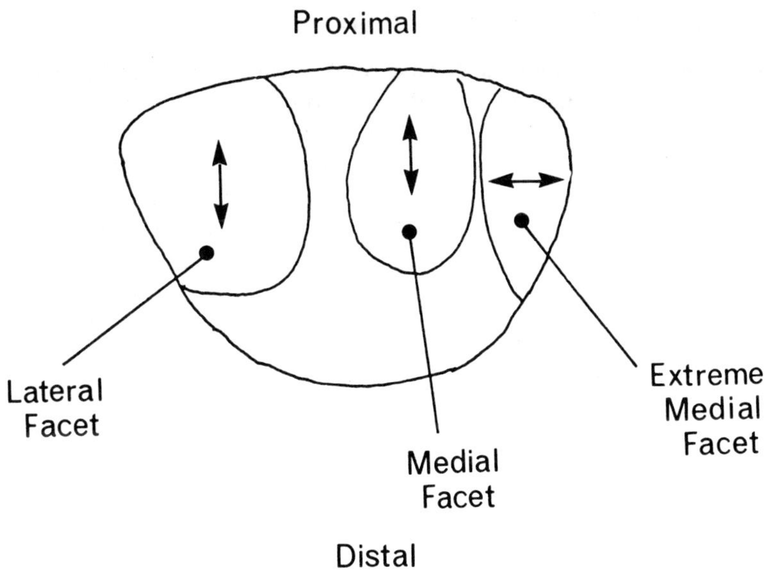

Fig.4.3 Diagram of the surface of a human left patella, defining the facets and indicating and directions of the preferred orientations for their collagen fibrils, in the plane of the articular surface.

the tibial plateaux of pigs and dogs (Yarker *et al.*, 1983, 1984) and from pig femoral head cartilage, where the deep zone appears to be calcified (Aspden & Hukins, 1981b).

There is also a preferred orientation for the collagen fibrils of the surface zone in the plane of the articular surface. The simplest way of deomonstrating this anisotropy is by introducing artificial splits in the cartilage by pricking it with a needle dipped in Indian ink (Meachim *et al.*, 1974). Because the mechanical properties of this surface are anisotropic, the resulting hole is not circular but appears as a crack. Cracks are more readily propagated parallel with than perpendicular to the collagen fibrils (Hukins, 1984). Thus the ink-filled 'split lines' can be identified with their directions of preferred orientation.

On the lateral and medial facets of the patella (defined in Figure 4.3) the lines tend to lie in a proximo-distal direction, i.e. this is the preferred direction for the collagen fibrils in this region of the patellar surface. However, on the extreme medial facet the lines, and hence the preferred fibril orientation, tend to lie in a medio-lateral direction. This same information has also been obtained, more quantitatively, from X-ray diffraction patterns of slices of human patellar cartilage (cut parallel to the plane of the surface) recorded with the beam normal to the plane of the surface (Aspden & Hukins, 1981a).

4.5 Efficiency of reinforcement by collagen

Since collagen fibrils provide tensile reinforcement, the mechanical properties of a tissue depend on the directions in which these fibrils are oriented. In particular, tensile reinforcement will be most effective if the fibrils are oriented in the directions of tensile stress. Consider a fibre-reinforced composite material in which all the fibres are parallel. If a tensile stress is applied at an angle ϕ to the axes of these fibrils, their effective stiffness is reduced by a factor of $\cos^4\phi$ assuming that the strains are sufficiently low for fibre reorientation to be negligible (Cox, 1952; Krenchel, 1964; Aspden, 1986). Our X-ray diffraction experiments indicate little collagen fibril reorientation during compression of articular cartilage, so that this low strain approximation appears reasonable.

Krenchel (1964) used this result to define an 'efficiency of reinforcement', η, for a fibre-reinforced composite material. He supposed that the reinforcing fibres could be considered to belong to M sets of parallel fibres; the mth set was oriented at an angle, ϕ_m, with respect to some direction within the material. The efficiency of reinforcement in this direction was then defined to be,

$$\eta = \sum_{m=1}^{M} f_m \cos^4 \phi_m \tag{4.7}$$

where f_m represents the fraction of the total number of fibres which belong to the mth set.

This result may be generalised to the case of a continuous distribution of fibre orientations represented by $g(\phi, \psi)$ as defined in eqn (4.5). By analogy with the derivation of eqn (4.3), the fraction of the total number of fibres which passes through the elemental area of surface, defined by the elemental angles $\delta\phi$ and $\delta\psi$ of Figure 4.1, is given by,

$$f_m = \frac{n(\phi, \psi)}{N} r^2 \sin \phi \, \delta\phi \, \delta\psi \tag{4.8}$$

The contribution of these fibres to the efficiency of reinforcement in the z-axis direction is then,

$$\begin{aligned} \delta\eta &= \frac{n(\phi, \psi)}{N} r^2 \sin \phi \cos^4 \phi \, \delta\phi \, \delta\psi \\ &= g(\phi, \psi) \sin \phi \cos^4 \phi \, \delta\phi \, \delta\psi \end{aligned} \tag{4.9}$$

given the definition of $g(\phi, \psi)$ in eqn (4.5). The total efficiency of reinforcement of the fibre distribution represented by $g(\phi, \psi)$ in this direction is given by integrating eqn (4.9) over the total angular range of ϕ and ψ, i.e.

$$\eta = \int_{\psi=0}^{2\pi} \int_{\phi=0}^{\pi/2} g(\phi, \psi) \sin \phi \cos^4 \phi \, d\phi \, d\psi \tag{4.10}$$

where the reason for the limits of integration has been explained in the derivation of eqn (4.3).

In order to use eqn (4.10) here, the z-axis is taken to be the direction of preferred orientation and cylindrical symmetry is assumed, as in the derivation of eqn (4.6). Then the orientation distribution function is no longer a function of ψ and eqn (4.10) can be integrated to yield,

$$\eta = 2\pi \int_{0}^{\pi/2} g(\phi) \sin \phi \cos^4 \phi \, d\phi \tag{4.11}$$

The application of eqn (4.11) may be illustrated by considering an assembly of fibrils with no preferred orientation; the result, which will be used below, has been obtained less rigorously by Krenchel (1964). For such an assembly the orientation distribution function, g, is independent of ϕ so eqn (4.6) may be written as,

$$2\pi \, g \int_{0}^{\pi/2} \sin \phi \, d\phi = 1$$

$$\Rightarrow g = 1/2\pi \tag{4.12}$$

The efficiency of reinforcement for this assembly is given by substituting the value for g from eqn (4.12) into eqn (4.11), with the result that,

$$\eta_o = \int_{0}^{\pi/2} \sin \phi \cos^4 \phi \, d\phi \tag{4.13}$$

Eqn (4.13) can be easily integrated to give,

$$\eta_o = 1/5 \tag{4.14}$$

Results for $g(\phi)$ from the three zones of articular cartilage, shown in Figure 4.2, may be substituted into eqn (4.11), to obtain the efficiency of reinforcement in the direction of preferred orientation. In both the surface and deep zones η is typically about 0.6; the intermediate zone is sufficiently disoriented for η to be assigned a value of 0.2 from eqn (4.14).

4.6 **Critical length**

There has to be some mechanism for stress to be applied to the collagen fibrils in articular cartilage which are appropriately oriented to provide tensile reinforcement. It appears that stress can be transferred to the fibrils by the ground substance in a way similar to that in which weak plastic fillers transfer stress to the fibres (glass, carbon, etc.) of synthetic composite materials (Hukins *et al.*, 1984). Stress transfer in these synthetic materials has been extensively investigated, both experimentally and theoretically (see, for example, Kelly, 1973; Agarwal & Broutman, 1980; Caddell, 1980).

Consider a fibre, surrounded by a weak gel, which is subjected to applied tension along its axis. The applied force may simply pull the fibre out of the gel. However, viscous forces will oppose the motion of the fibre. These viscous forces will increase as the length of the fibre increases, since they depend on its surface area. Thus if the fibres are sufficiently long and the gel is sufficiently viscous, the viscous forces may balance the applied tension and the system will be in equilibrium. If the system can reach equilibrium in this way, the fibres are able to provide effective reinforcement. Note that resistance to shear of the interface between a fibre and the gel may be provided by forces other than the purely viscous; in the case of connective tissues, like articular cartilage, they are likely to include the forces associated with specific chemical interactions between the proteoglycans of the ground substance and the surfaces of the collagen fibrils (Hukins & Aspden, 1985).

Fibres are considered to be sufficiently long enough to provide tensile reinforcement if their lengths are greater than the 'critical length', ℓ_c. For a fibre of radius, ϱ, surrounded by a gel which is capable of transferring a stress, τ, to unit area of the fibre, the critical length is given by,

$$\ell_c = \varrho\sigma_f/\tau \tag{4.15}$$

where σ_f is the fracture stress of the fibre (Kelly, 1973; Agarwal & Broutman, 1980; Caddell, 1980).

It appears that the collagen fibrils in articular cartilage have lengths which are as great as estimates of their critical value, and so are able to provide tensile reinforcement by this mechanism. The radius, ϱ, for collagen fibrils in articular cartilage is around 20 nm (see above). Tendon fractures at a stress of around 100 MN·m^{-2} (Elliott, 1965) and its mass consists of 30% collagen (Harkness, 1968), i.e. it has a value of 0.22 for β (see above); these observations suggest an estimate for the fracture stress of collagen fibrils, σ_f, of,

$$\sigma_f = 100/0.22 = 450\ \mathrm{MN{\cdot}m^{-2}} \tag{4.16}$$

Eqn (4.16) rests on the same assumptions as Eqn (4.1) (Caddell, 1980). A model for the creep behaviour of tendon indicates that has a value of 0.1 MN·m^{-2} (Hooley & Cohen, 1979). It is assumed here that articular cartilage has a similar value since it has a similar composition. Substituting these values for ϱ, σ_f and τ into eqn (4.16) yields a value of around 90 μm, for ℓ_c. Although the lengths of collagen fibrils in connective tissues have not been measured convincingly (Hukins, 1982), there is no doubt from published electron micrographs (Muir *et al.*, 1970) that they can be as long as this estimated critical length.

4.7 **Stability of articular cartilage**

Articular cartilage can only be stable if the internal swelling pressure exerted by its ground substance (see above) is balanced by a tension tangential to its surface. Since collagen fibrils are oriented roughly parallel to the surface, in the surface zone (Figure 4.2), they will be stretched by the internal pressure and so can provide reinforcement (Aspden & Hukins, 1981a); the restoring stress in the stretched fibrils balances the internal pressure (Hukins *et al.*, 1984). This hypothesis is consistent with the observation that the collagen fibrils in healthy articular cartilage do not display their normal crimped appearance (Gathercole & Keller, 1975) but that the crimps do appear when fragments are torn off in osteoarthritis (Ball, 1987).

A simple 'order of magnitude' calculation indicates that the structure of the surface zone enables it to withstand the internal swelling pressure (Hukins *et al.*, 1984). This calculation is based on the structure of femoral head cartilage whose surface zone resembles a thin-walled spherical pressure vessel enclosing the bulk tissue. The tensile stress, S, in a wall of thickness, t, required to balance a pressure, p, is then given by,

$$S = RP/2t \tag{4.17}$$

where R is the radius of the sphere (see, for example, Higdon *et al.*, 1976). Measurements from the femoral heads of mini-pigs provide approximate values of 0.3 mm for t (Aspden & Hukins, 1981b) and 2 cm for R (Hukins *et al.*, 1984). Taking a value for p of 0.3 $\mathrm{MN{\cdot}m^{-2}}$ (given, for human tissue, by Maroudas, 1976) then eqn (4.17) yields a value for S of 10 $\mathrm{MN{\cdot}m^{-2}}$.

Collagen fibrils in the surface zone can readily withstand a stress of this magnitude. Tendons fracture at stresses of about 100 $\mathrm{MN{\cdot}m^{-2}}$ and normally operate at no more than a quarter of this value (Elliott, 1965). Articular cartilage has a value of about 0.15 for β, while the value for tendon is about 0.22 (see above). Furthermore, η has a value of 0.6 in the surface zone of articular cartilage (see above) but is effectively 1 in stretched tendon where the collagen fibrils are oriented in the direction of applied tensile stress (Gathercole & Keller, 1975). Hence, the collagen fibrils of the surface zone would be expected to operate at a stress of,

$$25 \times 0.15 \times 0.6/0.22 = 10\ \mathrm{MN{\cdot}m^{-2}}$$

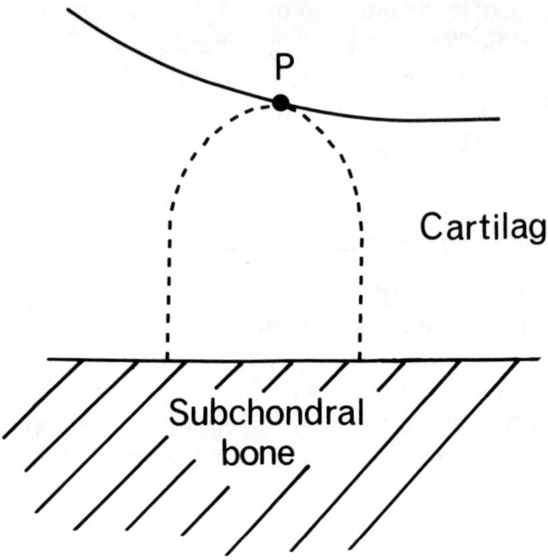

Fig.4.4 A point, P, on a concave surface of articular cartilage; there are collagen fibrils which lie, in the plane of this figure, along the curved locus indicated by the dotted line.

This estimate is of the same order as the estimated stress required to balance the swelling pressure. Furthermore, by analogy with tendon, the surface zone is expected to be able to withstand about four times this stress before it fractures.

This explanation for the mechanical stability can still apply when the cartilage surface is not spherical and, indeed, even to concave surfaces. The example of femoral head cartilage was chosen because it is easy to analyse semi-quantitatively.

Now consider a point P (Figure 4.4) on a concave surface of articular cartilage. In the surface zone, the preferred orientation of the collagen fibrils will still be parallel to the surface; fibrils in the deep zone have a preferred orientation which is perpendicular to this direction (Figure 4.2). Since the intermediate zone is a nearly randomly oriented assembly of fibrils (Figure 4.2), there will always be fibrils oriented to form a smoothly varying transition from the two extremes of the surface and deep zones. Thus there will be fibrils lying along the curved locus show in Figure 4.4; indeed, the slight peaks in $g(\phi)$ for the intermediate zone (Figure 4.2) correspond to the oblique orientation indicated here. This locus and corresponding loci in other planes enclose a volume of ground substance which exerts a swelling pressure. Provided that stress can be transferred to collagen fibrils which lie along these loci, they will be stretched by this pressure and so are appropriately oriented to provide tensile reinforcement. Thus articular cartilage could be considered as a system of microscopically small pressure vessels.

Support for this view is provided by two completely different kinds of experimental evidence. Firstly, unlike other cartilaginous tissues, excised samples of articular cartilage do not swell when placed in a medium of lower osmotic pressure, because their collagen fibril network is able to balance the increased internal pressure (Maroudas, 1979, 1980; Maroudas & Urban, 1980). Secondly, lacerating the cartilage surface of animals does not induce osteoarthritis (Carlson, 1957; Ghadially *et al.*, 1977; Mankin, 1982). Such lacerations would not lead to instability in a series of microscopically small, independent pressure vessels. However, injection of papain, which breaks down the ground substance interfering with its ability to function normally, destroys the balance of forces necessary for cartilage stability, leading to instability in the form of severe osteoarthritis (Murray, 1964; Bentley, 1972).

Collagen fibrils in the deep zone are oriented so that they tend to be stretched by any movement of the ground substance away from the underlying subchondral bone. Thus they tether the articular cartilage to bone and to the calcified cartilage above it, which has much the same pattern of collagen fibril orientations in the deep zone (Speer, 1982).

The pattern of preferred orientations in the plane of the articular surface of patellar cartilage (Figure 4.3) can be explained by the direction of contact stresses generated during flexion of the knee joint. During the normal range of flexion (in walking and running) the condyles of the femur move over the lateral and medial facets in a proximo-distal direction, i.e. in the direction of preferred orientation for their collagen fibrils. The extreme medial facet is only in contact with the femur (medial condyle) during extreme flexion of the knee. During extreme flexion the patella moves laterally, so that any contact stress on the medial facet would be in a lateral direction – once again in the direction of preferred orientation for the collagen fibrils.

4.8 **Osteoarthritis**

Osteoarthritis may be considered as mechanical instability of articular cartilage; it is classified as 'secondary', where a traumatic origin can be identified, or 'primary', where the aetiology is unknown (Ball, 1987). In the early stages of osteoarthritis the cartilage cracks and fragments. This degenerate cartilage is progressively destroyed, to the extent that the underlying bone is exposed in some regions of the joint surface. Progressive destruction of the cartilage is accompanied by formation of osteophytes, presumably in an attempt to stabilise an unstable joint by restricting its motion. It is not until this late stage that the disease can be detected radiologically (Ball, 1987).

Healthy articular cartilage is stable because its structure enables the swelling pressure of the ground substance to be balanced by tension in the collagen fibrils; disruption of this structure destroys the balance of forces – leading to mechanical instability. It is not therefore surprising that osteoarthritis is a progressive disease. If the structure of the surface zone of articular cartilage is sufficiently disturbed, its collagen fibrils will be unable to withstand the swelling pressure. This internal pressure will then further disturb the structure, so that the forces acting to disrupt the tissue experience positive feedback with the result that the system is unstable.

Experimental osteoarthritis has been induced in the knee joints of animals by cutting the cruciate ligaments or by partial removal of the meniscus; some obese strains of animals also develop the disease spontaneously (Adams & Billingham, 1982). In all these cases it appears that osteoarthritis develops because the joint has been subjected to an inappropriate pattern of loading. Secondary osteoarthritis in human knee joints can develop similarly as a result of damage to the ligaments (Jacobson, 1977; Fetto & Marshall, 1980), tears in the meniscus (McDaniel & Dameron, 1980) and surgical removal of the meniscus (Johnson *et al.*, 1974). In general, however, the aetiology of osteoarthritis is complicated and often obscure. Nevertheless, whatever the origin of the disease, articular cartilage is progressively disrupted because the balance of forces essential for its mechanical stability is destroyed.

Acknowledgements

We thank Dr Y. E. Yarker for collaboration in some of the experiments described here as well as the Science and Engineering Research Council, the Arthritis and Rheumatism Council and the Medical Research Council for support of various stages of this research.

References

Adams, M. E. & Billingham, M. E. J. (1982). Animal models of degenerative joint disease. *Current Topics in Pathology*, **71**, 265–97.

Agarwal, B. D. & Broutman, L. J. (1980). *Analysis and Performance of Fibre Composites*. Wiley, New York.

Armstrong, C. G., Bahrani, A. S. & Gardner, D. L. (1979). *In vitro* measurement of articular cartilage deformation in the human hip joint under load. *Journal of Bone and Joint Surgery*, **61A**, 744–55.

Aspden, R. M. (1986). Relation between structure and mechanical behaviour of fibre-reinforced composites at large strains. *Proceedings of the Royal Society*, **A406**, 287–98.

Aspden, R. M. & Hukins, D. W. L. (1979). Determination of the direction of preferred orientation and the orientation distribution function of collagen fibrils in connective tissues from high-angle X-ray diffraction patterns. *Journal of Applied Crystallography*, **12**, 306–11.

Aspden, R. M. & Hukins, D. W. L. (1981a). Collagen organisation in articular cartilage, determined by X-ray diffraction, and its relationship to tissue function. *Proceedings of the Royal Society*, **B212**, 299–304.

Aspden, R. M. & Hukins, D. W. L. (1981b). Calcification of the deep zone in pig femoral head cartilage. *Experientia*, **37**, 1333–4.

Ball, J. (1987). Aetiology and pathology of osteoarthrosis. In: *Copeman's Textbook of the Rheumatic Diseases*, 6th Edn., ed. J. T. Scott, pp. 821–45. Churchill Livingstone, Edinburgh.

Bayliss, M. T. (1984). Proteoglycans: structure and molecular organisation in cartilage. In: *Connective Tissue Matrix*, ed. D. W. L. Hukins, pp. 55–88. Macmillan, London.

Bentley, G. (1972). Papain-induced degenerative arthritis of the hip in rabbits. *Journal of Bone and Joint Surgery*, **53B**, 324–37.

Caddell, R. M. (1980). *Deformation and Fracture of Solids*, 2nd Edn., pp. 252–71. Prentice-Hall, Englewood Cliffs NJ.

Carlson, H. (1957). Reactions of rabbit patellary cartilage following operative defects. *Acta Orthopaedica Scandinavica*, Supplement 28.

Cox, H. L. (1952). The elasticity and strength of paper and other fibrous materials. *British Journal of Applied Physics*, **3**, 72–9.

Deas, H. D. (1952). The diffraction of X-rays by a random assemblage of molecules having partial alignment. *Acta Crystallographica*, **5**, 542–6.

Elliott, D. H. (1965). Structure and function of mammalian tendon. *Biological Reviews*, **40**, 392–497.

Fetto, J. F. & Marshall, J. L. (1980). The natural history and diagnosis of anterior cruciate ligament insufficiency. *Clinical Orthopaedics and Related Research*, **147**, 29–38.

Gathercole, L. J. & Keller, A. (1975). Light microscopic waveforms in collagenous tissues and their structural implications. In: *Structure of Fibrous Biopolymers*, eds. E. D. T. Atkins & A. Keller, pp. 153–87. Butterworth, London.

Ghadially, F. N., Thomas, I., Oryschak, A. F. & La Ronde, J. M. (1977). Long-term results of superficial defects in articular cartilage. A scanning electron microscope study. *Journal of Pathology*, **121**, 213–7.

Harkness, R. D. (1968). Mechanical properties of collagenous tissues. In: *Treatise on Collagen*, Vol. 2A, ed. B. S. Gould, 247–310. Academic Press, New York.

Higdon, A., Ohlsen, E. H., Stiles, W. B., Weese, J. A. & Riley, W. F. (1976). *Mechanics of Materials*, pp. 147–239. Wiley, New York.

Hooley, C. J. & Cohen, R. E. (1979). A model for the creep behaviour of tendon. *International Journal of Biological Macromolecules*, **1**, 123–32.

Hukins, D. W. L. (1982). Biomechanical properties of collagen. In: *Collagen in Health and Disease*, eds. J. B. Weiss & M. I. V. Jayson, pp. 49–72. Churchill Livingstone, Edinburgh.

Hukins, D. W. L. (1984). Collagen orientation. In: *Connective Tissue Matrix*, ed. D. W. L. Hukins, pp. 211–40. Macmillan, London.

Hukins, D. W. L. & Aspden, R. M. (1985). Composition and properties of connective tissues. *Trends in Biochemical Sciences*, **10**, 260–4.

Hukins, D. W. L., Aspden, R. M. & Yarker, Y. E. (1984). Fibre reinforcement and mechanical stability of articular cartilage. *Engineering in Medicine*, **13**, 153–6.

Jacobsen, K. (1977). Osteoarthrosis following insufficiency of the cruciate ligaments in man. *Acta Orthopaedica Scandinavica*, **48**, 520–6.

Jeronimidis, G. & Vincent, J. F. V. (1984). Composite materials. In: *Connective Tissue Matrix*, ed. D. W. L. Hukins, pp. 187–210. Macmillan, London.

Johnson, R. J., Kettlekamp, D. B., Clark, W. & Leaverton, P. (1974). Factors affecting late results after meniscectomy. *Journal of Bone and Joint Surgery*, **56A**, 719–29.

Kelly, A. (1973). *Strong Solids*. pp. 157–226. Clarendon Press, Oxford.

Krenchel, H. (1964). *Fibre Reinforcement*. Akademisk Forlag, Copenhagen.

Mankin, H. J. (1982). The reaction of articular cartilage to mechanical injury. *Journal of Bone and Joint Surgery*, **64A**, 460–6.

Maroudas, A. (1976). Balance between swelling pressure and collagen tension in normal and degenerative cartilage. *Nature*, **260**, 808–9.

Maroudas, A. (1979). Physicochemical properties of articular cartilage. In: *Adult Articular Cartilage*, 2nd Edn., ed. M. A. R. Freeman, pp. 215–90. Pitman Medical, Tunbridge Wells.

Maroudas, A. (1980). Physical chemistry of articular cartilage and the intervertebral disc. In: *The Joints and Synovial Fluid*, Vol. 2, ed. L. Sokoloff, pp. 240–93. Academic Press, New York.

Maroudas, A. & Urban, J. P. G. (1980). Swelling pressure of cartilaginous tissues. In: *Studies in Joint Disease*, Vol. 1, eds. A. Maroudas & E. J. Holborrow, pp. 87–116. Pitman Medical, Tunbridge Wells.

McDaniel, W. J. & Dameron, T. B. (1980) Untreated ruptures of the anterior cruciate ligament. A follow-up study. *Journal of Bone and Joint Surgery*, **62A**, 696–705.

Meachim, G., Denham, D., Emery, J. H. & Wilkinson, P. H. (1974). Collagen alignments and artificial splits at the surface of human articular cartilage. *Journal of Anatomy*, **118**, 101–18.

Meachim, G. & Stockwell, R. A. (1979). The matrix. In: *Adult Articular Cartilage*, 2nd Edn., ed. M. A. R. Freeman, pp. 1–67. Pitman Medical, Tunbridge Wells.

Muir, I. H. M. (1979). Biochemistry. In: *Adult Articular Cartilage*, 2nd Edn., ed. M. A. R. Freeman, pp. 145–214. Pitman Medical, Tunbridge Wells.

Muir, H., Bullough, P. & Maroudas, A. (1970). The distribution of collagen in human articular cartilage with some of its physiological implications. *Journal of Bone and Joint Surgery*, **52B**, 554–63.

Murray, D. G. (1964). Experimentally induced arthritis using intra-articular papain. *Arthritis and Rheumatism*, **7**, 211–9.

Myers, E. R., Armstrong, C. G. & Mow, V. C. (1984). Swelling pressure and collagen tension. In: *Connective Tissue Matrix*, ed. D. W. L. Hukins, pp. 161–86. Macmillan, London.

Speer, D. P. (1982). Collagenous architecture of the growth plate and perichondral ossification groove. *Journal of Bone and Joint Surgery*, **64A**, 399–407.

Swanson, S. A. V. (1979). Friction, wear and lubrication. In: *Adult Articular Cartilage*, 2nd Edn., ed. M. A. R. Freeman, pp. 415–460. Pitman Medical, Tunbridge Wells.

Woodhead-Galloway, J., Hukins, D. W. L., Knight, D. P., Machin, P. A. & Weiss, J. B. (1978). Molecular packing in elastoidin spicules. *Journal of Molecular Biology*, **118**, 567–78.

Yarker, Y. E., Aspden, R. M. & Hukins, D. W. L. (1983). Birefringence of articular cartilage and the distribution of collagen fibril orientations. *Connective Tissue Research*, **11**, 207–13.

Yarker, Y. E., Hukins, D. W. L. & Nave, C. (1984). X-ray diffraction studies of tibial plateau cartilage using synchrotron radiation. *Connective Tissue Research*, **12**, 337–43.

5

Temperature dependence of the mechanical properties of human dentine

D. C. Watts

5.1 Introduction

Many investigations of the mechanical properties of human dentine have been reported in the dental research literature; these studies have been carefully reviewed by Braden (1976) and by Waters (1980). Much previous work was concerned with measurements at ambient temperatures and in earlier papers the measurement temperature was often unspecified. However, unlike the slightly less mineralised material, compact bone, dentine is frequently – albeit transiently – exposed to a range of temperatures (approximately 0–60°C) in the oral environment. Yet study of temperature-dependent behaviour has been largely neglected. This information is important both for dental science and restorative treatment and also in terms of structure/property relationships in bone. The temperature coefficients of the mechanical properties of bone have been extensively studied by Bonfield & Li (1968).

Dentine is generally regarded as much less anisotropic in its properties than bone. Nevertheless there is a pronounced anisotropic structural feature: the presence of cylindrical channels, the dentinal tubules. The tubule diameter and concentration per unit area decrease in the direction of the outer surface. Diameters decrease from 4 μm to 1 μm and the concentration from 7.5 to 2 per 100 μm^2. At an average or central location with 5 tubules per 100 μm, of diameter 2 μm, this gives a tubule 'volume' of 16% of the dentine. The organic substance between the tubules, within which most of the mineral is dispersed, consists mainly of collagen fibrils ca. 0.3 μm thick. These appear to be arranged in planes perpendicular to the tubules.

The purpose of the present chapter is to summarise the state of prior knowledge of dentine biomechanics and to review some recent investigations on temperature coefficients over the clinically relevant range of

0–80°C, (Rasmussen & Patchin, 1984; El Mowafy & Watts, 1986; Watts *et al.*, 1987). The compressive stress–strain behaviour of dentine, including derived properties such as elastic modulus and proportional limit, is given particular consideration. Compressive properties are intrinsically important and such measurements are more experimentally amenable than studies of tension. The newer techniques of fracture mechanics are also included in this survey. The measurement of these parameters enables comparisons to be made with bone and with direct restorative materials.

5.2 **Stress–strain and fracture properties of dentine**

The earliest reported investigations into the compressive properties of human dentine are those of Black (1895). He obtained a value for the compressive strength close to the average reported by later workers. Since that time there have been a number of studies (Peyton *et al.*, 1952 ; Craig & Peyton, 1958; Stanford *et al.*, 1958, 1960). Other modes of mechanical deformation have also been employed. Thus Tyldesley (1959) and Renson (1968) used bending modes and Bowen & Rodriguez (1962) and Lehman (1967) studied the strength and elasticity of dentine in direct tension, while Hannah (1970) obtained values for tensile strength from diametral compression. More recently the compressive viscoelastic properties of dentine have been investigated by means of stress–relaxation measurements (Duncanson & Korostoff, 1975; Korostoff *et al.*, 1975).

For relatively brittle materials, such as calcified tissue, resistance to fracture is best quantified in terms of the parameters of fracture mechanics. The powerful theoretical and experimental approaches of fracture mechanics have been extensively applied to engineering materials (Ewalds & Wanhill, 1984; Kinloch & Young, 1983). Fracture toughness measurements have been made for dental polymethylmethacrylate (Causton 1975; Hargreaves, 1981), dental amalgam (Roberts *et al.*, 1978; Cruickshanks-Boyd & Lock, 1981; Lloyd & Adamson, 1985) and tooth-coloured restorative materials (Lloyd & Mitchell, 1985). The fracture toughness of human enamel has been determined with an indentation procedure (Hassan *et al.*, 1981). The most extensive studies of the fracture mechanics of tooth tissue have been conducted by Rasmussen *et al.* (1972, 1976) and by Rasmussen & Patchin (1984). These latter investigators utilized a work-of-fracture (W_f) approach, originally introduced by Tattersall & Tappin (1966), which has also been applied to bone (Piekarski, 1970). The material's work-of-fracture is determined from the ratio of the total work expended during fracture to the projected normal area of the two fractured surfaces.

In addition to this work-of-fracture parameter there are two other parameters which are regarded as of central importance in linear elastic fracture mechanics. These are the plane strain fracture toughness (K_{ic}) and the critical strain energy release rate (G_{ic}). These quantities have been extensively measured for bone (Wright & Hayes, 1977; Behiri & Bonfield, 1984). Several specimen designs have been used in these fracture-toughness studies on bone, including the compact tension (CT) geometry. This design has now been used for dentine (Watts *et al.* 1987).

5.3 **Experimental methods**

5.3.1 *Specimen preparation*

Recently extracted permanent lower molar teeth were employed as the source material for dentine. After extraction the teeth were initially stored for 24 h in buffered formal saline, pH 7, to inhibit bacterial and fungal growth and prevent demineralisation. Although this medium may produce a change in water content (Waters, 1980), through subsequent storage in fresh refrigerated distilled water the water content of the dentine should restabilise. Furthermore, although formaldehyde may cause cross-linking of the proteinaceous material within the dentine, the net change in the elastic properties of the dentine as a whole is likely to be small.

Dentine specimens of rectangular cross-section were produced by sectioning the tooth crowns. A standard micro-slicing machine was used with a thin diamond disc. A constant cooling water flow was provided at the cutting field and a linear cutting speed of 1 $mm \cdot min^{-1}$ was used to minimise any temperature rise in the dentine during sectioning. Test specimens were polished carefully under water on the flat surface of a white stone (Aloxite Superfine, No. 201A, Carborundum Co., Manchester, UK). Prepared specimens were checked under the microscope and any defective ones were discarded. The overall specimen dimensions were measured by means of a micrometer and numbered on both sides before storing in water until testing.

For uniaxial compressive measurements, the length, width and thickness were typically 3.00 × 2.50 × 1.25 mm, respectively, with the axial (length) direction perpendicular to the tubule orientation.

For fracture-toughness measurements, 'compact tension' (CT) test pieces, conforming to British Standard 5447 (1977), were prepared (Figure 5.1) having approximate rectangular dimensions of 4.60 × 4.50 × 1.60 mm. One sample only was obtainable from each tooth. A central notch of specific length (a) was then made in each specimen using the diamond disc along a marked guideline. This stage was of prime importance, since notch length is highly critical. According to the British Standard, a/W must be in the range 0.45–0.55, where W is the net width of the specimen. These dimensions were measured with a travelling microscope and fell within the acceptable range. Two cylindrical holes, 0.80 mm in diameter, were

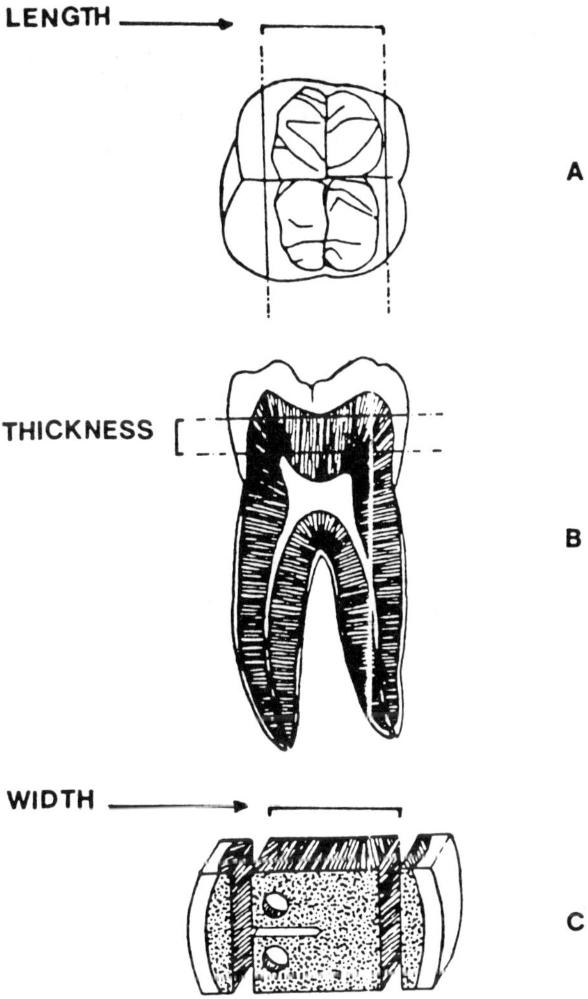

Fig. 5.1 Successive stages in the preparation of compact tension test specimens of dentine from a molar tooth.

drilled in each specimen, to provide the means of attachment for tensile loading using stainless steel wire of diameter 0.75 mm.

5.3.2 *Temperature control*
During specimen testing, each sample group was maintained at thermal equilibrium in water at preset temperatures, ± 0.1°C. For compressive measurements the five thermostatically-controlled temperatures were 0, 23, 37, 50 and 80°C.

For fracture-toughness measurements, a small plastic container was used to hold water around the specimen at preset temperatures during each 15 s test interval. Testing was conducted at 0, 37 and 60°C. As subsequently noted, the resultant data were relatively temperature invariant. For tests at 0°C, a mixture of ice and water was placed in the container and temperature was constantly monitored by means of a digital thermometer with a thermocouple placed in close contact with the specimen surface.

5.3.3 *Mechanical loading*

Mechanical loads were applied to the specimens at a cross-head speed of 1 $cm \cdot min^{-1}$, using a Universal testing machine (Model 50 TSB, RDP Howden Ltd., Leamington Spa, UK). Analogue load/deflection curves were recorded together with digital recording of peak loads. The compressive loads were applied uniaxially, parallel to the long axes of specimens, and thus perpendicular to dentinal tubule orientation, until fracture occurred. For the CT fracture specimens, loads were applied in tension.

5.3.4 *Fractography*

Following the fracture of each specimen type, the fragments were retained. These were viewed by optical and scanning electron microscope to ascertain the fracture pattern. The orientation of the applied stress in the CT fracture specimens resulted in a fracture surface parallel to the tubule orientation. To confirm compliance with the British Standard, each CT specimen was reassembled and examined under the microscope to confirm that the inclination of the fracture plane in relation to the notch plane was within 10°.

5.3.5 *Analysis of data*

The compressive load/deflection data were converted into engineering stress–strain data and minor corrections were made to allow for machine compliance. Modulus of elasticity, proportional limit, ultimate compressive strength, and the energy quantities resilience and toughness were calculated from the data. Analysis of variance and linear regression analysis methods were applied to each of these quantities as measured at the different temperatures.

Plane strain fracture toughness (K_{ic}) and critical strain energy release rate (G_{ic}) were calculated using the following equations:

$$K_{ic} = P \cdot Y_2 / B \cdot (W)^{1/2} \quad (5.1)$$

and,

$$G_{ic} \approx K_{ic}^2 / E \quad (5.2)$$

where P is load required to fracture specimen, Y_2 is a tabulated function of a/W (British Standard 5447, 1977), B is specimen thickness, W is specimen net width, and E is the modulus of elasticity of the material under the condition of the test.

5.4 **Results**

5.4.1 *Compressive behaviour of dentine (0–80°C)*

Representative stress–strain curves for dentine samples at three temperatures are given in Figure 5.2. These show the Hookean elastic behaviour at lower stress levels, and the reduction in modulus which occurs with increasing temperature. Beyond the proportional limit, non-linear deformation continued up to the point of fracture. The quantities derived from the stress–strain data are presented in Table 5.1. Each of these quantities was subjected to one way analysis of variance (ANOVA), with respect to temperature, and highly significant differences ($p < 0.0001$) were found in all cases, despite the fact that ANOVA does not take into account the ordering of the data with respect to temperature. Thus, for example, in the case of the modulus (E), $F_{(data)} = 17.08$, compared with $F_{0.0001}$ $(4,30) = 8.54$.

Following ANOVA, linear regression analysis was successfully applied to all of the quantities, with the exception of toughness. The F-test was also applied to the analysis of the residuals. In each case F (residuals) was less than $F_{0.05}$ $(3,30) = 2.92$. This confirmed the null hypothesis of the linear fit in each case. Details of the linear regression analysis are given in Table 5.2 and in the following equations for modulus (E), proportional limit (PL), compressive strength (CS) and resilience (R), respectively:

$$E\ (\mathrm{GN{\cdot}m^{-2}}) = 15.55 - 0.0734\ (T\ °\mathrm{C}) \qquad (5.3)$$

$$PL\ (\mathrm{MN{\cdot}m^{-2}}) = 198.2 - 0.8673\ (T\ °\mathrm{C}) \qquad (5.4)$$

$$CS\ (\mathrm{MN{\cdot}m^{-2}}) = 298.4 - 1.0988\ (T\ °\mathrm{C}) \qquad (5.5)$$

$$R\ (\mathrm{MJ{\cdot}m^{-3}}) = 1.265 - 0.0051\ (T\ °\mathrm{C}) \qquad (5.6)$$

As can be seen from Table 5.1, the toughness decreased from 0 to 37°C and then remained essentially constant up to 80°C. Specimen failure occurred through lines determined by maximum shear stresses, approximately 45° to the axial direction.

5.4.2 *Fracture-toughness parameters for dentine (0–60°C)*

Table 5.3 records individual values obtained for K_{ic} and G_{ic} from examined specimens at the measurement temperatures. Values of the modulus of elasticity of dentine used to calculate G_{ic} were 15.20, 13.26

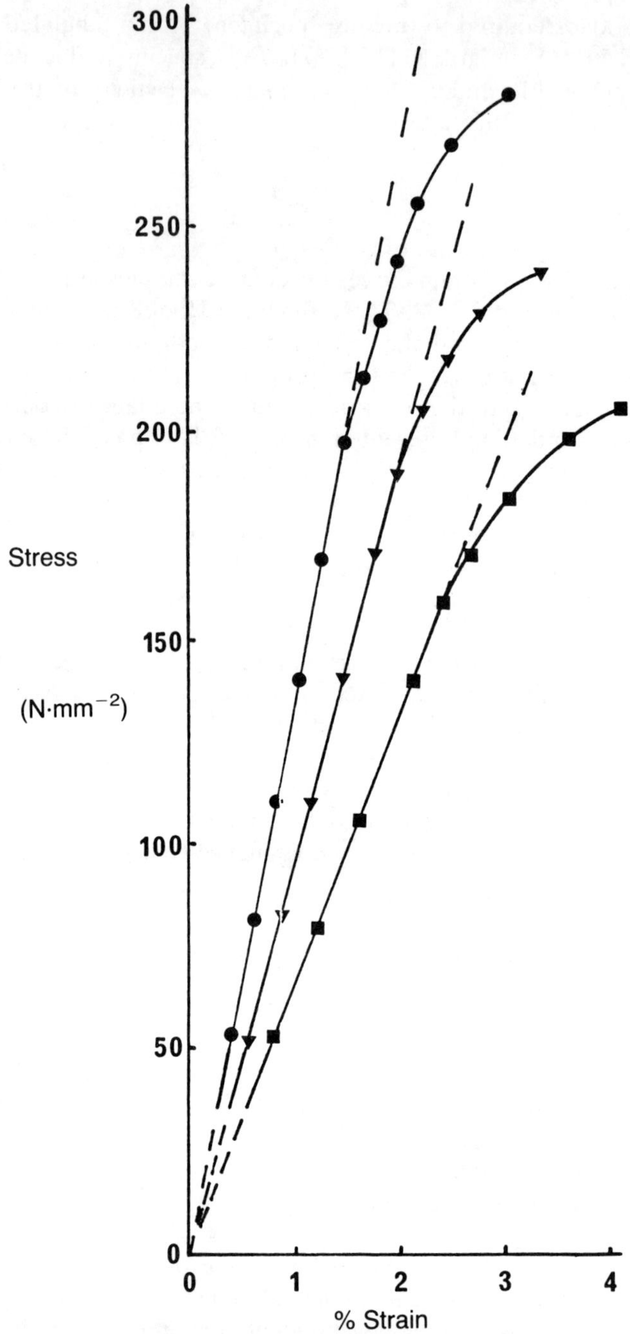

Fig. 5.2 Stress–strain curves for dentine specimens in uniaxial compression at 0°C (●), 37°C (▼), and 80°C (■).

Table 5.1 *Uniaxial compression data for dentine.*

Parameter	Temperature 0°C	23°C	37°C	50°C	80°C
Modulus of elasticity (E); $GN{\cdot}m^{-2}$	15.20 (1.6)	13.94 (1.5)	13.26 (1.8)	12.06 (1.2)	9.35 (0.8)
Proportional limit (PL; $MN{\cdot}m^{-2}$)	198.8 (7.4)	178.3 (8.0)	165.0 (17.0)	155.0 (10.0)	129.3 (8.5)
Compressive strength (CS; $MN{\cdot}m^{-2}$)	297.5 (12.2)	270.9 (7.0)	260.0 (19.0)	247.1 (14.6)	207.9 (9.6)
Resilience (R; $MJ{\cdot}m^{-3}$)	1.30 (0.25)	1.14 (0.34)	1.03 (0.21)	1.00 (0.24)	0.89 (0.30)
Toughness (T; $MJ{\cdot}m^{-3}$)	6.87 (0.87)	6.23 (0.45)	4.82 (0.53)	4.98 (0.52)	5.19 (0.29)

Values are means with SD in parentheses; seven samples per group.

Table 5.2 *Linear regression analysis of mechanical properties of dentine as a function of temperature.*

Property	Temperature coefficient (A)	Coefficient of determination (r^2)	F (residuals)
Modulus (E)	−0.0734, $GN{\cdot}m^{-2}{\cdot}°C^{-1}$	0.9772	0.0429
Proportional limit (PL)	−0.8673, $MN{\cdot}m^{-2}{\cdot}°C^{-1}$	0.9993	0.0059
Compressive strength (CS)	−1.0988, $MN{\cdot}m^{-2}{\cdot}°C^{-1}$	0.9929	0.0660
Resilience (R)	−0.0051, $MJ{\cdot}m^{-3}$	0.9523	0.0209
Toughness (T)	—	0.5871	—

Table 5.3 *Fracture-toughness parameters of dentine compact tension specimens at three temperatures.*

	Temperature 0°C	37°C	60°C
K_{ic} ($MN{\cdot}m^{-1.5}$)	3.112 (0.498)	3.132 (0.194)	2.964 (0.145)
G_{ic} ($J{\cdot}m^{-2}$)	652 (220)	742 (137)	790 (110)

Values are means with SD in parentheses.

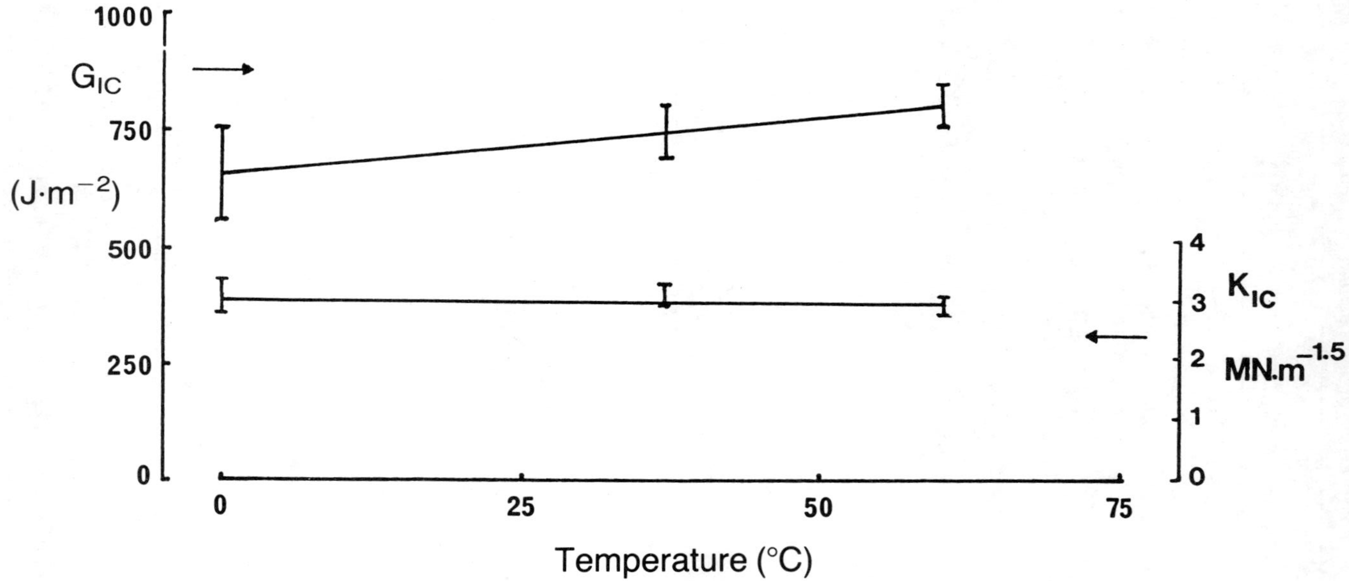

Fig. 5.3 Mean values of fracture toughness (K_{ic}) and critical strain energy release rate (G_{ic}) as a function of temperature for human dentine.

and 11.15 GN·m^{-2} at 0, 37 and 60°C, respectively. The standard deviations of the G_{ic} values for each specimen group were calculated from eqn (5.2) taking into account the quadratic sum of the original fractional errors (Taylor, 1982).

Mean values of K_{ic} and G_{ic} as a function of temperature are plotted in Figure 5.3. Analysis of variance showed no significant difference between K_{ic} values at the three test temperatures at the 0.05 significance level. Also no significant difference was found between G_{ic} values at the three test temperatures and the same level, $F_{(2,19,\ \alpha=0.05)} = 3.52$. However ANOVA does not take into account the ordering of the data with respect to temperature. Accordingly the data for both K_{ic} and G_{ic} were analysed by linear regression. Examination of the residuals confirmed that K_{ic} was invariant with temperature at the 0.05 level. By contrast, G_{ic} fitted the following equation, $0.01 < p < 0.05$:

$$G_{ic} = 2.31\ (T\ °\mathrm{C}) + 653.2 \qquad (5.7)$$

Thus there is statistically significant, but not highly significant evidence for slight temperature-dependence of G_{ic}. Since G_{ic} values were calculated utilising modulus (E) values, eqn (5.2), then the temperature-dependence of G_{ic} may be a consequence of the above-mentioned temperature-dependence of E. Mean values obtained over the measured temperature range were as follows:

$$K_{ic} = 3.08\ (0.33)\ \mathrm{MN{\cdot}m^{-1.5}}$$

$$G_{ic} = 722\ (150)\ \mathrm{J{\cdot}m^{-2}}$$

Figure 5.4 shows scanning electron microscope view of a typical fracture surface, where the direction of crack propagation is from left to right. The crack has propagated with a crack front aligned 'parallel' to the tubules and the fracture surface shown is also parallel to the dentinal tubules. However there is slight evidence for some lateral movement of the crack front perpendicular to the dentinal tubules.

5.5 **Discussion**

5.5.1 *Compressive behaviour*

Dentine has been shown to exhibit viscoelastic as well as elastic properties (Duncanson & Korostoff, 1975; Korostoff *et al.*, 1975). Hence the possibility of a viscoelastic contribution to the present data should not be entirely discounted. This raises the question of the exact definition of modulus, which is normally understood to be a purely elastic quantity. Craig & Peyton (1958) reported a relatively high value for the 'static' modulus of dentine, in the range 16.6–18.5 GN·m^{-2}, having subtracted from the strain

Fig. 5.4 Scanning electron microscope view of a typical fracture surface in human dentine. The direction of crack propagation is from left to right. Bars represent 10 μm.

(through which the modulus is calculated) a contribution arising from anelastic (i.e. retarded elastic) deformation. Elsewhere in the literature, somewhat lower values (11–13 GN·m^{-2}) are reported – presumably for ambient temperatures (Waters, 1980). These have evidently been uncorrected for any anelastic effects.

In the present work, a relatively high rate of strain, 3.3 min^{-1}, has been applied to the dentine specimens in order to minimise anelastic effects and hence to approach true elastic modulus values. Nevertheless this rate is still very low in comparison with ultrasonic frequencies. The values obtained for modulus, compressive strength and proportional limit at 23°C are consistent with previous data in the literature, as collated and compared by Waters (1980).

The temperature dependence of the modulus of dentine is a statistically significant negative linear relationship. This decrease with increasing temperature parallels a similar trend measured for compact bone by Bonfield & Li (1968) and Bonfield & Tully (1982). Indeed there is close agreement obtained between the value found here for the temperature coefficient of modulus: −0.0734 GN·m^{-2} per °C (Table 5.2) and that reported for cortical bone: −0.07 GN·m^{-2} per °C (Bonfield & Tully, 1982), over the range 0–41°C, at ultrasonic frequencies and −0.08 GN·m^{-2}

per °C (Bonfield & Li, 1968), over the range −59 to 25°C, based on direct stress–strain measurements.

The attempt may be made to analyse the elastic properties of dentine and bone in terms of the composite structure of the materials. At the simplest level, dentine may be regarded as a binary composite of collagen and calcium hydroxyapatite (HAP). Since HAP is chemically stable below 400°C, there is no significant change in the modulus value of 114 $GN{\cdot}m^{-2}$, between 0 and 80°C (Katz & Ukraincik, 1971). By contrast, collagen may exhibit temperature-dependence of modulus in this region (Bonfield & Li, 1968). The elastic modulus of collagen was determined by Currey (1965) as 1.24 $GN{\cdot}m^{-2}$. However, Lees & Davidson (1977) suggest that for cross-linked collagen the higher value of 3.75 $GN{\cdot}m^{-2}$ is more appropriate. The application of composite models to hard mineralised tissues was reviewed by Katz (1971). From this work it appears that the simplest model for interpreting the 'lower bound' behaviour is that of the Reuss or uniform stress–cumulative strain model, for which the total modulus (E), in terms of volume fractions (V), is:

$$E = (E_{HAP} \cdot E_c)/[V_c \cdot E_{HAP} + V_{HAP} \cdot E_c] \tag{5.7}$$

where subscript c denotes collagen.

Taking $V_{HAP} = 0.45$ and $V_c = 0.30$ and ignoring the volume fraction of water, this gives the result for E of 11.9 $GN{\cdot}m^{-2}$. It is likely that more refined models, taking into account the hierarchical structural organisation of dentine, will achieve a closer correlation between theoretical predictions and empirical data. At present, the decrease in the modulus, and other compressive properties, of dentine with temperature may be attributed either to a reversible decrease in the modulus of collagen with temperature or to a possible weakening in the bonding between collagen and hydroxyapatite micro-crystals. The extent of reversibility or irreversibility of these effects requires further investigation. However, Bonfield & Tully (1982) found that the modulus changes in cortical bone were completely reversible in the range 17–41°C.

With respect to the 'ultimate' properties of proportional limit and compressive strength, it has been suggested by Korostoff *et al.* (1975) that strains exceeding 0.6% produce permanent changes in dentine, even including the collapse of dentinal tubules. Under such high stress conditions it may be speculated that the effect on the properties of anisotropic structured features will be minimised. Certainly the typical failure modes are consistent with isotropic behaviour. Fracture occurred along planes determined by directions of maximum shear stresses rather than along either the direction of the matrix fibres – which are parallel to the axial loading – or the direction of the dentinal tubules.

5.5.2 *Fracture behaviour*

The small volume of human dentine available for the production of samples having a simple geometry places a severe constraint on the design of test specimens. Nevertheless the compact-tension geometry, although difficult to fabricate on a small scale, yields data with quite small scatter. Rasmussen & Patchin (1984) report large standard deviations for most of their work-of-fracture specimen groups. The grand average value for fracture toughness (K_{ic}) is midway in the range (0.23–6.56) tabulated by Bonfield & Behiri (1982) for cortical bone. The average value of G_{ic} obtained is at the lower end of the range reported by these same workers for bovine bone. It should be noted that the values of G_{ic} reported here are calculated via eqn (5.2) from the values of K_{ic} and modulus (E). Since the latter quantity could vary somewhat with rate of measurement, it may be that the G_{ic} values should be recalculated using a ‘high frequency’ value for the modulus when this becomes available. The temperature-invariant fracture-toughness (K_{ic}) value should not be confused with the temperature-dependent toughness quantity, (Table 5.1).

For plane strain conditions, the Critical Strain Energy Release Rate is given by:

$$G_{ic} = K_{ic}^2 \cdot (1-\nu^2)/E \qquad (5.9)$$

where ν is Poisson’s ratio. Renson & Braden (1975) quoted a value of 0.12 for this ratio. However, according to Waters (1980), if the original data quoted by Renson & Braden are correct, their value for Poisson’s ratio has been miscalculated and should be 0.23 which is much closer to the value of 0.31 obtained by Gilmore *et al.* (1969) for bovine dentine. If the value of ν is taken as 0.23, then the term $(1-\nu^2)$ is 0.947 which is approaching unity. Hence follows eqn (5.2).

Recent investigations of the fracture mechanics of bone have been concerned with the rate-dependence of the fracture-toughness, K_{ic}. The larger specimen sizes available for bone enable the crack velocity to be measured directly (Behiri & Bonfield, 1984). This was not possible in the present work. However there have been few investigations to date of the rate-dependence of fracture parameters for dental materials (Roberts *et al.*, 1977).

Temperature variations in the teeth occur during normal function and also in water-cooled cavity preparation (Brown *et al.*, 1972). However the results indicate the absence of any pronounced temperature-dependency of fracture properties, at least up to 60°C. This is fairly consistent with the findings of Rasmussen & Patchin, although detailed numerical comparison of K_{ic} and G_{ic} with work of fracture is difficult.

The present study is limited to one specimen orientation as far as the investigation of anisotropic effects relative to tubule orientation is concerned. Three basic orientations appear possible: one perpendicular and two in the parallel plane. Rasmussen *et al.* (1976) looked at two of the orientations (perpendicular and one of the parallel) while this study is concerned with the third. Rasmussen & Patchin (1984) conclude that dentine is mildly anisotropic under controlled, as distinct from catastrophic fracture. They also note that dentine anisotropy is less in terms of strength than in terms of work of fracture. Consequently, we may expect some variation of K_{ic} and G_{ic} with sample orientation. The electron microfractography, (Figure 5.4), illustrates the propagation of the crack front generally parallel to the tubule direction; that is, with the crack growth direction perpendicular to the tubules. This means that the crack front line is constantly passing through a succession of tubules. If the crack tip is effectively blunted by each tubule that it crosses, the round tubules may be contributing to the moderate level of fracture-toughness exhibited by dentine rather than necessarily being a source of weakness. The stability of tubule geometry with temperature (0–60°C) may then explain the invariance of K_{ic} in this range. However Rasmussen & Patchin (1984) also found little variation of work-of-fracture with temperature. Since a blunting effect would most likely not have been a factor for their orientations (Rasmussen *et al.*, 1976) it may not be a factor for this one. Figure 5.4 shows some evidence of lateral movement of the crack front perpendicular to the tubule direction. The matrix surrounding the tubules consists of calcified layers of collagen oriented approximately perpendicular to them (Kramer, 1951). Rasmussen & Patchin (1984) suggest that fracture along the collagen fibrils, i.e. along the 'grain', meets less resistance than does fracture perpendicular to the 'grain' of the collagen fibrils.

The fracture toughness (K_{ic}) of dentine may thus vary to some extent with the orientation of the tubules and with the crack velocity. However the value obtained under the conditions of the present experiment may be briefly compared with those of available direct restorative materials. From the compilation of Lloyd (1983) it is apparent that the currently available restorative materials are of a lower fracture toughness. Dental cements are in the range 0.1–0.5 $MN{\cdot}m^{-1.5}$, amalgams in the range 1.0–1.6 and composites in the range 0.6–2.0. Human dental enamel also has a low fracture toughness: within the range 0.7–1.27 $MN{\cdot}m^{-1.5}$, according to Hassan *et al.* (1981).

References

Behiri, J. C. & Bonfield, W. (1984). Fracture mechanics of bone – The effects of density, specimen thickness and crack velocity on longitudinal fracture. *Journal of Biomechanics*, **17**, 25–34.

Black, G. V. (1895). An investigation into the physical characters of the human teeth in relation to their diseases and to practical dental operations. *Dental Cosmos*, **37**, 353–421, 469–84, 553–71, 637–61 and 737–57.

Bonfield, W. & Li, C. H . (1968). The temperature-dependence of the deformation of bone. *Journal of Biomechanics*, **1**, 323–9.

Bonfield, W. & Behiri, J. C. (1982). Fracture mechanics of compact bone. In: *Biomedical Engineering. I. Recent Developments*, ed. S. Saha, pp. 343–7. Pergamon Press, Oxford.

Bonfield, W. & Tully, A. E. (1982). Ultrasonic analysis of the Young's modulus of cortical bone. *Journal of Biomedical Engineering*, **4**, 23–7.

Bowen, R. L. & Rodriguez, M. S. (1962). Tensile strength and modulus of elasticity of tooth structure and several restorative materials. *Journal of the American Dental Association*, **64**, 378–87.

Braden, M. (1976). Biophysics of the tooth. In: *Frontiers of Oral Physiology*, Vol. 2, ed. Y. Kawamura, pp. 1–37. S. Karger, Basel.

British Standard 5447 (1977). *Plane strain fracture toughness of metallic materials*. British Standards Institution, London.

Brown, W. S., Jacobs, H. R. & Thompson, R. E. (1972). Thermal fatigue in teeth. *Journal of Dental Research*, **51**, 461–7.

Causton, B. E. (1975). Fracture mechanics of dental polymethylmethacrylate. *Journal of Dental Research*, **54**, 339–45.

Craig, R. G. & Peyton, F. A. (1958). Elastic and mechanical properties of human dentin. *Journal of Dental Research*, **37**, 710–8.

Cruickshanks-Boyd, D. W. & Lock, W. R. (1981). A comparison of fracture toughness test methods for dental amalgam. *Journal of Dental Research*, **60**, 1148.

Currey, J. D. (1965). Anelasticity in bone and echinoderm skeletons. *Journal of Experimental Biology*, **43**, 279–92.

Duncanson, M. G. & Korostoff, E. (1975). Compressive viscoelastic properties of human dentin: 1. Stress–relaxation behaviour. *Journal of Dental Research*, **54**, 1207–12.

El Mowafy, O. M. and Watts, D. C. (1986). Fracture toughness of human dentin. *Journal of Dental Research*, **56**, 677–81.

Ewalds, H. L. and Wanhill, R.J.H. (1984). *Fracture Mechanics*. Edward Arnold, London.

Gilmore, R. S., Pollack, R. P. & Katz, J. L. (1969). Elastic properties of bovine dentine and enamel. *Archives of Oral Biology*, **15**, 787–96.

Hannah, C. McD. (1970). The Tensile Properties of Human Enamel and Dentine, *IADR British Division, Progr. and Abstracts*, No. 113.

Hargreaves, A. S. (1981). The effect of the environment on the crack initiation toughness of dental polymethylmethacrylate. *Journal of Biomedical Materials Research*, **15**, 757–68.

Hassan, R., Caputo, A. A. & Bunshah, R. F. (1981). Fracture toughness of human enamel. *Journal of Dental Research*, **60**, 820–7.

Katz, J. L. (1971). Hard tissue as a composite materials – I. Bounds on the Elastic Behaviour. *Journal of Biomechanics*, **4**, 455–73.

Katz, J. L. & Ukraincik, K. (1971). On the anisotropic elastic properties of hydroxyapatite. *Journal of Biomechanics*, **4**, 221–7.

Kinloch, A. J. & Young, R. J. (1983). *Fracture behaviour of polymers*. Applied Science Publishers, London.

Korostoff, E., Pollack, S. R. & Duncanson, M. G. (1975). Viscoelastic properties of human dentin. *Journal of Biomedical Materials Research*, **9**, 661–74.

Kramer, I.R.H. (1951). Distribution of collagen fibrils in the dentine matrix.

British Dental Journal, **91**, 1–7.

Lees, S. & Davidson, C. L. (1977). The role of collagen in the elastic properties of calcified tissues. *Journal of Biomechanics*, **10**, 473–86.

Lehman, M. L. (1967). Tensile strength of human dentin. *Journal of Dental Research*, **46**, 197–201.

Lloyd, C. H. (1983). Resistance to fracture in posterior composites. *British Dental Journal*, **155**, 411–4.

Lloyd, C. H. & Adamson, M. (1985) The fracture toughness (K_{ic}) of amalgam. *Journal of Oral Rehabilitation*, **12**, 59–68.

Lloyd, C. H. & Mitchell, L. (1985). The fracture toughness of tooth-coloured restorative materials. *Journal of Oral Rehabilitation*, **11**, 256–72.

Peyton, F. A., Mahler, D. B. & Hershahnov, B. (1952). Physical properties of dentin. *Journal of Dental Research*, **31**, 366–70.

Piekarski, K. (1970). Fracture of bone. *Journal of Applied Physics*, **41**, 215–23.

Rasmussen, S. T., Heuer, A. H., Patchin, R. E. & Scott, D. B. (1972) Fracture properties of enamel and dentin. *IADR Progr. and Abstracts 51*, No. 862.

Rasmussen, S. T., Patchin, R. E. Scott, D. B. & Heuer A. H. (1976). Fracture properties of human enamel and dentin. *Journal of Dental Research*, **55**, 154–64.

Rasmussen, S. T. & Patchin, R. E. (1984). Fracture properties of human enamel and dentin in an aqueous environment. *Journal of Dental Research*, **63**, 1362–8.

Renson, C. E. (1968). Elastic and mechanical properties of dentine. *IADR British Division Progr. and Abstracts*, No. 79.

Renson, C. E. & Braden, M. (1975). Experimental determination of the rigidity modulus, Poisson's ratio and elastic limit in shear of human dentine. *Archives of Oral Biology*, **20**, 43–7.

Roberts, J. C., Powers, J. M. & Craig, R. G. (1977). Fracture toughness of composites and unfilled resins. *Journal of Dental Research*, **56**, 748–53.

Roberts, J. C., Powers, J. M. & Craig, R. G. (1978). The fracture toughness and critical strain energy release rate of dental amalgam. *Journal of Materials Science*, **13**, 965–71.

Stanford, J. W., Paffenberger, G. C., Kampula, J. W. & Sweeney, W. T. (1958). Determination of some compressive properties of human enamel and dentin. *Journal of the American Dental Association*, **57**, 487–95.

Stanford, J. W., Weigel, K. V., Paffenberger, G. C. & Sweeney, W. T. (1960). Compressive properties of hard tooth tissues and some restorative materials. *Journal of the American Dental Association*, **60**, 746–51.

Tattersall, H. G. & Tappin, G. (1966). The work of fracture and its measurement in metals, ceramics and other materials. *Journal of Materials Science*, **1**, 296–301.

Taylor, J. R. (1982). *An Introduction to Error Analysis*, pp. 52–9. Oxford University Press, Oxford.

Tyldesley, W. R. (1959). Mechanical properties of human dental enamel and dentine. *British Dental Journal*, **106**, 269–78.

Waters, N. E. (1980). Some mechanical and physical properties of teeth. In: *The Mechanical Properties of Biological Materials*, eds. J. F. V. Vincent & J. D. Currey, pp. 99–135. Cambridge University Press, Cambridge.

Watts, D. C., El Mowafy, O. M. & Grant, A. A. (1987). Temperature-dependence of compressive properties of human dentin. *Journal of Dental Research*, **66**, 29–32.

Wright, T. M. & Hayes, W. C. (1977). Fracture mechanics parameters for compact bone. *Journal of Biomechanics*, **10**, 419–30.

6

Mechanical properties of vertebral cancellous bone

K. Short

6.1 **Introduction**

In order to produce a desired effect within a system by external manipulation, it is necessary to understand the mechanisms which govern its behaviour, including the mechanisms which operate at the lowest level upon which all others depend.

This philosophy can be applied to most fields of research, and is applicable as much to medical science as it is to structural mechanics, in that the extent of knowledge in both fields has been increased by study of the underlying mechanisms which govern the behaviour of the system as a whole.

In medicine, research conducted into biology at the cellular level has contributed greatly to the development of treatment at the systemic level, while structural analysis is carried out using advanced numerical techniques which consider the behaviour and interaction of discrete finite elements resulting from partitioning of the whole structure into smaller parts.

The work presented here is intended to demonstrate how the application of the advanced technology of structural analysis can be applid to a load-bearing anatomical structure in order to provide a greater understanding of the way in which the structure behaves *in vivo*.

The anatomical structure in question is the human vertebral centrum, which forms part of the spinal column, a major load-bearing structure of the body and the primary source of back pain. Injury to the spine can lead to a wide range of conditions from discomfort and pain to paralysis and, in some cases, death. It is important to realise that although only a small proportion of back-pain patients progress to the stage where surgical intervention is necessary, the total cost to the National Health Service, the Departments of Health and Social Security and British industry resulting from treatment costs, sickness benefit and loss of earnings, is in excess of 1000 million pounds sterling per year (Tesh, 1986).

It is desirable, therefore, to minimise the occurrence and duration of each back pain incident by the development of new methods of conservative treatment and through improved surgical techniques. This can only be achieved by an increased understanding of the mechanical behaviour of the spine and through an appreciation of the role of individual anatomical structures in governing this behaviour.

This chapter is concerned with an investigation into the mechanical behaviour of the vertebral centrum in its function as an integral part of the spinal column. The technique employed involved the simulation of the physical structure of the centrum, including accurate representation of the macro-structure, using the finite element method.

6.2 **Anatomy**

The spine, or vertebral column, is a strong, flexible load-bearing segmented rod, extending from the base of the cranium to the pelvic girdle. It serves to support the trunk and the head, to protect the spinal cord, and as a point of attachment for the ribs, ligaments, muscles and other soft tissue structures of the back.

The spine is made up of a succession of comparitively rigid bony vertebral bodies (vertebrae) and soft intervertebral discs, the flexibility of the spine originating from the deformability of the discs, which allow the relative angular displacement of successive vertebrae. The posterior elements of the vertebral bodies form the vertebral canal, which encloses the spinal cord, the most important neural pathway in the body. It is this dual function of the spine in load-bearing and protection of the spinal cord which makes it one of the most vulnerable structures in the body. If the spine should fail in its load-bearing capacity, then damage to the spinal cord or associated nerves may ensue causing pain, paralysis or death. It is important, therefore, that the structural mechanisms of spine and the role of specific anatomical structures within the spine are understood.

The basic repeating structural unit of the spine is commonly referred to as a spinal motion segment (SMS), and the work presented here is concerned with the centrum of the vertebral body, which is an integral part of the anterior portion of the SMS. SMS is the name given to an intervertebral disc, its two neighbouring vertebral bodies, and the soft tissue structures associated with the posterior elements of the vertebral bodies. Each vertebra is composed of a centrum and posterior elements, the centrum being a cylindrical mass of cancellous bone which is surrounded laterally by a thin cortical shell of dense bone. It is bounded in man, superiorly and inferiorly, by vertebral endplates which possess a bony layer and a cartilaginous layer. Due to the ability of the disc to deform under load, the SMS can be thought of as a polycentric multi-axial joint.

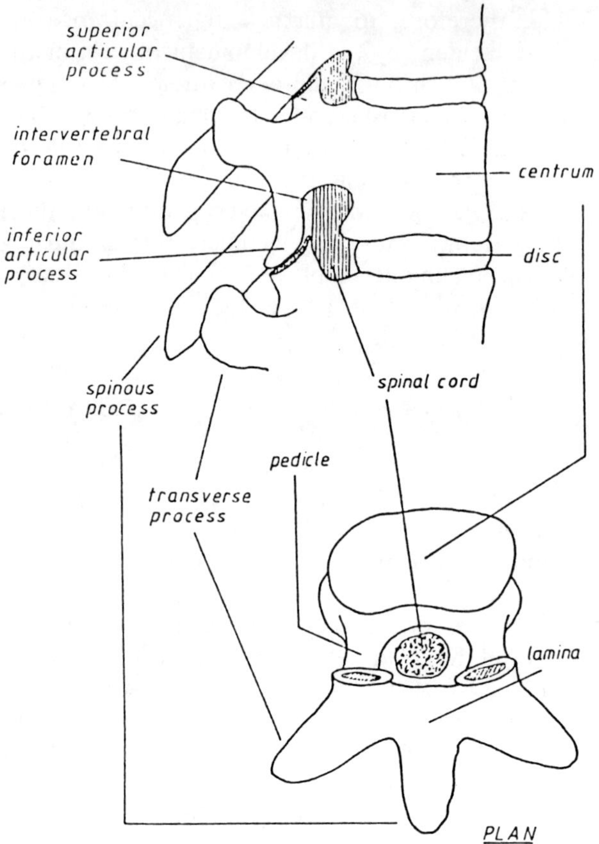

Fig. 6.1 Anatomy of the vertebrae.

The posterior elements include the pedicles and lamina, forming the vertebral canal which encloses the spinal cord, the transverse and spinous processes which serve as points of attachment for soft tissue structures, and superior and inferior articular processes which articulate with neighbouring vertebrae above and below. The intervertebral disc is composed of a central, soft, hydrophilic gel-like nucleus (nucleus pulposus), which is enclosed by a fibrous sleeve (annulus fibrosus). The annulus is made up of numerous fibrous collagenous layers, each consecutive layer possessing alternate fibre orientation.

When loaded in axial compression via the endplates of neighbouring vertebral bodies, the disc acts as a thick-walled pressure vessel. The nucleus is placed in a state of hydrostatic compressive stress which exerts a pressure loading on the endplates and annulus. The annulus acts as a containment structure, its fibres being subjected to tensile forces originating from hoop stresses which are developed in resisting the radial

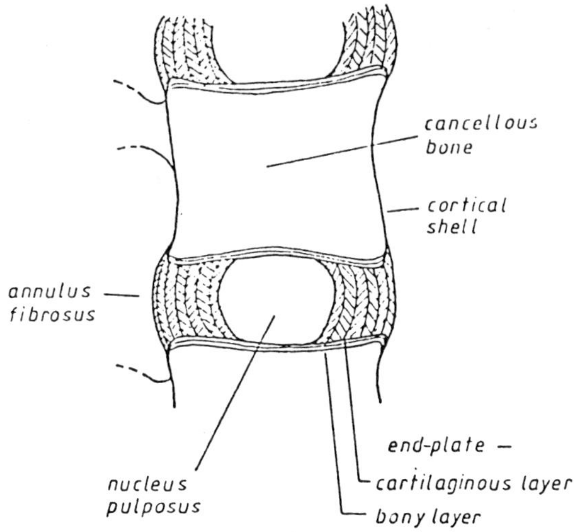

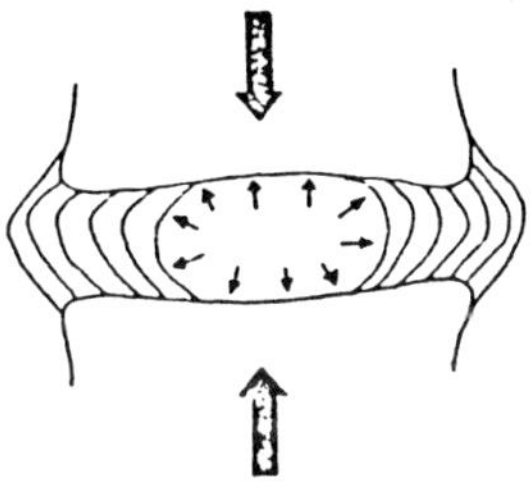

Fig. 6.2 Section through anterior SMS.

component of the hydrostatic pressure in the nucleus. The annulus also provides resistance to axial compressive loads, but to a much lesser extent than the nuclear mechanism, and therefore a non uniform distribution of load is developed across the vertebral endplate.

The spine can be considered as two separate columns: the anterior column made up of vertebral centra and discs; and the posterior column made up of bony posterior elements and associated soft tissue structures. The anterior column is normally in state of axial compression due to the weight of the trunk, head and upper extremities which it supports, although the articular processes of the posterior elements do contribute to resisting this compressive load but to a much lesser extent. The posterior column, on the other hand is normally in a state of tension, especially during physiological movements such as lifting, due to tensile forces in the soft tissues between the posterior elements originating from the supraspinous and interspinous ligaments, muscle groups and the action of the thoracolumbar fascia (Tesh, 1986). Torsional loading is resisted partly by

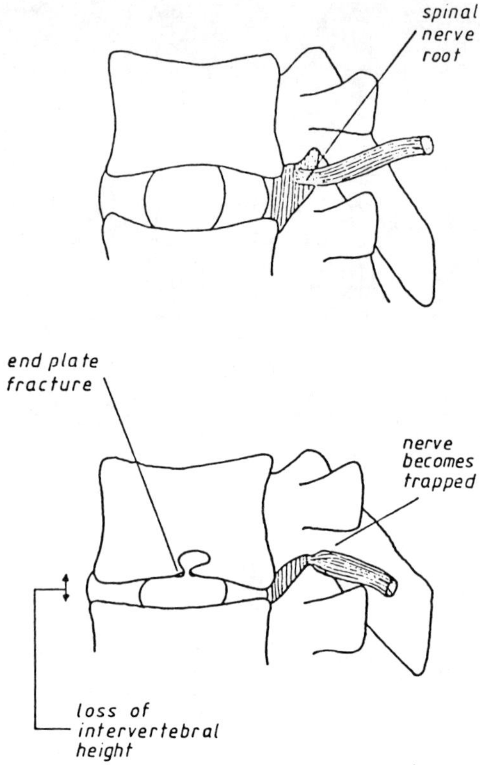

Fig. 6.3 Failure of the SMS.

the annular fibres of the disc and by interaction of the articular processes of the posterior elements.

The centrum resists compression via the geometrical structure of the honeycomb-like cancellous bone. This material has a high strength to weight ratio to enable the centrum to accomodate the large static and dynamic compressive loads to which it is subjected to *in vivo*. The above description is intended as a basic guide to the relevant anatomy, and a more comprehensive account can be found in any of the classic texts or from Panjabi & White (1971).

6.3 **Failure of the SMS**

Failure of a load-bearing structure may be taken as meaning that the structure has been loaded to a point at which damage has occurred and the structure can no longer resist the applied loads. In anatomical load-bearing structures however, failure may be interpreted differently.

For example, damage may occur to a structure in the body which produces no immediate pain or discomfort, so that the injury may go undetected. It can be argued that failure has not occurred since no detrimental effects have ensued. However, the structural damage which may have occurred can alter the balance of forces within the system leading to more extensive damage at a later date when symptoms may then appear. Therefore the consequences of non-catastrophic structural failure in terms of pathological anatomy must be considered in the analysis of anatomical structures.

An illustration of this scenario can be found when considering low back pain; a majority of back pain and referred pain associated with back injury arises from pressure being put on the spinal cord or its associated nerves. A large proportion of this pain is caused by what is commonly known as a 'slipped disc', which occurs when material from the nucleus herniates through a rupture in the annulus and impinges on the spinal cord or nerve root causing pain.

However, another mechanism which produces similar symptoms can be caused by a decrease in disc height, resulting in a decrease of the intervertebral space, when nuclear material herniates into the cancellous bone mass of a vertebral body via a fracture in the intervening vertebral endplate. When combined with the normal translations of successive vertebral bodies due to the physical movements which are part of everyday life, this can result in entrapment of the spinal nerve root producing neuritis which is characterised by pain along the length of the nerve or its branches. The most common manifestation of this condition is sciatica, where pain occurs in the low back and down the back of the leg due to pressure on the sciatic nerve root where it emerges from the lumbar region of the spine.

This type of failure may also be responsible for the formation of Schmorl's nodes (Hilton *et al.*, 1976), which are calciferous nodules found in the cancellous bone of the centrum adjacent to the endplates (and elsewhere). The significance of Schmorl's nodes in the centrum is not fully understood, but Hilton suggests that the type found near the endplate are due to endplate lesions occurring during adolescence, and may predispose the spine to degeneration in later life.

This localised failure of the cancellous bone adjacent to the endplate, leading to fracture of the endplate and ingress of nuclear material, may be responsible for long-term development of back pain which may not easily be attributable to the incident which caused the intitial damage.

6.4 Vertebral cancellous bone

Bone exists within the mature human body in two main forms: compact (dense) bone, commonly known as cortical bone, and cancellous (trabecular or spongy) bone. Compact bone occurs mostly in the diaphyses of the long bones of the skeleton such as the femur, in plate-like bones such as the cranium, and as shell-like containment structures which enclose a mass of cancellous bone, as in the femoral head and vertebral body.

Although the shape and thickness of compact bone may vary in different anatomical structures, the material itself is essentially a composite of organic and mineral content. The relative proportions of the two phases may vary according to the anatomical site of the structure.

Cancellous bone is composed of a matrix of bony elements called trabeculae, arranged with some degree of regularity in a honeycomb-like three-dimensional structure, and exists in different forms throughout the body. The geometrical arrangement of the trabeculae varies according to the geometry and load system associated with the structure. In many instances, such as the femoral head, the trabeculae are arranged according to the pattern of the stress field which is developed in the structure due to physiological loading, (Singh *et al.*, 1972). Moreover, if this pattern is altered due to a change in the applied loads, by surgery for example, then the trabecular structure exhibits a tendency to remodel itself according to the new stress field. Therefore cancellous bone is not a permanent structure but rather a dynamically changing system of trabeculae which attempts to accomodate the loads applied to it in the most efficient manner by constant remodelling and replacement of bone material.

Compact bone structures are not difficult to observe as they generally occur at the gross level, while the geometrical trabecular arrangement of cancellous bone is not easy to observe due to the very much smaller scale of the trabeculae and the difficulties involved in gaining access to cancellous bone within anatomical structures. Consequently, there are conflicting reports in the literature of the macro-structure of cancellous bone, including descriptions of plate-like and strut formations, together with a scarcity of quantitative information. A source of this confusion arises because of the commonly made, but incorrect assumption that the structure of cancellous bone is universal throughout the body.

6.4.1 *Method of investigation*

A significant proportion of the work described here involved an investigation of the macro-structure of vertebral cancellous bone. The full protocol involved in the preparation and investigation of specimens is beyond the scope of this chapter and a full account can be found in the thesis from which this work is derived (Short, 1986); however the basic method will be outlined.

Cadaveric lumbar spines were obtained from which individual vertebrae were dissected. Specimens of vertebral centra were then prepared by removal of discs, posterior elements and other gross soft-tissue structures. The centra were then immersed in a tissue removal solution under the action of heat, which acted to remove all remaining soft tissue including intra-trabecular material comprised mostly of vertebral bone marrow. This procedure was developed so that damage to the trabecular structure was minimised. After dehydration, the specimens were embedded in polymethylmethacrylate (PMMA) prior to sectioning with the aid of a diamond-edged geological circular saw, sections being of the order of 2 mm thick. The PMMA was then dissolved away from the sections by immersion in chloroform and the sections prepared for scanning electron microscopy (SEM) or macrophotography. The embedding process ensured that no artifacts or damage were introduced by the sectioning process.

Generally, the structure of the vertebral cancellous bone was found to be that of an open-celled matrix of beam-like trabeculae of circular cross-section with lengths and cross-sectional diameters of the order of 1.0 mm and 0.1 mm, respectively. Near the endplates, the structure is fairly regular and tightly packed, with the trabeculae arranged in an approximately orthogonal matrix. This type of structure can be modelled in two dimensions as a quadrilateral open-celled foam. Away from the endplates, especially in the central region of the centrum, the structure is less regular and more open. This type of structure can be modelled in two dimensions as a hexagonal open-celled foam, whilst the endplates and cortical shell can be modelled in section by continuous beams. The quantitative data necessary for generation of a representative model was obtained by analysis of the SEM and photographic images using dedicated data collection software incorporated into a projection/digitising tablet system.

The concept of modelling cancellous bone as a foam is not new; Ashby (1970) stated that natural cellular materials, including cancellous bone, could be modelled in this way and analysed theoretically to produce equations of stiffness and strength, while Gibson (1985) applied the same technique specifically to cancellous bone. The work presented here extends Ashby's theory to representation of individual trabeculae by finite beam elements in order that a stress analysis of a complete anatomical structure, including the macro-structure of cancellous bone, may be carried out using the finite element method (FEM).

6.5 **Mechanical properties of cellular materials**

Ashby (1970) explained how a cellular material (foam), when compressed, will exhibit a stress–strain relationship with three distinct phases. At low strains, the foam shows a linear elastic response followed by a plateau which represents continuing deformation at a constant stress. Finally the curve becomes asymptotic to a constant strain value independent of increases in stress. These three phases can be explained in terms of the individual cells which make up the bulk material.

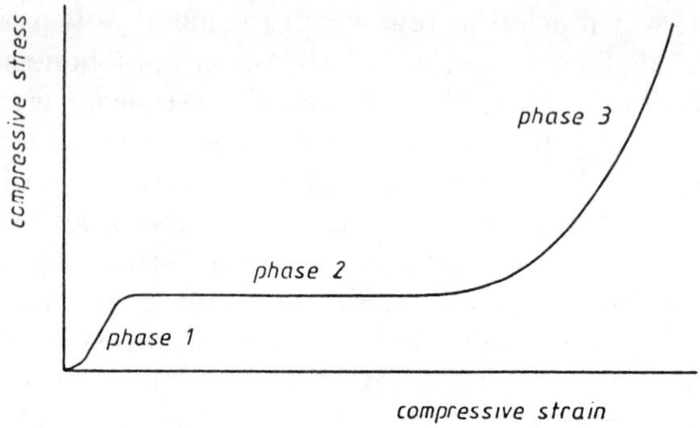

Fig. 6.4 Mechanical behaviour of a cellular solid.

The concepts presented here are a two-dimensional simplification of the real three-dimensional structure, made in order to accomodate limitations of time, resource and practicality. However, justification of this simplification may be made by considering, in the case of the vertebral centrum, that the centrum is a series of two dimensional slices, each acting independently under its own equivalent line load. Since cellular materials exhibit zero Poisson effect in compression, the consequences of lateral spreading of adjacent slices can be ignored. The contribution of neighbouring slices can therefore be considered to be confined to shear resistance in the vertical plane, and in the absence of significant physiological shear stresses of this type within the centrum these effects can also be ignored for the purposes of this model. Therefore a two-dimensional approach can be used in describing the mechanisms involved in the three-phase stress–strain response curve.

When a foam is loaded in compression, the first type of deformation which occurs is due to bending of the cell walls (beams). The amount of deflection produced by the applied force is dependent on the second moment of area, the length and material properties of the beams. This linear elastic phase is limited to small strains (typically 5% in compression), and therefore some other mechanism must be responsible for the phase two plateau region.

There are several possible deformation modes for the phase two response including plastic yielding or brittle fracture, however Townsend *et al.* (1975) concluded that inelastic buckling of trabeculae loaded in axial compression was the most likely failure mode. In the same paper, Townsend combined the classical Euler equation for elastic buckling with the Engresser equation to give the following relationship;

$$s = g \frac{E}{(L/r)^2} \tag{6.1}$$

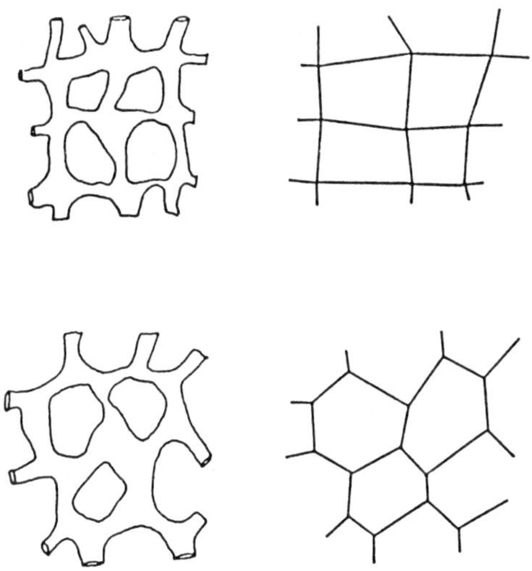

Fig. 6.5 Quadrilateral and hexagonal structure.

where, s = critical buckling stress; E = Young's modulus; r = radius of gyration which = the square root of the second moment of the area (I) cross-sectional area (A); L = length; L/r = slenderness ratio; and g = a constant dependent upon beam end conditions and deviation from ideal Euler behaviour.

The effect of buckling of the trabeculae on the overall stress–strain behaviour is to produce the plateau of constant stress at increasing strain. This behaviour can be described as a phase of energy absorption by the cellular structure, and is utilised in the design of energy absorption systems, such as crash barriers, where foams are used extensively. Ideally then, this phase is initiated when the critical buckling stress for individual beams is reached. Once the beams begin to buckle, deformation increases at a constant load accompanied by an increase in the strain energy of the deformed structure.

As the beams buckle, a point will be reached when the cell walls crush together, usually accompanied by beam fracture. Obvously any further deformation of the cellular structure is impossible since it no longer posesses any coherence. The stress–strain curve reflects this consolidation by exhibiting a sharp increase in stress with negligible change in strain.

6.6 Stress analysis

The basic finite element model of the centrum used for the analysis was composed entirely of beam and spring elements, with the trabecular

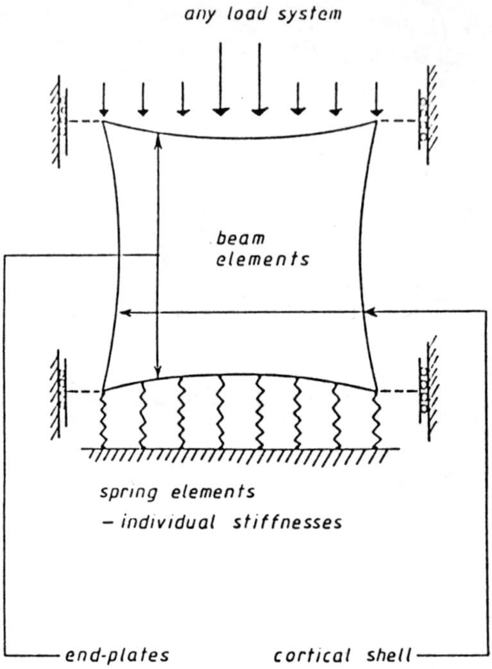

Fig. 6.6 General model.

structure of the vertebral cancellous bone being represented by a matrix of beam elements, each trabecula being composed of two two-noded beams in order to achieve a mid-point node. The cortical shell and vertebral endplates were modelled by continuous beams made up from a series of conventional beam elements while the whole structure was supported via the inferior endplate by a series of spring elements. The support conditions attempted to represent the characteristic response of the inferior intervertebral disc, and the stiffnesses of the springs could be varied independently in order to examine the effects of various support conditions. The beam elements of the cancellous bone were arranged according to two regions: the endplate region where the arrangement was of quadrilateral cells, and the central region where the arrangement was of hexagonal cells.

A dedicated mesh generator, GENEL, was developed for the task of generating the large amount of data required to describe the finite element model. This program uses fundamental parameters as input data, thus enabling manipulation of the mesh by the user with the minimum of data preparation effort, and the program produces graphical and numerical output. The numerical output is in the format of a bulk data file suitable for use with the NASTRAN finite element code. The input data used by

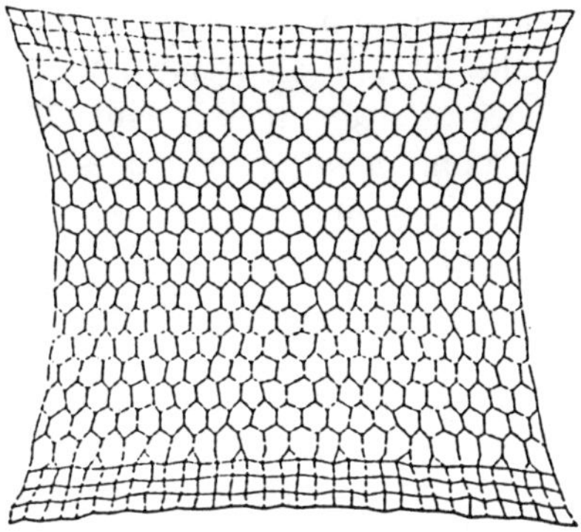

Fig. 6.7 Typical mesh.

GENEL consist of geometric and material parameters which govern features such as beam dimensions, cell geometry and packing, gross dimensions of the mesh, and other characteristics of the model including a randomisation routine which introduces a degree of irregularity into the mesh in order to produce a structure more closely resembling that of cancellous bone. A fuller description of GENEL and other specifically written software can be found in Short (1986). The analysis of the model was performed using the MacNeal–Schwendler/NASTRAN finite element program mounted on the Science and Engineering Research Council's Rutherford Appleton Laboratories computing facility accessed via the Joint Academic Network communications gateway and the University of Strathclyde's mainframe computers.

Due to the large number of elements used in the model (approximately 3000), and the consequent amount of output data produced by NASTRAN, an automated post-processing program was developed called STRESSPLOT. Essentially, this program plotted deflected shape and calculated maximum buckling stress within each individual element. This value was then compared with the value of the critical buckling stress derived from eqn (6.1) for each element and used as a criterion for failure on an element-by-element basis. A point to note is that STRESSPLOT calculates values for g in eqn (6.1) for each element separately, using a polynomial equation fitted to Townsend's (1975) graph of g versus slenderness ratio. The graphical results produced by STRESSPLOT took advantage of colour plots to provide greater clarity; the typical results presented here are monochrome derivations of the original results.

There were two objectives in performing analyses of the model; to validate the behaviour of the model against experimental data, and to investigate the structural mechanics of the vertebral centrum. Existing experimental data concerning the testing of cancellous bone was of limited use in validating the model since much of this work was performed using samples excised from the original anatomical structure, providing comparative numerical values of strength, but with little relationship to the behaviour of cancellous bone *in situ* within the structure under physiological loading. Such tests show marked inconsistency between tests and authors, and include attempts at defining strength using techniques such as indentation testing which are not applicable to cellular materials due to the absence of Poisson's effect in compression. It should be emphasised that the criteria for evaluation and comparison of properties for cellular materials cannot be assumed to be identical to those used for conventional continuum materials, especially in a complex anatomical situation. Therefore although results based on these techniques may be useful for comparative purposes, they were of little use in validation of the model. Similarly, previous mechanical test data of vertebral bodies were not suitable for validation because of the variations between test conditions used by those workers; the conditions specified in this work being a good simulation of those thought to exist *in vivo*.

A programme of mechanical testing was therefore carried out using specimens of human vertebral centra derived from cadaveric spines. Loading of the specimens was achieved using an Instron hydraulic testing machine acting on a compression rig designed specifically for this work which held the specimen. Within the rig the specimen was mounted between two silastic J-RTV silicone spacers, which in turn were mounted between steel loading platens. The spacers and platens were used to represent the compliance of the discs and were profiled in order to reproduce the non-uniform load distribution across the endplates described *in vivo* by Nachemson (1960). The theory behind calculating the profiles used the fact that the stiffness of the rubber at a particular position is inversely proportional to the thickness of the rubber at that point. The required load distribution for a healthy disc, based on values of intra-discal pressures measured by Nachemson, was that load applied to the endplate by the nucleus was twice the applied by the annulus as a percentage of that total load, assuming that nucleus is in contact with the central third of the endplate (Short, 1986).

The specimens were loaded at a constant cross-head speed of $2\,\text{mm}\cdot\text{min}^{-1}$, with the applied load being measured via a load cell incorporated into the cross-head. The load-deformation response of the specimen was obtained by subtraction of the preconditioned response curve of the rubber spacers from that of the specimen plus spacer system. The resulting curves showed the phase 1 and phase 2 response characteristics of a classical cellular material, as would be expected, and were in qualitative agreement with previous curves such as those obtained by Kazarian (1977). From the resultant curves it was possible to derive values of phase 1 linear elastic stiffness, and the magnitude of the phase 2 failure (plateau) load, which could then be used for validation of the model.

6.7 **Results**

The gross dimensions of the finite element models were chosen to agree approximately with those of the physical specimens used in the mechanical testing programme, while the dimensions of the beam elements, used to represent the trabecular structure, the cortical shell and the vertebral endplates, were derived from the data collection work described previously, although because of the randomisation of the mesh the lengths of the beams varied about a mean value. Young's modulus and mass density for the trabecular bone, cortical shell and bony endplates were assumed to be those of dense bone and were taken from the results of ultrasonic non-desctructive tests performed by Bonfield & Tully (1982), whilst an arbitrarily fixed value of Poisson's ratio was used throughout since its effect on the global solution was small. Typical values used were:

Young's modulus = $23\,000\ \mathrm{N\cdot mm^{-2}}$
Mass density = $2E{-}6\ \mathrm{kg\cdot mm^{-3}}$
Poisson's ratio = 0.1

Under loading and restraint conditions similar to those used in the mechanical tests, the finite element model was analysed at progressively increasing load magnitudes. This iterative analysis was accompanied by removal of elements deemed to have buckled, according to the criteria described previously at intermediate load levels, before proceeding with the next iteration. The results of the analyses produced deformation characteristics of the classical phase 1 and phase 2 response curve of a cellular material, giving excellent qualitative agreement with experimental curves. Quantitatively, model predictions of phase 1 stiffness and phase 2 load magnitude were compared after compensating for the absence of a third dimension in the model (Short, 1986).

Table 6.1 *Typical results of compression tests of centra.*

Specimen	Height (mm)	Width (mm)	Phase 1 stiffness $\mathrm{kN\cdot mm^{-1}}$	Phase 2 load (kN)
Male (26 yrs)	28	34	0.7	1.75
Male (26 yrs)	28	36	0.5	1.78
FE model	30	33	0.4	1.78

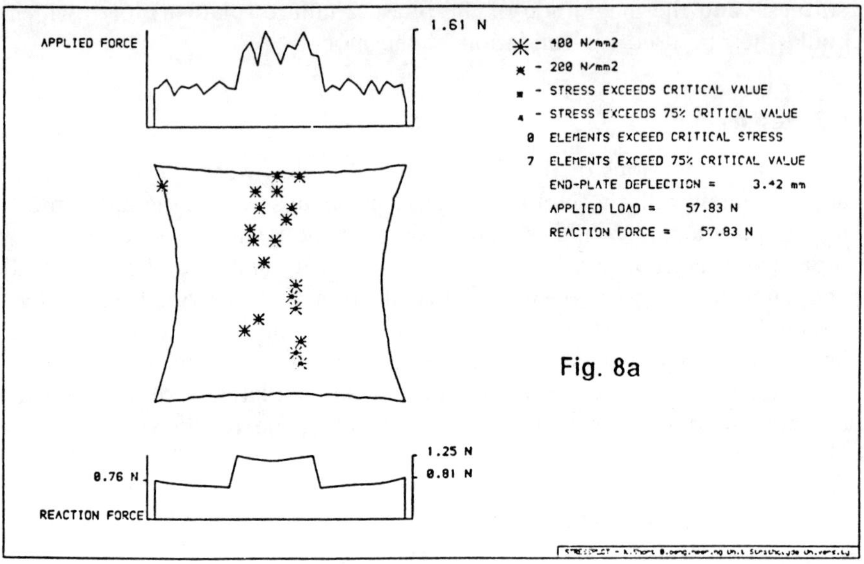

APPLIED FORCE
1.61 N
- 400 N/mm2
- 200 N/mm2
- STRESS EXCEEDS CRITICAL VALUE
- STRESS EXCEEDS 75% CRITICAL VALUE
0 ELEMENTS EXCEED CRITICAL STRESS
7 ELEMENTS EXCEED 75% CRITICAL VALUE
END-PLATE DEFLECTION = 3.42 mm
APPLIED LOAD = 57.83 N
REACTION FORCE = 57.83 N
1.25 N
0.81 N
0.76 N
REACTION FORCE

Fig. 8a

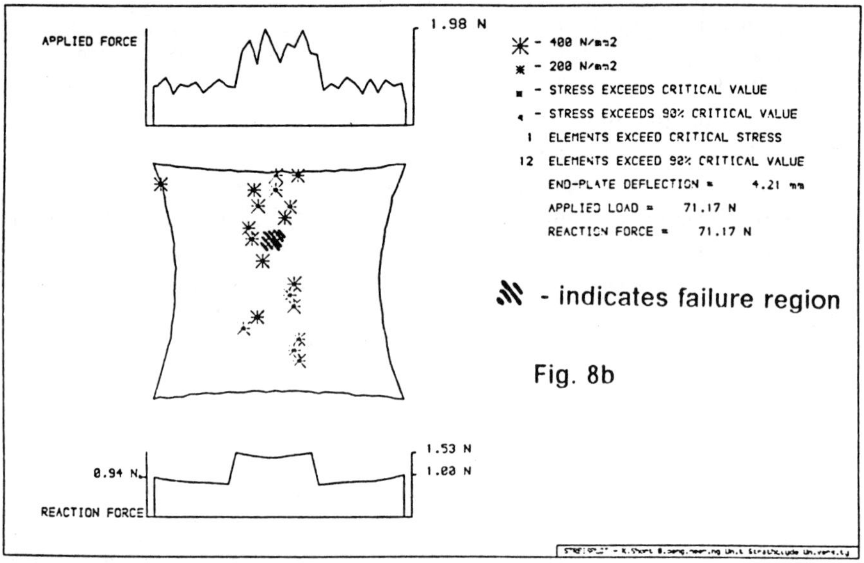

APPLIED FORCE
1.98 N
- 400 N/mm2
- 200 N/mm2
- STRESS EXCEEDS CRITICAL VALUE
- STRESS EXCEEDS 90% CRITICAL VALUE
1 ELEMENTS EXCEED CRITICAL STRESS
12 ELEMENTS EXCEED 90% CRITICAL VALUE
END-PLATE DEFLECTION = 4.21 mm
APPLIED LOAD = 71.17 N
REACTION FORCE = 71.17 N
- indicates failure region
1.53 N
1.00 N
0.94 N
REACTION FORCE

Fig. 8b

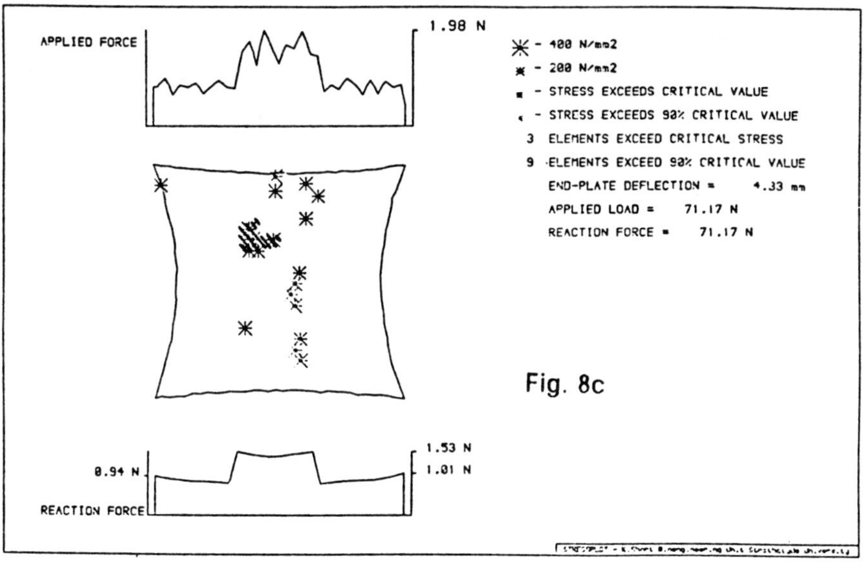
APPLIED FORCE
1.98 N
- 400 N/mm2
- 200 N/mm2
- STRESS EXCEEDS CRITICAL VALUE
- STRESS EXCEEDS 90% CRITICAL VALUE
3 ELEMENTS EXCEED CRITICAL STRESS
9 ELEMENTS EXCEED 90% CRITICAL VALUE
END-PLATE DEFLECTION = 4.33 mm
APPLIED LOAD = 71.17 N
REACTION FORCE = 71.17 N
1.53 N
0.94 N
1.01 N
REACTION FORCE

Fig. 8c

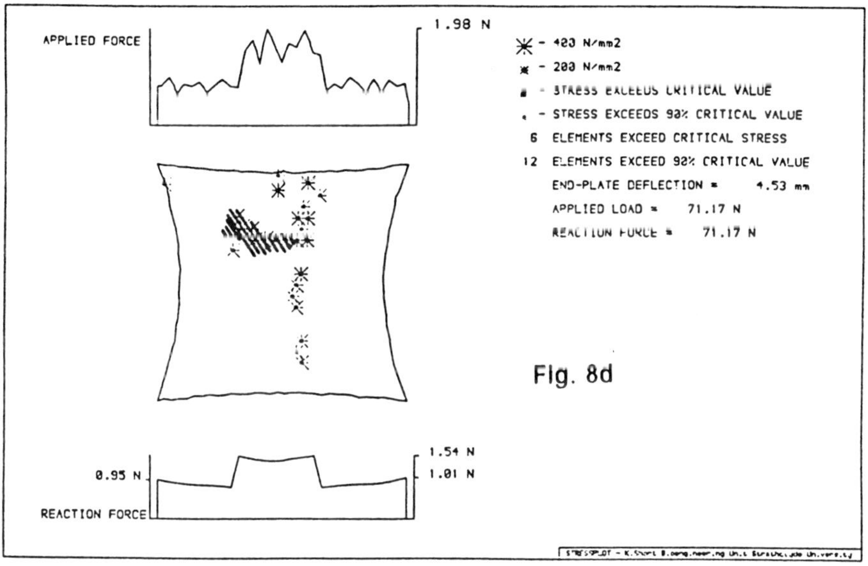
APPLIED FORCE
1.98 N
- 400 N/mm2
- 200 N/mm2
- STRESS EXCEEDS CRITICAL VALUE
- STRESS EXCEEDS 90% CRITICAL VALUE
6 ELEMENTS EXCEED CRITICAL STRESS
12 ELEMENTS EXCEED 90% CRITICAL VALUE
END-PLATE DEFLECTION = 4.53 mm
APPLIED LOAD = 71.17 N
REACTION FORCE = 71.17 N
1.54 N
0.95 N
1.01 N
REACTION FORCE

Fig. 8d

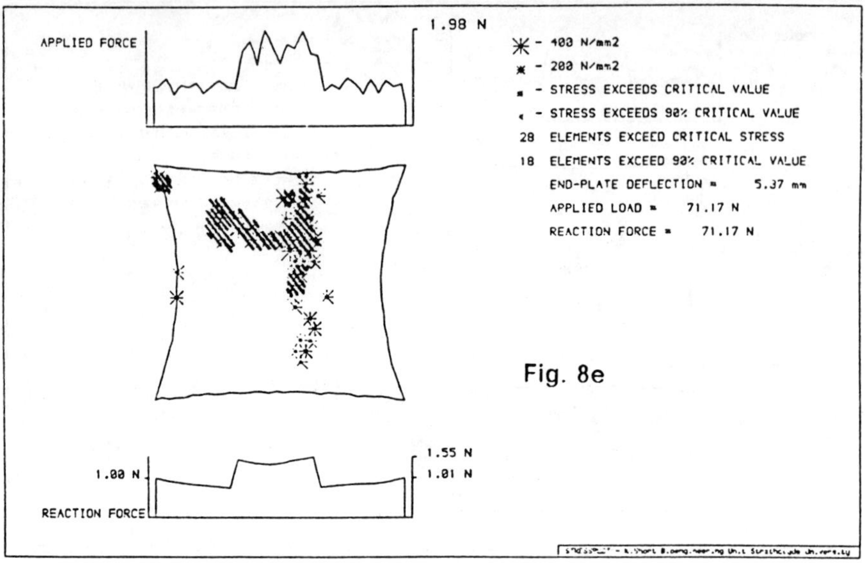

1.98 N
APPLIED FORCE
- 400 N/mm2
- 200 N/mm2
- STRESS EXCEEDS CRITICAL VALUE
- STRESS EXCEEDS 90% CRITICAL VALUE
28 ELEMENTS EXCEED CRITICAL STRESS
18 ELEMENTS EXCEED 90% CRITICAL VALUE
END-PLATE DEFLECTION = 5.37 mm
APPLIED LOAD = 71.17 N
REACTION FORCE = 71.17 N
1.55 N
1.00 N
1.01 N
REACTION FORCE

Fig. 8e

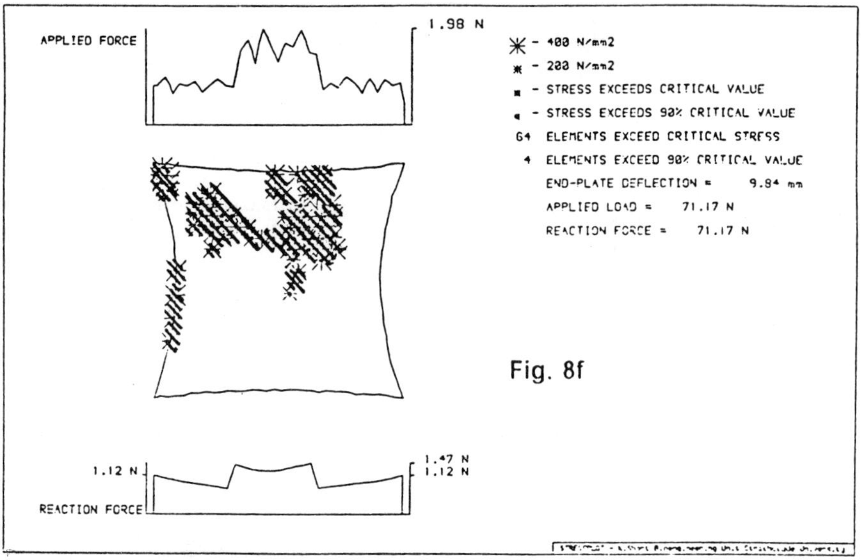

1.98 N
APPLIED FORCE
- 400 N/mm2
- 200 N/mm2
- STRESS EXCEEDS CRITICAL VALUE
- STRESS EXCEEDS 90% CRITICAL VALUE
64 ELEMENTS EXCEED CRITICAL STRESS
4 ELEMENTS EXCEED 90% CRITICAL VALUE
END-PLATE DEFLECTION = 9.84 mm
APPLIED LOAD = 71.17 N
REACTION FORCE = 71.17 N
1.47 N
1.12 N
1.12 N
REACTION FORCE

Fig. 8f

It was concluded from the experimental results that the model provided a good simulation of the human vertebral centrum within the constraints of the validation criteria. Therefore there is a high probability that the model will behave in a similar way to a physical specimen when subjected to conditions which are beyond the scope of validatory evidence. In fact this assumption underlies the technique of mathematical modelling which uses simple situations for validation purposes in order to predict the effects of complicated or unexpected situations which cannot be validated directly.

Analysis of the model, under axial loading similar to that associated with normal disc loading *in vivo*, predicted that at some elevated load level the area of trabecular failure within the cancellous matrix would progressively increase, whilst maintaining a constant load at the level at which initial failure occurred. The results suggested that the extent of the failure area depended on the duration of loading at the critical level and predicted characteristic zones of trabecular disruption. There was an agreement between model behaviour at an intermediate load step and clinical observation of load patterns *in vivo* caused by trauma associated with vertical falls described as crush–cleavage fractures (Lindahl *et al.*, 1983). Consideration of the series of analyses suggested that maintenance of the initial failure load produces a cascade effect of trabecular failure with the total area of failure accelerating with duration of load. However, in situations where that total applied load was reduced after initiation failure within the cancellous bone early in the iterative analysis, the model showed that the failure region did not enlarge even with the application of significant proportions of the original failure load. This was accompanied by a redistribution of stress within the trabecular matrix away from the damage area to structurally intact cancellous bone illustrated by changes in the stress distribution throughout the model.

When the above analysis was extended by continuing application of the load to a state of advanced failure, the load transmission properties of the

Fig. 6.8 Progressive failure of the model: * shows where stress exceeds 90% of critical value; cross hatching indicates failure region. (a) No elements exceed critical stress; seven elements exceed 75% critical value. Endplate deflection = 3.42 mm; applied load = 57.83 N; reaction force = 57.83 N. (b) One element exceeds critical stress; 12 elements exceed 90% critical value. Endplate deflection = 4.21 mm; applied load = 71.17 N; reaction force = 71.17 N. (c) Three elements exceed critical stress; nine elements exceed 90% critical value. Endplate deflection = 4.33 mm; applied load = 71.17 N; reaction force = 71.17 N. (d) Six elements exceed critical stress; 12 elements exceed 90% critical value. Endplate deflection = 4.53 mm; applied load and reaction force both = 71.17 N. (e) 28 elements exceed critical stress; 18 elements exceed 90% critical value. Endplate deflection = 5.37 mm; applied load and reaction force both = 71.17 N. (f) 64 elements exceed critical stress; four elements exceed 90% critical value. Endplate deflection = 9.84 mm. Applied load and reaction force both = 71.17 N.

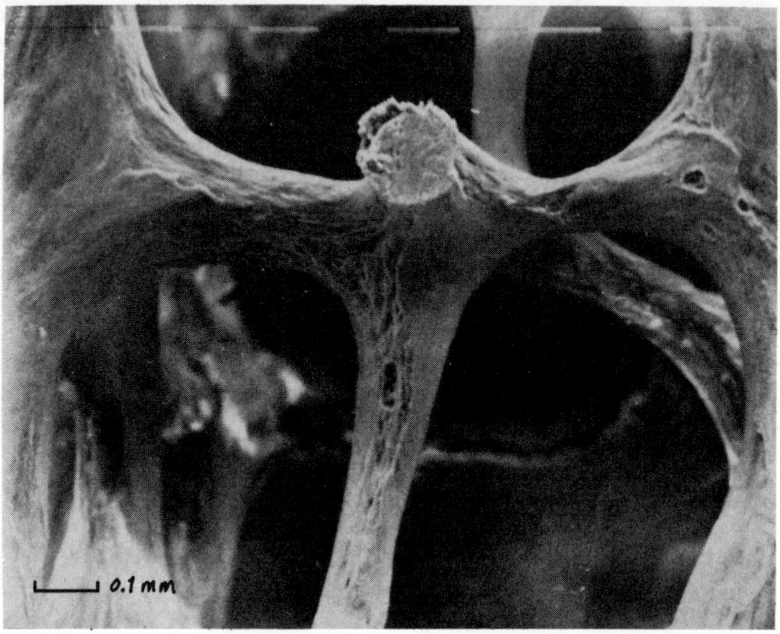

Fig. 6.9a Scanning electron micrograph of human vertebral cancellous bone.

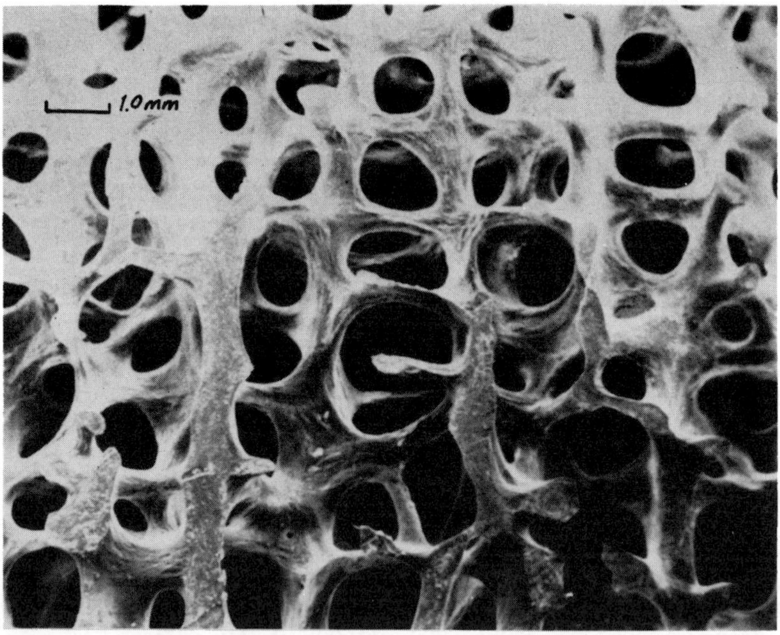

Fig. 6.9b Scanning electron micrograph of human vertebral cancellous bone

model were found to change so that the load applied by the centrum to the inferior disc became redistributed away from the nuclear region in favour of the lateral margins (annular region) of the endplate. It is suggested that such a redistribution may be detrimental to the nutrition of the disc by affecting the mechanical pumping action, which has been described as an active transportation mechanism for nutrients to the nucleus via the end-plates during cyclical axial compression of the spine during physiological movements such as locomotion.

Investigation of the contribution of the cortical shell in resisting axial compressive load was carried out by duplicating a conventional analysis on a model with cortical shell finite elements removed. Comparison of the results showed that approximately 10% of the total load was transmitted via the cortex, which supports the theory that the majority of the compressive strength of the centrum is derived from the cellular trabecular matrix and that the role of the cortex lies in containment of vertebral bone marrow and other secondary functions. However, it should be noted that the containment of bone marrow under high compressive strain rates may be important in dynamic loading of the centrum due to the viscosity of the marrow and its difficulty in flowing laterally under high strain rates. In this case a hydrostatic compressive stress within the marrow may contribute to the overall resistance to deformation (Kazarian, 1977).

Analyses were performed with varying degrees of redistribution of applied compressive load, away from the nuclear region of the endplate to the annular regions, in order to simulate disc degeneration where the nucleus of the intervertebral disc is unable to develop its full potential hydrostatic pressure and therefore cannot transmit its normal level of axial load, due to dehydration of the nuclear material. Under these loading conditions, the model predicted a change in distribution of stress within the trabecular matrix, compared with normal disc loading, away from the central region towards the lateral extremities of the cancellous bone. It can be argued that such a depression in stress levels may promote osteoporosis, which is essentially demineralisation, and hence loss of strength in bone, by the following mechanism.

In healthy bone, the deposition of new bone material in the cancellous matrix is promoted by cyclical loading of the trabeculae (Burr *et al.*, 1983), therefore removal of a proportion of the normal physiological load from a region of cancellous bone is likely to upset the balance between tissue removal and deposition to result in a net decrease in bone tissue. This can lead to formation of regions of osteoporosis within the cancellous bone. By extending this hypothesis, an argument can be put forward whereby degeneration of the intervertebral discs at a particular level in the spine may encourage degeneration of discs at other levels through promotion of osteoporosis in intervening centra. This is because the loading applied to a healthy disc by an osteoporotic centrum will tend to produce the effects described previously, where applied loads were shifted away from the nuclear (osteoporotic) region to the annular region of the endplate with the accompanying detrimental effect on disc pumping.

A series of analyses was performed in which the model was supported inferiorly by a simulation of a bone chip of the type used in surgical interbody fusion. These analyses were speculative in nature and were intended to give an indication of future possibilities. The width of the graft was varied in each analysis and the resulting stress distributions showed that a centrally located graft of approximately four-fifths of the width of the endplate was best at simulating the stress distribution associated with normal disc support and loading. Moreover, the results showed susceptibility to gross shear failure of the centrum at the lateral margins of the graft, which suggests that the avoidance of sharp edges when fashioning the graft material is important in limiting the type of stress concentrations observed at the graft edges.

6.8 **Discussion**

The results presented here are in a much summarised form and reference to the thesis from which they were extracted should be made for a full and proper description. With hindsight, the analysis method could be improved to provide a more elegant solution by the introduction of large displacement, post-buckling and non-linear considerations which were not available at the time. Moreover, an input of clinical expertise into any future work would be necessary in order to achieve physically significant results rather than those obtained from academically chosen boundary conditions.

It is hoped that this chapter, whilst not providing a detailed and conclusive solution to the problem, will encourage future work to be carried out using a similar modelling philosophy and help point the way towards the development of useful mathematical tools to assist the clinician by enabling biomechanical analyses to be carried out for existing and novel applications.

References

Ashby, M. F. (1983). The metallurgical properties of cellular solids. *Metallurgical Transactions A*, **14A**, 1755–69.

Bonfield, W. & Tully, A. E. (1982). Ultrasonic analysis of the Young's modulus of cortical bone. *Journal of Biomedical Engineering*, **4**, 23–27.

Burr, D. B., Martin, R. B. & Martin, P.A. (1983). Low extremity loads stimulate bone formation in the vertebral column: Implications for osteoporosis *Spine*, **8**, 681–6.

Gibson, L. J. (1985). The mechanical behaviour of cancellous bone. *Journal of*

Biomechanics, **5**, 317–28.

Hilton, R. C., Ball, J. & Benn, R. T. (1976). Vertebral end-plate lesions (Schmorl's nodes) in the dorsolumbar spine. *Annals of Rheumatic Disease*, **35**, 127–32.

Kazarian, L. & Graves, G. A. (1977). Compressive strength characteristics of the human vertebral centrum. *Spine*, **1**, 1–14.

Lindahl, S., Willen, J., Wordwall, A. & Irstam, L. (1983). A 'new' thoracolumbar unstable fracture. *Spine*, **6**, 559–69.

Nachemson, A. (1960). Lumbar intradiscal pressure. *Acta Orthopaedica Scandinavica*, Suppl. XLIII.

Panjabi, M. M. & White, A. A. (1980). *Clinical Biomechanics of the Spine*. Lippincott, Toronto.

Singh, M., Riggs, B. L., Beabout, J. W. & Jowsey, J. (1972). Femoral trabecular index for evaluation of spinal osteoporosis. *Annals of Internal Medicine*, **77**, 63–7.

Short, K. (1986). A finite element model of the human vertebral centrum. PhD Thesis, Bioengineeing Unit, University of Strathclyde.

Tesh, K. M. (1986). The abdominal muscles and vertical stability. PhD Thesis, Bioengineering Unit, University of Strathclyde.

Townsend, P. R., Rose, K. M. & Radin, E. L. (1975). Buckling studies of single human trabeculae. *Journal of Biomechanics*, **8**, 199–201.

Further reading

Cowin, S. C. (1986). Wolff's Law of trabecular architecture at remodelling equilibrium. *Journal of Biomechanical Engineering*, **108**, 83–8.

Dyson, E. D., Jackson, C. K. & Whitehouse, C. K. (1983). Scanning electron microscope studies of human trabecular bone. *Nature*, **225**, 957–9.

Lin, P. M. (ed.) (1982). *Posterior Interbody Fusion*. Charles C. Thomas, Springfield, Illinois.

Zienkiewicz O. C. (1980). *The Finite Element Method*, (3rd Edition). McGraw-Hill, London.

7

Correlation between apatite crystal orientation in bone and stress trajectories: a preliminary study

A. L. Yettram

7.1 **Introduction**

A method developed by Bacon *et al.* (1979a) uses neutron diffraction to measure the preferred orientation of the apatite crystals in a bone structure. They have applied the method to various bones in the human body, *viz*. the scapula (Bacon *et al.*, 1979b), the vertebrae (Bacon *et al.*, 1979c), the mandible (Bacon *et al.*, 1980), the femur (Bacon & Griffiths, 1985) and the bones of the foot (Bacon *et al.*, 1984). Furthermore, they have described the method in general terms, and also their hypothesis that the crystal orientation corresponds to stress trajectories, in a further article (Bacon *et al.*, 1981)

Bacon *et al.* have tested their theory against simplified stress analysis models for some of their cases. These tended to be from photoelastic experiments, such as those of Oxnard (1973) with respect to the scapula, and the basic proposition appeared to be valid. The main virtue in the use of apatite crystal orientation measurements as an indication of principal stress trajectories over, say, the familiar trabecular theory, lies in the better certainty with which the former can be determined compared to the possible vagueness which can enter into the visual interpretation of patterns of trabeculae. Knowledge of the major principal stress directions in bone is clearly of great potential importance in that if the muscle force values and directions which correspond to them can be set up, they can then be used with some confidence when artificial replacements for bone are being designed.

The work of Bacon *et al.*, although so widely publicised, suprisingly has aroused little response or interest from stress analysts in the orthopaedic field. With so much work being carried out at present in terms of finite element analysis (FEA) of orthopaedic structures, the crystal orientation concept could be used as an important check against any particular FEA

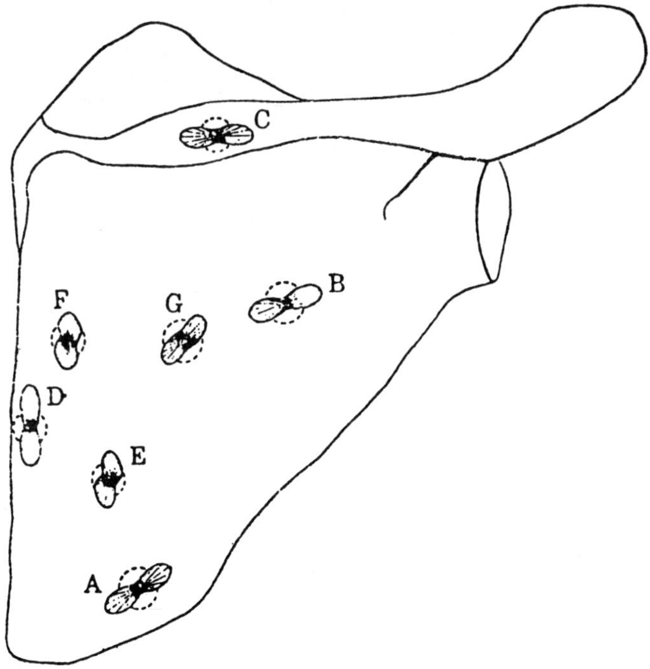

Fig. 7.1 Sites A–G on scapula with apatite crystal alignment indicated. (From Bacon, G. E., Bacon, P. J. & Griffiths, P. K., 1979b; reprinted by permission of *Proceedings of the Royal Society, London*, **B204**, 359.)

models. The work presented here is a preliminary investigation into the usefulness of these procedures. As such the FEA modelling was fairly simple. For example, the models were assumed to be completely isotropic and to behave in a linearly elastic manner. The Pafec FEA package (Henshell, 1984) was used throughout. Two situations were dealt with and the main results are reported below.

7.2 The scapula

The scapula is one of the most complicated bones in the human skeleton and is subjected to a great variety of forces from attached muscles during elevation of the shoulder girdle. The orientation of the apatite crystals at several sites A to G, as presented by Bacon *et al.* (1979b), are shown in Figure 7.1.

A finite element model of the scapula was constructed by embedding an actual bone into a block of Araldite, sectioning it in a predetermined way, as shown in Figure 7.2, photographing and enlarging each slice to a

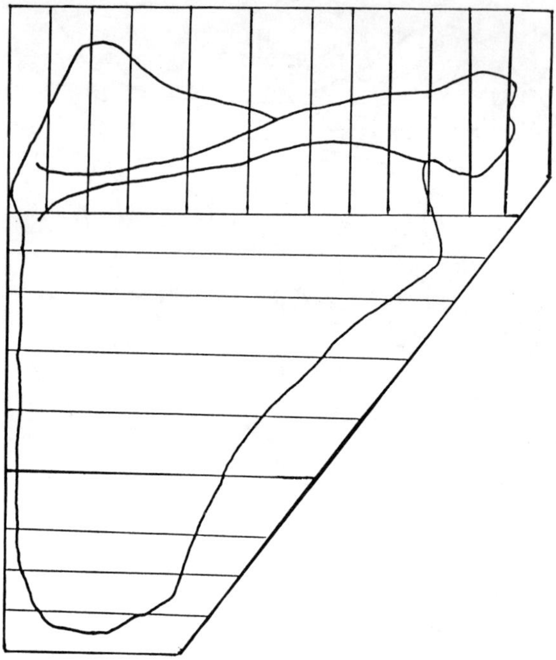

Fig. 7.2 Arrangement for slicing scapula embedded in a block of Araldite.

set scale, selecting nodal points on each section, digitising them into a computer file and then adding the element topology. Although the scapula consists of an outer layer of cortical bone of varying thickness and an inner core of cancellous bone, for simplicity the model was here considered as homogeneous throughout with average elastic properties (modulus of elasticity = 13.8 $kN \cdot mm^{-2}$; and Poisson's ratio = 0.3). Twenty-noded brick and 15-noded wedge elements were used resulting in 110 elements with 838 nodes. The mesh is shown in Figure 7.3.

Oxnard (1973), MacConaill & Basmajian (1969) and Dvir & Berme (1978) all suggest that the principal motion of the scapula is a rotation about the point of attachment of the lower trapezius muscle at the scapular spine, this point being known as 'the root of the spine' (RS). This results in the upper trapezius and serratus anterior muscles forming a couple that rotates the scapula upward. The first load case considered here used this simplified approach for 90° of elevation with the force in the former being 180 N and in the latter 100 N, both lying in the mean plane of the scapula. Restraints were applied at the RS point along all axes and sufficient further restraints were added to prevent rigid body movement. The results at some

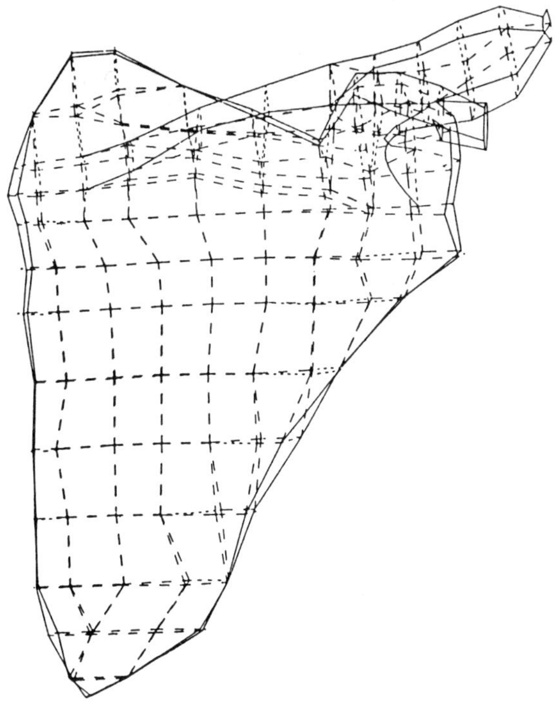

Fig. 7.3 Finite element mesh for stress analysis of scapula.

of the selected sites showed fair agreement with the results of Bacon *et al.* (1979b), at others the correlation was less good. An investigation into the reactive forces at the assumed constraints showed them to be excessively high, overwhelming the applied forces.

In a second load case, a set of self-equilibrating forces was applied to the model with restraint only to provide against rigid body movement. A just sufficient number of muscle forces were applied so that their values could be obtained from static equilibrium conditions in terms of the prescribed muscle forces given by Inman *et al.* (1944). This case is shown in Figure 7.4, where the letters A to E represent the insertions and lines of action of the deltoid, upper trapezius, levator scapulae and upper serratus anterior, lower trapezius and lower serratus anterior muscles, respectively. *R* represents the point of reaction at the gleno-humeral joint. It is assumed that the reaction here is totally in the z-direction. The situation is, of course, three-dimensional and although the scapula lies generally in the zx-plane, there are components of force in all three directions. The values of the force components used in the set are as in Table 7.1.

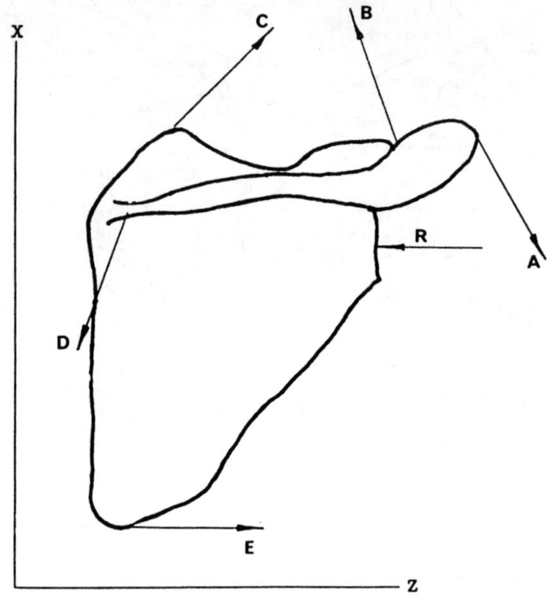

Fig. 7.4 Configuration of system of forces applied to scapula.

The results for this load case are shown in Figure 7.5 as principal stress vectors from the mid-plane of the structure at the seven selected sites in Figure 7.1. Generally this second load case, which represents a self-equilibrating set of muscle forces, provides stress trajectories which correlate better with the crystal orientations. This in itself is not proof of the original hypothesis but could be considered as supporting the possibility of its validity.

Table 7.1 *Force system assumed to act on scapula.*

Muscle	Location in Figure 7.4	*x*-component (N)	*y*-component (N)	*z*-component (N)
Deltoid	A	−106.0	−40.0	61.5
Upper trapezius	B	83.5	60.0	−30.5
Levator scapulae and upper serratus anterior	C	57.5	−6.0	57.5
Lower trapezius	D	−35.0	−10.0	−12.5
Lower serratus anterior	E	0.0	−4.0	46.5
(Reaction)	R	0.0	0.0	−122.5

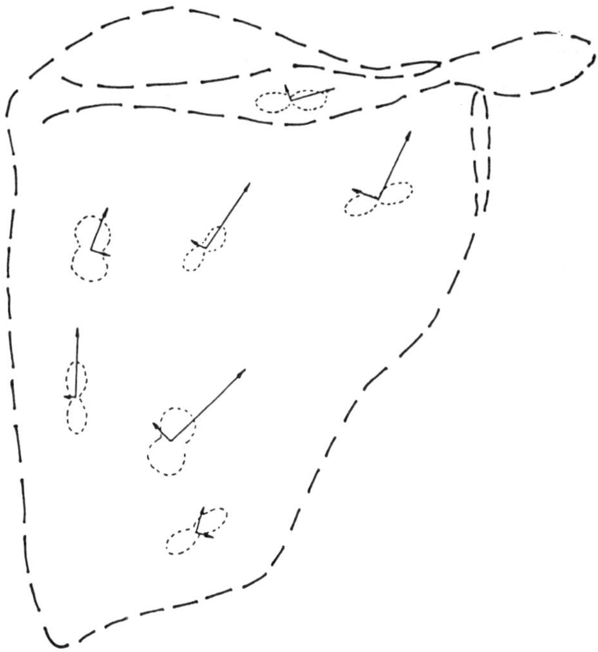

Fig. 7.5 Principal stress vectors at sites A–G in scapula overlaid on crystal orientation plots.

7.3 **The mandible**

The crystal orientations for the mandible given by Bacon *et al.* (1980) are shown in Figure 7.6. An FEA model of the half-mandible was constructed by a similar process to that employed for the scapula. In the mandible the cortical bone is quite thin and the cancellous bone is extremely open. Hence this bone was modelled using triangular, six-noded flat plate elements for the cortical bone while the cancellous bone was ignored. From symmetry only one half of the mandible was modelled, the sectioning being as shown in Figure 7.7. The model, shown in Figure 7.8, resulted in 1497 elements with 776 nodes.

The load case considered for this analysis follows that used by Barbenel (1972) and Arvikar & Seireg (1975). Their data are then transformed into a set of four muscle forces active in the process of biting with a set occlusal force, as shown in Figure 7.9. Restraint is applied in the form of a 'ball and socket' support at the temporo-mandibular joint and also in the x-direction at all the nodes on the plane of symmetry. The letters A to D represent the insertions and the lines of action of the temporalis, external pterygoid, internal pterygoid and masseter muscles respectively. E represents the

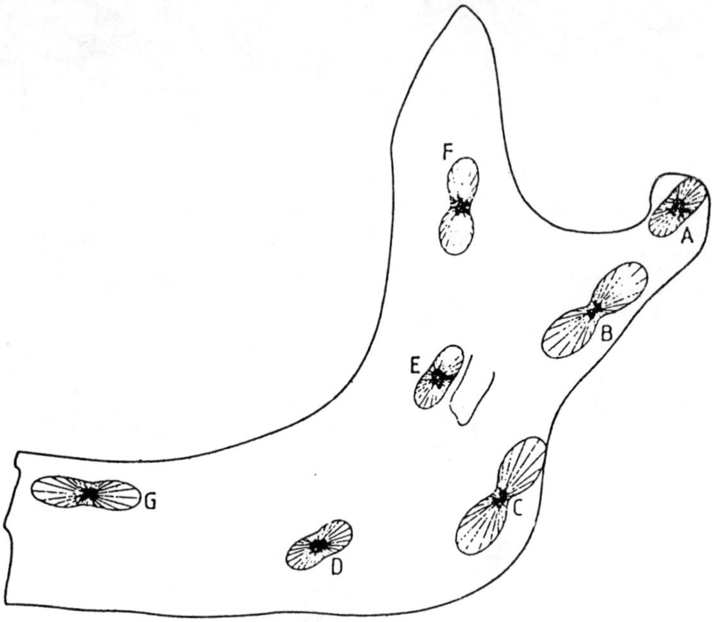

Fig. 7.6 Sites A–G on mandible with apatite crystal alignment indicated. (From Bacon, G. E., Bacon, P. J. & Griffiths, P. K., 1980; reprinted by permission of *Journal of Biomechanics*, **13**, 726.)

occlusal force and R_1 the reaction at the temporo-mandibular joint. The values of the force components used in the set are given in Table 7.2.

Figure 7.10 shows the corresponding vector plots of principal stress direction compared to those proposed by Bacon *et al.* (1980) at their measured sites of Figure 7.6. Overall the correlation in direction is reasonably good. The principal stresses in the body of the mandible are generally directed along its length while those in the ramus are mainly directed toward the temporo-mandibular joint.

7.4 **Conclusion**

In the late nineteenth century Wolff (1892) published his work on bone structure and form, *The Law of Bone Transformation*. In it he formulated the principle which has become known as Wolff's Law, that is that 'the form of bone is determined by its mechanical function'. In other words the external shape and the internal structure of a bone is arranged in such a way as to produce an optimum structure, one which will transmit the applied forces within an acceptable range of stress and with the minimum amount of material. Wolff's Law has been generally accepted since its

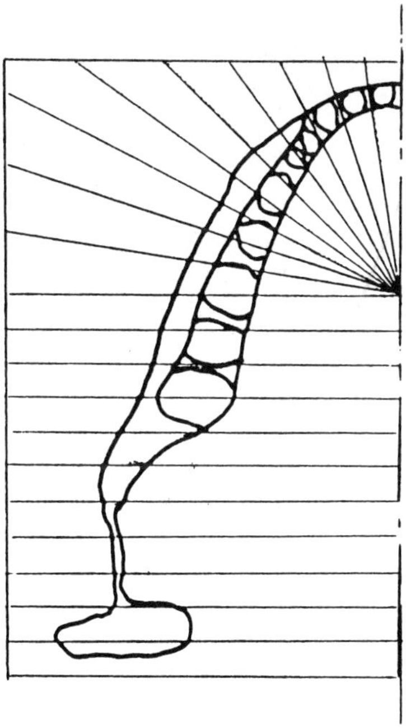

Fig. 7.7 Arrangement for slicing semi-mandible embedded in a block of Araldite.

inception. Koch (1917), in considering the femur, argued that it was not necessary to know every possible stress state in a bone to discover whether it is designed for mechanical efficiency. In line with this, he analysed the stress distribution in the proximal end of the femur and, ignoring the effect of muscle force, confirmed that the stress trajectories in the proximal femur corresponded with the trabeculae in the cancellous bone.

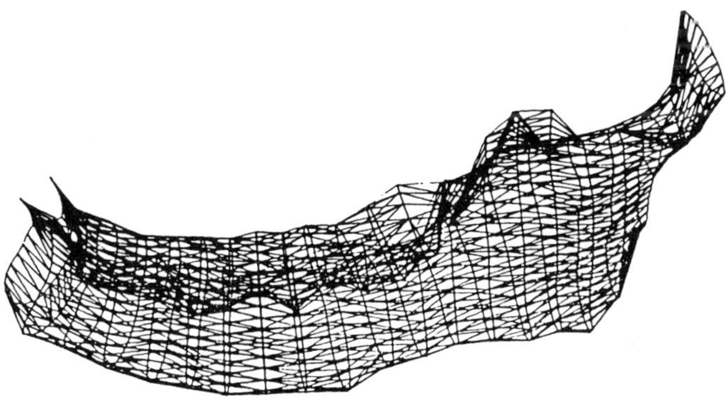

Fig. 7.8 Finite element mesh for stress analysis of semi-mandible.

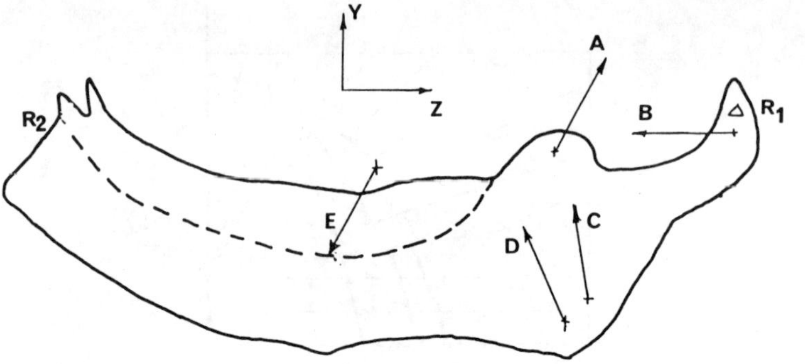

Fig. 7.9 Configuration of system of forces applied to semi-mandible.

In the case of long load-bearing bones, such as the femur, it may well be that the dominant stress state is determined by an easily defined loading system such as is enforced through articulations at the ends. For other bones, such as those considered earlier, the 'designed for' loading pattern is not so obvious. Also the trabecular pattern cannot be easily interpreted and so may be of little help in determining what such a pattern might be. It is here that the crystal orientation mapping of Bacon *et al.* (1981) could be most useful in applied orthopaedics.

The preliminary finite element study here generally supports the hypothesis of Bacon *et al.* (1981) with regard to apatite crystal orientation indicating preferred stress direction. The FEA models were fairly simple and further refinement is required before the experimental values can be used to determine the corresponding distribution of muscle forces definitively. This would involve an iterative process of adjustment and readjustment to produce good correlation at all the measured sites.

Table 7.2 *Force system assumed to act on mandible.*

Muscle	Location in Figure 7.9	*x*-component (N)	*y*-component (N)	*z*-component (N)
Temporalis	A	−3.04	14.19	−7.77
External pterygoid	B	3.54	−1.54	8.25
Internal pterygoid	C	0.01	0.01	0.01
Masseter	D	−0.90	6.26	4.10
Occlusal force	E	0.00	−21.28	2.04
Reaction at temporo-mandibular joint	R_1	−0.26	2.36	−6.63
Reaction at mid-plane of mandible	R_2	0.65	0.00	0.00

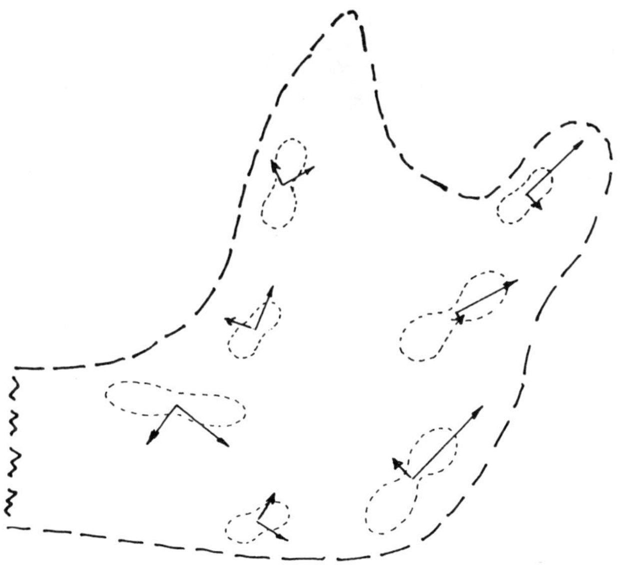

Fig. 7.10 Principal stress vectors at sites A–G in mandible overlaid on crystal orientation plots.

Acknowledgement

The author acknowledges the assistance of K. Culmer and C. Dunn in the above work.

References

Arvikar, R. & Seireg, A. (1975). A three-dimensional model for analysis of the human jaw during mastication. *Proceedings of the 28th ACEMB Conference*, p. 184.

Bacon, G. E., Bacon, P. J. & Griffiths, R. K. (1979a). The orientation of apatite crystals in bone. *Journal of Applied Crystallography*, **12**, 99–102.

Bacon, G. E., Bacon, P. J. & Griffiths, R. K. (1979b). Stress distribution in the scapula studied by neutron diffraction. *Proceedings of the Royal Society of London*, **B204**, 355–62.

Bacon, G. E., Bacon, P. J. & Griffiths, R. K. (1979c). Neutron diffraction studies in the lumbar vertebrae. *Journal of Anatomy*, **128**, 277–83.

Bacon, G. E., Bacon, P. J. & Griffiths, R. K. (1980). Orientation of apatite crystals in relation to muscle attachment in the mandible. *Journal of Biomechanics*, **13**, 725–9.

Bacon, G. E., Bacon, P. J. & Griffiths, P. K. (1981). The unseen stresses on the human frame. *New Scientist*, **26**, 796–9.

Bacon, G. E., Bacon, P. J. & Griffiths, P. K. (1984). A neutron diffraction study of the bones of the foot. *Journal of Anatomy*, **139**, 265–73.

Bacon, G. E. & Griffiths, P. K. (1985). Texture, stress and age in the human femur. *Journal of Anatomy*, **143**, 97–101.
Barbenel, J. C. (1972). The biomechanics of the temporomandibular joint: a theoretical study. *Journal of Biomechanics*, **5**, 251–6.
Dvir, Z. & Berme, N. (1978). The shoulder complex in elevation of the arm. *Journal of Biomechanics,* **11**, 219–5.
Henshell, R. D. (1984). *Pafec Theory*. Pafec Ltd, Nottingham.
Inman, V. T., Dec, J. B., Saunder, M. & Abbott, L. C. (1944). The function of the shoulder joint. *Journal of Bone and Joint Surgery,* **26**, 14–20.
Koch, J. C. (1917). The laws of bone architecture. *American Journal of Anatomy*, **21**, 177–298.
MacConaill, M. A. & Basmajian, J. V. (1969). *Muscles and Movements*. Williams and Wilkins, Baltimore MD.
Oxnard, C. E. (1973). *Form and Pattern in Human Evolution*. University of Chicago Press, Chicago.
Wolff, J. (1892). *Das Gesetz der Transformation der Knochen*. Quarto, Berlin.

8

Structure, function and mechanical failure of the meniscus

R. M. Aspden and D. W. L. Hukins

8.1 **Introduction**

The menisci of the knee joint are two crescents of fibrocartilage with a curved wedge-shaped cross-section that fit the space between the almost flat tibial plateau and the rounded femoral condyles. The shapes and ligamentous attachments of the menisci in position on the tibial plateau and a typical cross-section through a meniscus are shown in Figure 8.1. The main function of the meniscus is to bear compressive load (Seedhom, 1976, 1979). Mechanical tests, performed *in vitro*, suggest that well over 50% of the load acting on the knee joint is transmitted through the intact menisci (Krause *et al.*, 1976; Seedhom & Hargreaves, 1979; Ahmed & Burke, 1983).

A mechanism for load-bearing by the meniscus was first suggested by Fairbank (1948). He supposed that as the meniscus is compressed it is squeezed radially outwards, generating a hoop strain around the circumference, and that this motion is resisted by the strong ligamentous attachments. In this chapter recent evidence is reviewed which suggests that the fibrocartilaginous structure and overall shape of the meniscus enable it to support an applied load in this way. Furthermore, the approach used to relate the structure of the meniscus to its function suggests how it might fail.

8.2 **Composition**

Like other connective tissues, the fibrocartilage of meniscus is a composite material (Hukins & Aspden, 1985). It consists of a weak, highly hydrated glycosaminoglycan gel which is reinforced by collagen fibres. Glyco-

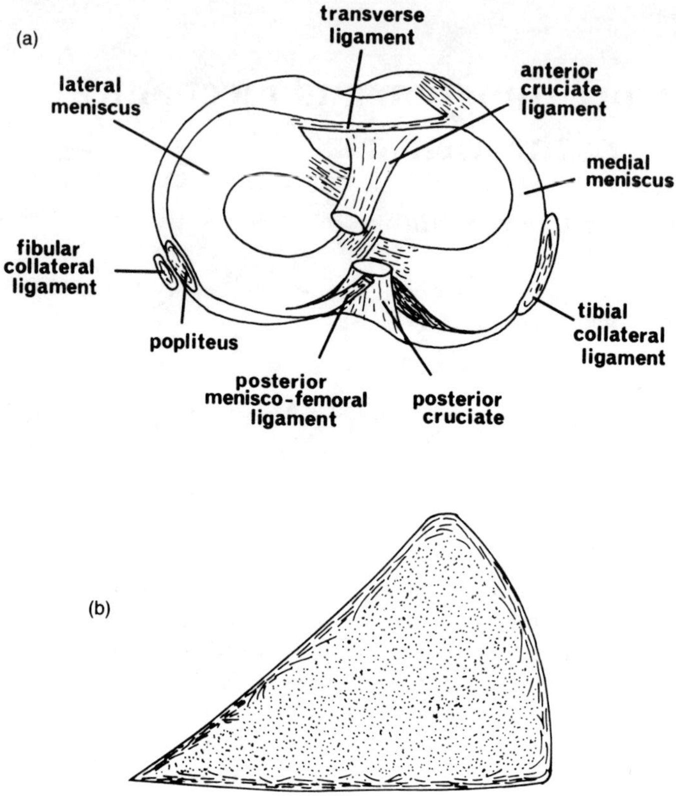

Fig. 8.1 Gross anatomy of the menisci and related structures: (a) superior view of the left tibial plateau; (b) typical cross-section of a meniscus showing the preferred orientations of the collagen fibres.

saminoglycans are linear polysaccharide molecules which contain a large number of sulphate and carboxylate groups, i.e. they are polyanions (Bayliss, 1984). These anions attract small cations such as Ca^{2+} and Na^{+} to balance their negative charges which leads to an imbalance in the partition of the cations between the tissue and surrounding body fluids. The higher concentration of ions within the tissue produces a higher osmotic pressure than in the surroundings and attracts water from them into the tissue resulting in a tendency to swell (Maroudas & Urban, 1980). The swelling of the gel is resisted by the collagen fibres which are strong in tension but weak in compression and flexion (Hukins, 1982). These fibres can only reinforce the gel providing they are arranged so that the restoring force generated in the stretched fibres balances the applied compression (Hukins, 1984).

8.3 **Collagen fibre orientations**

X-ray diffraction (Aspden *et al.*, 1985) and polarised light microscopy (Bullough *et al.*, 1970; Aspden *et al.*, 1985) have been used to determine the direction of preferred orientation of the collagen fibres and the distribution of orientations about this preferred direction in dog, pig and human menisci. X-ray diffraction patterns were recorded from blocks of tissue, about 1 mm thick, whose position and orientation with respect to the intact meniscus were carefully recorded. These blocks were fixed in formol saline and kept moist by enclosing them in an airtight cell while each diffraction pattern was recorded. It was assumed that fixation in formol saline did not affect the orientation of fibres in meniscus because fibre orientations in annulus fibrosus of the intervertebral disc are unchanged by this procedure (Hickey & Hukins, 1979). Each X-ray diffraction pattern was analysed to yield the direction of preferred orientation and an orientation distribution function for the fibres in the block of tissue from which it was recorded. The orientation distribution function describes the probability of finding a fibre at a given angle to the preferred direction (Aspden & Hukins, 1979). The results show that the fibres are oriented predominantly circumferentially, i.e. around the curve of the meniscus. The orientation distribution function, which may be characterised by its full width at half its maximum height, has a full width typically of about 35° for these circumferential fibres. The distribution is approximately Gaussian and this width corresponds to a standard deviation of about 15° (Aspden *et al.*, 1985). There is also a thin surface region, about 100 μm thick in the human, in which the fibres are oriented preferentially around the wedge-shaped cross-section. The width of the orientation distribution function in this region is about 60°, corresponding to a standard deviation of about 25° (Aspden *et al.*, 1985). The fibres are therefore much less highly aligned in the surface region than in the bulk tissue.

Polarised light microscopy confirmed the main features of these results and clearly demonstrated collagen fibres oriented predominantly circumferentially in the bulk tissue, with a thin surface region in which the fibres were oriented perpendicularly to this. The collagen fibres could be seen to be crimped, which is a common feature in many connective tissues (Gathercole & Keller, 1975).

This pattern of fibre orientations, shown schematically in Figure 8.1, is consistent with the mechanism for the function of the meniscus suggested by Fairbank (1948). The circumferential collagen fibres are oriented in just the direction to resist the hoop stress generated by loading the tissue in compression. The surface zone will help to bind the bulk fibres together, rather like the binding round the end of a rope, so that the fibres are not pushed apart by compression. A model has been developed based on this structure that demonstrates the generation of a hoop strain under load, and also suggests that normal compressive loading can lead to the most commonly observed injuries sustained by the meniscus (Aspden, 1985).

8.4 **Model**

The simplest model of the meniscus in which the shape and mechanical properties can be varied systematically is a toroid with a constant wedge-shaped cross-section (Aspden, 1985). The dimensions of the toroid and the range of cross-sections used corresponded to those measured from different parts of adult human menisci. The outer radius, 20 mm, and the height at the crest, 10 mm, were maintained constant and the inner radius varied from 5 to 12 mm, thus altering the steepness of the wedge. The toroid was considered to stand on a compliant base of modulus 5 $MN{\cdot}m^{-2}$, to represent the underlying articular cartilage of the tibial plateau, and a compressive load was applied along the axis of cylindrical symmetry. The loads applied in the model varied from 350 to 3500 N; the lower figure, about half body weight, is estimated to be a typical static load for a standing man assuming most of the load to be transmitted by the menisci (Seedhom, 1979); the maximum load, approximately five times body weight, is the average maximum tibio-femoral force normal to the tibia measured during squatting by Dahlkvist *et al.* (1982). The applied load was assumed to be transmitted in such a way that it always acted perpendicularly to the upper surface of the toroid.

A finite element representation of this model was developed which, because of the axial symmetry, needed only to divide a cross-section into triangular elements. Each element therefore was itself a toroid with a constant triangular cross-section. The first model had 33 elements and 28 nodes (Aspden, 1985) but this has subsequently been increased to 90 elements and 64 nodes. Triangular elements enabled the model to approximate closely the shape of the meniscus. A finite element program was written for the BBC microcomputer (Acorn Computers, Cambridge) as described by Rockey *et al.* (1983). Though this is only a small computer it was adequate to analyse the simple shape described and programs were written to take account of the non-linear nature of the mechanical properties. The number of nodes was restricted by the memory size of the computer but the results did not change significantly on increasing the number of nodes from 28 to 64, indicating that this was not a serious limitation.

The mechanical properties of the tissue appear in the program in the form of an elasticity matrix for each element. Because of the anisotropy of meniscus the elements of this matrix were found by calculating the stiffness tensor and putting this into its matrix form (Nye, 1957). In the absence of any known, reliable measurements of the mechanical properties of the meniscus, these matrix elements were initially calculated on the assumption that meniscus behaves like articular cartilage in compression in the plane of a cross-section, and like tendon in extension around the circumferential direction. More recent tensile testing of thin slices of meniscus has provided a better estimate of the mechanical properties in the circumferential direction (Wilkins, 1987). However, these measurements were obtained from sections of meniscus only 50 μm thick and so may not properly represent the mechanical behaviour of the bulk tissue. As the mechanical properties of the tissue are not reliably known, any stresses calculated in the material can only be estimates of the stresses actually present. Because of this the results are expressed as strains and no attempt is made to calculate stresses or apply failure criteria that rely on them. However it has recently been suggested that strains are of more importance in this sort of functional modelling than are stresses (Hukins *et al.*, 1989). Uncertainty regarding the mechanical properties does not invalidate the conclusions drawn from the model; the range of values used will almost certainly include some typical values for this sort of tissue and trends become obvious that depend on the relative values of the stiffnesses used rather than on their absolute value.

The combinations of Young's modulus and shear modulus in the plane of the cross-section used in the most recent analyses using this model are shown in Table 8.1. Poisson's ratio was assigned the values 0.2 or 0.4. These parameters assume that the meniscus behaves like articular cartilage in compression in the transverse plane and are in the range of typical values for articular cartilage quoted in the review by Kempson (1979).

The tensile properties of the meniscus around the circumference were originally derived (Aspden, 1985) from published data on tendon and more recently from measurements by Wilkins (1987). The stress–strain relationship for tendon may be expressed in the following form:

$$\text{stress} = k\,(\text{strain})^2$$

Table 8.1 *The combinations of values for the elastic moduli assigned to fibrocartilage and used in the model for the meniscus in the plane of the cross-section.*

Young's modulus ($MN \cdot m^{-2}$)	Shear modulus ($MN \cdot m^{-2}$)
10	5
10	2
10	1
5	2
5	1
5	0.5
2	1
2	0.5
2	0.2

where $k = 6.25 \times 10^8$ N·m^{-2} (Eldon, 1968). Comparison of the proportions of collagen in meniscus, about 70% of the dry weight (Eyre & Muir, 1975), and tendon, about 90% of the dry weight (Lowry *et al.*, 1941), leads to a value of k for meniscus of about $6.25 \times 10^8 \times (7/9) = 4.86 \times 10^8$ N·m^{-2}. The slope of this curve was used in the first model to provide a value for the stiffness of the material in the circumferential direction as the tissue was loaded. This was compared in more recent modelling with values obtained from the experimental data of Wilkins (1987). Fitting a cubic polynomial to his measurements of stress, σ, as a function of strain, e, yields the following expression:

$$\sigma = 1.22 \times 10^6\, e - 1.66 \times 10^7\, e^2 + 1.57 \times 10^9\, e^3 \qquad (8.1)$$

and the gradient of this was used as the strain dependent modulus. This is considerably less stiff than the tendon-like model for the circumferential stiffness but the results show that this did not significantly affect the pattern of strains generated by loading.

The load on the model was applied in fixed increments; at each increment the hoop strain at the centroid of each element was calculated and used to determine the appropriate values for the circumferential elements of the stiffness matrix for the next load increment. For each total load applied, the displacement of the nodes and the values of the hoop, radial, vertical and shear strains were calculated at the centroid of each element. The final state of stress of the meniscus was not calculated, due to the uncertainty regarding the mechanical properties of the material mentioned previously. Even if these properties were better known, the viscoelastic nature of such a tissue makes the state of stress strongly dependent on the loading history. The non-linear, time-dependent mechanical properties of connective tissues means that both theoretically and practically strain is a more useful concept than stress (Hukins *et al.*, 1987).

8.5 **States of strain calculated for the model**

The pattern of strains developed within the toroid were strongly dependent on the cross-sectional shape and only weakly on the mechanical properties of the fibrocartilage. For a given cross-sectional shape the magnitudes of the strains were dependent on the mechanical properties. The pattern and location of the three most important strains are shown in Figure 8.2. These diagrams are partly schematic, to show the consistent pattern of strains developed, but are based on the results obtained for the flattest wedge-shape with a Young's modulus E = 10 MN·m^{-2}, shear modulus G = 5 MN·m^{-2}, and a Poisson's ratio of 0.2 in the transverse plane. The consistency with which various strains were present is notable: if, for instance, a radial tensile strain was developed in a particular cross-section

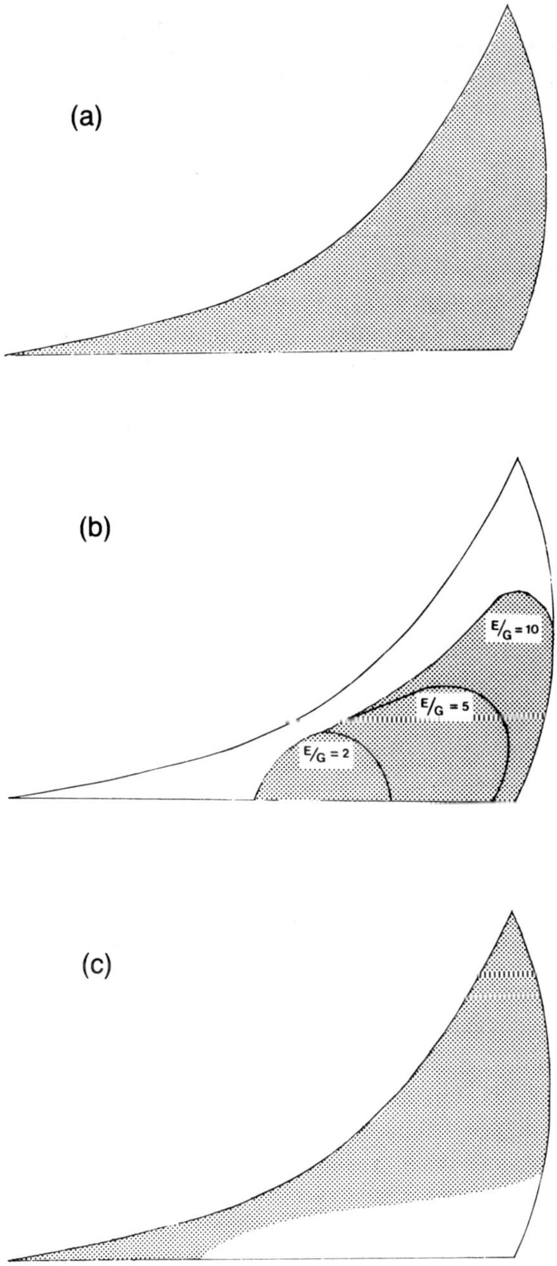

Fig. 8.2 The three most important strains generated by compressive loading and their locations: (a) circumferential tensile strain; (b) regions of radial tensile strains for various values of E/G; (c) shear strains, positive shear shaded.

then it was always present in the region shown in Figure 8.2 for all values of the stiffness parameters used. The area it covered depended on the relative values of E and G but it always appeared first on the inferior surface of the meniscus as shown.

As expected, a circumferential tensile strain was developed across most of the the cross-section for all shapes and elastic moduli of the meniscus, Figure 8.2a. This strain is in exactly the same direction as that in which the collagen fibres can provide reinforcement (Aspden *et al.*, 1985) and confirms that the meniscus bears a compressive load by generating a circumferential tension as suggested by Fairbank (1948). This strain was less than 0.04 for all values of elastic moduli used and for both of the stress–strain relations used in the hoop direction. This strain is of interest because of the similarity between meniscus and tendon. Both contain highly aligned, crimped fibres of type I collagen (Bullough *et al.*, 1970; Aspden *et al.*, 1985; Eldon, 1968) which do not acquire their maximum stiffness until the crimp is removed at a strain of about 0.04. It has been reported that tendons which are subjected to strains in excess of this value remain strained after the applied stress is removed (Rigby *et al.*, 1959). This residual strain may represent damage to the tissue (Hukins, 1982) although it is not certain that the tendons do not slowly return to their original length. Whatever the interpretation, such high strains are expected to be actually or potentially hazardous because any further application of stress will be to an effectively prestrained material.

The pattern of radial strains is shown in Figure 8.2b. The extent of the area of tensile strain is very dependent on the cross-sectional shape of the toroid, being greatest for the flattest wedge-shaped and vanishing for the steepest. For the flattest wedge-shape the region of radial tensile strain was always present and the area of the cross-section it covered was dependent on the ratio of Young's modulus/shear modulus, E/G, and Poisson's ratio. The area was smallest for the smallest value of E/G and, starting at the tibial surface, it extended into the bulk of the material as E/G increased as shown in Figure 8.2b. Keeping E/G constant and increasing Poisson's ratio had the same effect as might be expected.

A typical pattern of shear strains developed in the toroid is shown in Figure 8.2c. This pattern was almost independent of cross-sectional shape and mechanical properties. A region of positive shear strain always occupied the upper portion of the toroid with a region of negative shear strain adjacent to the tibial plateau. The effect of these strains is that the upper portion will attempt to move outwards while the lower portion is trying to move inwards. The boundary between the two regions is not a unique, definable plane but will almost certainly move up and down slightly in the tissue depending on the loads imposed. This boundary region, therefore, will be subject to strains first in one direction then in the other.

Repeated loading, of the order of a million times a year over a lifetime, may then lead to fatigue in the region between these two conflicting strains.

An alternative way of illustrating these different strains is by plotting the principal strains (see, for instance, Timoshenko & Goodier, 1982). Knowing the strain components e_x, e_y and γ_{xy} at a point, where e_x and e_y are the direct strains and γ_{xy} is the engineering shear strain, enables the direct strain, e, and the shear strain, γ, to be found for any direction, ϑ, at that point from:

$$e = e_x\cos^2\vartheta + e_y\sin^2\vartheta + \gamma_{xy}\cos\vartheta\sin\vartheta \tag{8.2}$$

$$\gamma = \gamma_{xy}(\cos^2\vartheta - \sin^2\vartheta) + 2(e_y - e_x)\sin\vartheta\cos\vartheta \tag{8.3}$$

The angle ϑ can be chosen so that the shear strain, γ, becomes zero and in this case,

$$\begin{aligned}\gamma_{xy}/(e_x - e_y) &= 2\sin\vartheta\cos\vartheta/(\cos^2\vartheta - \sin^2\vartheta) \\ &= \tan 2\vartheta\end{aligned} \tag{8.4}$$

From this equation it may be seen that there are two perpendicular directions along which the shear strain is zero and these are called the principal directions. The strains along these directions are the principal strains and represent the greatest and least values of strain, e, as a function of ϑ. The maximum values of shear strain are at 45° to the principal directions. The principal strains shown in Figure 8.3 are those corresponding to the strain diagrams of Figures 8.2b and 8.2c. The region of radial tensile strain can be clearly seen corresponding to Figure 8.2b. The directions of maximum shear strain bisect the strains shown and it can be seen that these change their orientation in passing from the bottom right of Figure 8.3 into the bulk of the tissue.

8.6 **Mechanical failure**

The regions of shear and radial tensile strains generated by compression of the material coincide strikingly with the most common injuries sustained by the meniscus. The pathology of the meniscus is described in detail by Smillie (1978) as the result of 10 000 meniscectomies performed between 1940 and 1974. Of the menisci excised 86% were for tears, of which 71% occurred in the medial meniscus and 29% in the lateral. The remainder were for problems without a direct mechanical aetiology such as cysts and congenital abnormalities. A diagrammatic representation of the predominant tears is shown in Figure 8.4 with their frequency of occurence. It may be seen that 50% of all the tears are horizontal tears involving the posterior segment of the medial meniscus (Figure 8.4e) and 25% are longitudinal tears affecting the same region (Figures 8.4a and 8.4b). For comparison

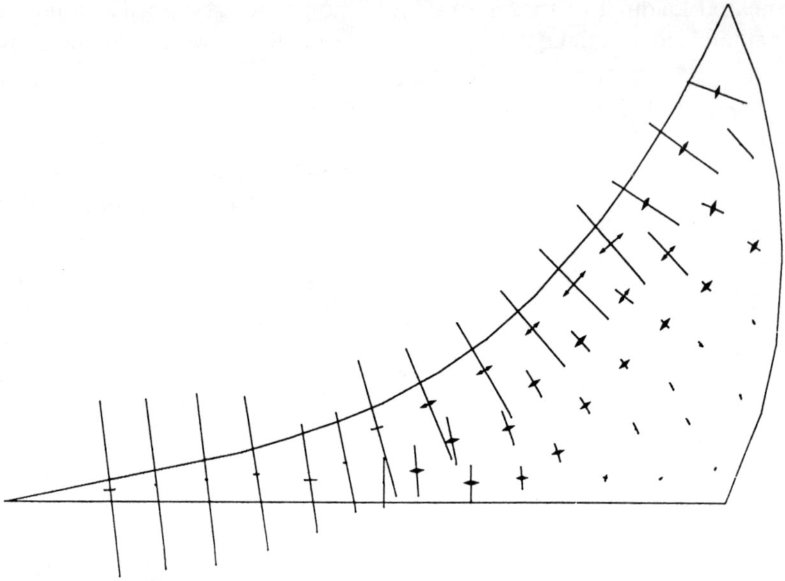

Fig. 8.3 Principal strains in the plane of the cross-section calculated for the same meniscus as Figure 8.2.

the frequency of similar tears of the lateral meniscus are 2% and 3%, respectively.

The horizontal tear is exactly that which would be caused by a shearing motion between the upper and lower portions of the meniscus. The model shows that this motion can be produced by compressive loading of the meniscus and that the greater the load the greater the strains produced. The posterior segment of the medial meniscus, which has the flattest cross-section, will experience the greatest loads and thus the greatest strains. Repeated loading and unloading may cause fatigue in the region of the boundary between the different strains leading to a lesion of the form shown in Figure 8.4d and eventually to that of Figure 8.4e. The increasing incidence of this lesion with age (Smillie, 1978) supports this hypothesis. The biochemical composition in the tissue adjacent to the tear has a reduced ratio of collagen to chondroitin sulphate compared to that in the normal or the remaining intact tissue, caused by an increase in chondroitin sulphate (Peters & Smillie, 1971, 1972). Rather than being a nutritional defect (Smillie, 1978) this change could be an attempt by the cells to repair the damaged tissue and to respond to the loading demands being placed upon it. Chondroitin sulphate is a glycosaminoglycan with a relatively large number of sulphate groups which give it a high fixed charge density. Increasing the fixed charge density will increase the swelling pressure

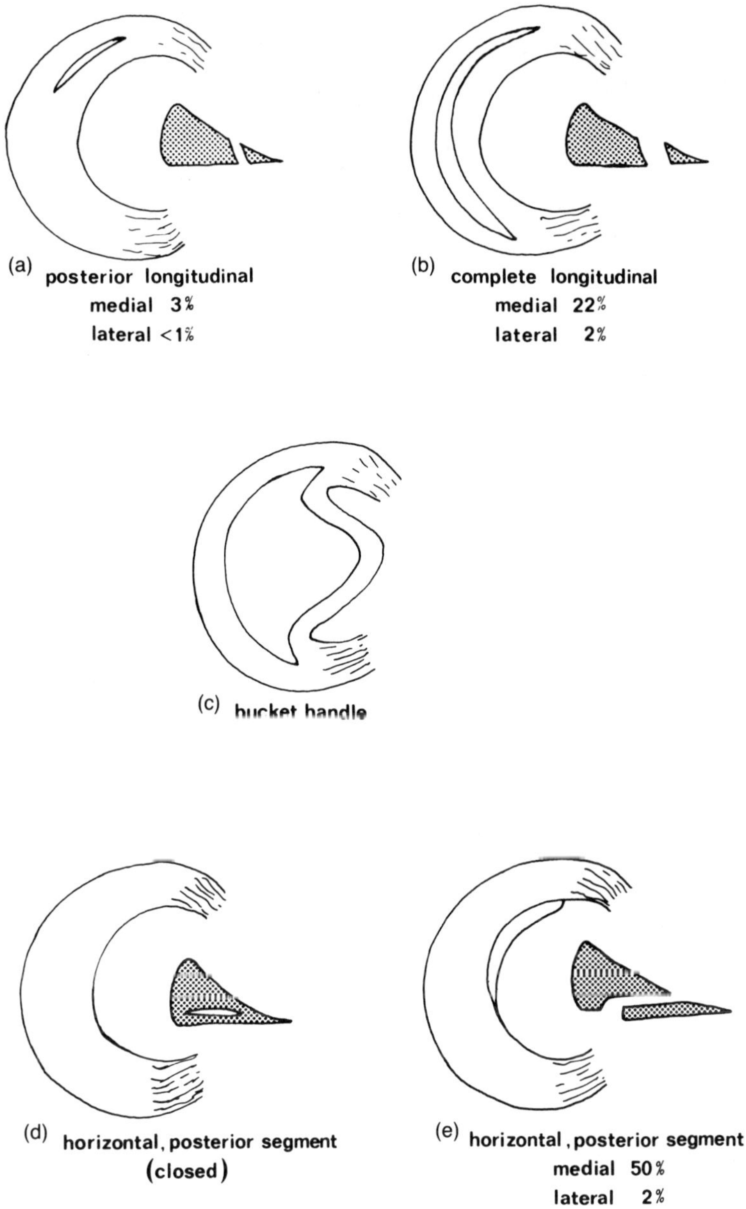

Fig. 8.4 Schematic representation of the most common meniscus tears and their frequency of occurrence. (Reprinted by permission of the Council of the Institution of Mechanical Engineers from *Engineering in Medicine*, **14**, 121, 1985).

which in turn may increase the compressive stiffness of the tissue. This may, therefore, be an attempt by the damaged tissue to maintain its load bearing capacity.

The longitudinal tears (Figures 8.4a and 8.4b) are caused by radial tensile strains which, because they are perpendicular to the predominant fibre direction, lead to separation of thc fibrcs and hence a tear. The model shows that the generation of these strains is strongly shape-dependent, being greatest for the flattest wedge-shaped cross-section. The model also shows that these strains are greatest at the tibial surface and spread from there into the bulk material depending on the mechanical properties, as shown in Figure 8.2. This is in agreement with observations of early stages of tearing being on the inferior surface of the meniscus (Smillie, 1978). These results again suggest that the posterior segment of the medial meniscus is most at risk due to its flatter cross-section. Failure in this case is not due to fatigue but will depend on the magnitude of the strains produced and therefore a large impulsive load presents the most danger to the meniscus. This is demonstrated by the incidence of this lesion which occurs predominantly in active, young adults and is generally a consequence of trauma (Smillie, 1978). That direct compression can cause this damage to the tissue differs from the traditional view that some form of twisting or rotating action is required (e.g. Shakespeare & Rigby, 1983; Smillie, 1978). Rotation and twisting require an axis to be defined about which this motion can take place and unfortunately little attention has been paid to this in descriptions of this injury. Also frictional forces in the joint are very small (Swanson, 1979) and so it would be difficult to transmit shearing forces to the meniscus by any rotation of the femur relative to the tibia. The model suggests, though, that what will cause damage to the meniscus is any motion that results in the relatively immobile medial meniscus being compressed between the femur and the tibia. This could well be caused by some 'twisting' motion but it is the resulting increase in the compressive load in the posterior medial compartment of the knee that produces the damage and not shearing or torsional forces directly.

8.7 **Conclusions**

This simple model for the meniscus shows that a circumferential strain is generated to balance an applied compressive load. This stress is in the direction of the preferred orientation of the collagen fibres and it confirms the load-bearing mechanism proposed for the meniscus. It also provides an explanation for the most commonly observed injuries, those of horizontal and longitudinal tears of the medial meniscus. It shows that lesions can be produced by pure compressive loading, transmission of shear forces to the meniscus is not required, and that because of its flatter cross-section the posterior segment of the medial meniscus is especially vulnerable.

Acknowledgements

We thank Dr Y. E. Yarker for collaboration in the measurement of collagen fibre orientations in the meniscus. Financial support was provided by the Medical Research Council and the Arthritis and Rheumatism Research Council.

References

Ahmed, A. M. & Burke, D. L. (1983). *In vitro* measurement of static pressure distribution in synovial joints. Part I: tibial surface of the knee. *Journal of Biomechanical Engineering*, **105**, 216–25.

Aspden, R. M. (1985). A model for the function and failure of the meniscus. *Engineering in Medicine*, **14**, 119–22.

Aspden, R. M. & Hukins, D. W. L. (1979). Determination of the direction of preferred orientation and the orientation distribution function of collagen fibrils in connective tissues from high-angle X-ray diffraction patterns. *Journal of Applied Crystallography*, **12**, 306–11.

Aspden, R. M., Yarker, Y. E. & Hukins, D. W. L. (1985). Collagen orientations in the meniscus of the knee joint. *Journal of Anatomy*, **140**, 371–80.

Bayliss, M. T. (1984). Proteoglycans: structure and molecular organisation in cartilage. In: *Connective Tissue Matrix*, ed. D. W. L. Hukins, pp. 55–88. Macmillan, London.

Bullough, P. G., Munuera, L., Murphy, J. & Weinstein, A. M. (1970). The strength of menisci of the knee as it relates to fine structure. *Journal of Bone and Joint Surgery*, **52B**, 564–70.

Dahlkvist, N. J., Mayo, P. & Seedhom, B. B. (1982). Forces during squatting and rising from a deep squat. *Engineering in Medicine*, **11**, 69–76.

Eldon, H. R. (1968). Physical properties of collagen fibres. *International Review of Connective Tissue Research*, **4**, 283–348.

Eyre, D. R. & Muir, H. (1975). The distribution of different molecular species of collagen in fibrous, elastic and hyaline cartilage of the pig. *Biochemical Journal*, **151**, 595–602.

Fairbank, T. J. (1948). Knee changes after meniscectomy. *Journal of Bone and Joint Surgery*, **30B**, 666–70.

Gathercole, L. J. & Keller, A. (1975). Light microscopic waveforms in collagenous tissues and their structural implications. In: *Structure of Fibrous Biopolymers*, eds. E. D. T. Atkins & A. Keller. Butterworth, London.

Hickey, D. S. & Hukins, D. W. L. (1979). Effects of methods of preservation on the arrangement of collagen fibrils in connective tissue matrices: an X-ray diffraction study of annulus fibrosus. *Connective Tissue Research*, **6**, 223–8.

Hukins, D. W. L. (1982). Biomechanical properties of collagen. In: *Collagen in Health and Disease*, eds. J. B. Weiss & M. I. V. Jayson, pp. 49–72. Churchill Livingstone, London.

Hukins D. W. L. (1984). Collagen orientation. In: *Connective Tissue Matrix*, ed. D. W. L. Hukins, pp. 211–40. Macmillan, London.

Hukins, D. W. L. & Aspden, R. M. (1985). Composition and properties of connective tissues. *Trends in Biochemical Sciences*, **10**, 260–4.
Hukins, D. W. L., Aspden, R. M. & Hickey, D. S. (1989). Strain relates connective tissue properties to joint movement *Clinical Biomechanics*, 4, 3–4.
Kempson, G. E. (1979). Mechanical properties of articular cartilage. In: *Adult Articular Cartilage* (2nd Edn), ed. M. A. R. Freeman, pp. 333–414. Pitman Medical, Tunbridge Wells.
Krause, W. R., Pope, M. H., Johnson, R. J. & Wilder, D. G. (1976). Mechanical changes in the knee after meniscectomy. *Journal of Bone and Joint Surgery*, **58A**, 599–605.
Lowry, O. H., Gilligan, D. R. & Katersky, E. M. (1941). The determination of collagen and elastin in tissue, with results obtained in various normal tissues from different species. *Journal of Biological Chemistry*, **139**, 795–804.
Maroudas, A. & Urban, J. P. G. (1980). Swelling pressures of cartilaginous tissues. In: *Studies in Joint Disease*, vol. 1, eds. A. Maroudas & E. J. Holborrow, pp. 86–117. Pitman Medical, Tunbridge Wells.
Nye, J. F. (1957). *Physical Properties of Crystals*. Oxford University Press, Oxford.
Peters, T. J. & Smillie, I. S. (1971). Studies on chemical composition of menisci from the human knee joint. *Proceedings of the Royal Society of Medicine*, **64**, 261–2.
Peters, T. J. & Smillie, I. S. (1972). Studies on the chemical composition of the menisci of the knee joint with special reference to the horizontal cleavage lesion. *Clinical Orthopaedics and Related Research*, **86**, 245–52.
Rigby, B. J., Hirai, N., Spikes, J. D. & Eyring, H. (1959). The mechanical properties of rat-tail tendon. *Journal of General Physiology*, **43**, 265–83.
Rockey, K. C., Evans, H. R., Griffiths, D. W. & Nethercot, D. A. (1983). *The Finite Element Method*. Granada Publishing, St. Albans.
Seedhom, B. (1976). Loadbearing function of menisci. *Physiotherapy*, **62**, 223–6.
Seedhom, B. (1979). Transmission of load in the knee joint with special reference to the role of the menisci. Part I: anatomy, analysis and apparatus. *Engineering in Medicine*, **8**, 207–19.
Seedhom, B. & Hargreaves, D. J. (1979). Transmission of load in the knee joint with special reference to the role of the menisci. Part II: experimental results, discussion and conclusions. *Engineering in Medicine*, **8**, 220–8.
Shakespeare, D. T. & Rigby, H. S. (1983). The bucket handle tear of the meniscus. A clinical and arthrographic study. *Journal of Bone and Joint Surgery*, **65B**, 383–90.
Smillie, I. S. (1978). *Injuries of the Knee Joint* (5th Edn.). Churchill Livingstone, Edinburgh.
Swanson, S. A. V. (1979). Friction, wear and lubrication. In: *Adult Articular Cartilage* (2nd Edn.), ed. M. A. R. Freeman, pp. 415–60. Pitman Medical, London.
Timoshenko, S. P. & Goodier, J. N. (1982). *Theory of Elasticity* (3rd Edn.), pp. 17–24. McGraw-Hill, Tokyo.
Wilkins, R. D. (1987). Tensile properties of the human knee joint meniscus in relation to composition. *Ph D Thesis*, University of Sydney, Australia.

9

Contact friction effects in flexural strength tests of alumina for orthopaedic implants

D. C. Fricker

9.1 **Introduction**

The tensile strength of ceramic materials is usually measured by manufacturing and testing to failure specimens made using the same material and processes as components. Because the compressive strengths of dense ceramics are so much greater than their tensile strengths, failure due to compressive stresses is virtually impossible. Thus the structural safety of ceramic components depends on the material's tensile strength, amongst other things.

The International Standard ISO 6474 (1981) specifies the physical, chemical and mechanical properties of alumina for biomedical use and gives corresponding test methods to be employed. Alumina is, at present, the most common ceramic for implants in the body. ISO 6474 specifies that tensile strength is to be determined using 3-point flexural fracture tests of a batch of specimens of the alumina material and that the arithmetic mean nominal failure stress be calculated to give a value for 'flexural strength'. The ISO test conditions require that each specimen surface be typical of critical surfaces of components. For the alumina universal heads attached to metal femoral joint prostheses of relevance to this study, the critical component surfaces are either as-fired (the crown of the tapered bore) or else ground (the taper); thus specimens for 3-point bending should either have as-fired surfaces or, more practically, representatively ground surfaces where fracture occurs. ISO 6474 specifies two alternative specimen geometries, *viz*. prismatic 4.5 ($\pm$0.5) $\times$ 4.5 ($\pm$0.5) mm square-section bars (see Figure 9.1 in section 9.3.1) or square bars 'waisted' to a minimum central diameter 3 ($\pm$0.1) mm by a large (unspecified) radius circular arc (see Figure 9.2 in section 9.3.2). In both cases, the support span ℓ = 25 mm.

A previous paper (Fricker, 1987) considered the statistical nature of ceramic specimen strengths and, using Weibull failure probability statistics, showed that the choice between these two alternative specimen geometries significantly affects flexural strength results. Results from the bending tests of the specimens considered here were analysed, and surface roughness and grain size results were also given. This study, however, confined itself to non-statistical analyses determined by maximum stress values alone.

This chapter describes work prior to the flexural testing of the alumina specimens using fixed loading and support rods to guarantee the loading rod to be equispaced between the support rods. It considers the effects of (inevitable) friction at these contacts on the flexural strength results. Stanley *et al.* (1976) have previously shown that contact friction in 3- or 4-point bending tests of rectangular or circular section prismatic bars reduces nominal bending stresses, so that flexural strengths calculated neglecting friction are always overestimates. For waisted bars, here, an additional effect of friction is to prevent the axial self-centralisation of the specimen with respect to the loading points; ISO 6474 shows a symmetrical loading arrangement. The importance of accurate axial positioning of waisted bars is also examined. It is assumed throughout this chapter that the axis of

Table 9.1 *Notation.*

$b, d, D, \ell, r, R, t, x, \phi$	Dimensions defined in Figures 9.1 and 9.2
e	Maximum stable axial positioning eccentricity of the minimum diameter section from the load point on waisted bars
F	Total transverse applied force
F_1, F_2	Transverse reactions at the two supports for waisted bars
W	Total force applied normal to waisted bars
θ	Angle defining any transverse section of waisted bars
θ_c	Value of θ corresponding to the critical section of greatest bending stress in waisted bars
μ	Coefficient of friction
σ	Maximum tensile bending stress at any section of a waisted bar
$\hat{\sigma}$	Maximum tensile bending stress anywhere in a bar
σ_a	Uniform axial stress
σ_{nom}	Nominal maximum bending stress calculated as per ISO 6474 neglecting contact friction

the loading rod is *exactly* midway between those of the two support rods (see Figures 9.1 and 9.2) and that the rod axes are perpendicular to the longitudinal axis of the specimen.

The notation used in this work is set out in Table 9.1.

9.2 **Previous coefficient of friction results**

Because the small number of alumina specimens available precluded separate friction tests, typical values for the coefficient of friction μ were assumed based on the previous friction test results given in Table 9.2. Friction tests 1 and 2 (Fessler & Fricker, 1989) concerned the fixation of alumina universal heads on Co-Cr-Mo alloy cone taper spigots and it can be seen that the different surface roughnesses did not influence the coefficient of static friction. Friction tests 3 to 7 (Fessler & Fricker, 1984a) concerned a polished steel rod sliding against different surface finishes of reaction-bonded silicon nitride (RBSN) and hot-pressed silicon nitride (HPSN) with different lubricants. It was found that the coefficients of sliding friction equalled the initial static values.

Five batches of bend specimens from four manufacturers were tested (Fricker, 1987) and their contact surface roughnesses had been measured to vary from CLA $R_a \cong 0.05\ \mu m$ (one batch with polished waists) to CLA $R_a = 1.8\ \mu m$ (a ground finish everywhere), whilst the profile ground, hardened steel loading and support rods had roughnesses $R_a = 0.22\ \mu m$. Since the μ values in Table 9.2 for silicon nitride/steel unlubricated agree with the value for alumina/Co–Cr–Mo unlubricated, it is assumed below that the values $\mu = 0.21$ or $\mu = 0.05$ apply for all alumina bending specimen/steel loading and support rod contacts, when respectively unlubricated or lubricated using a PTFE tape interlayer. PTFE tape has previously been used in RBSN bar and disc bending tests (Fessler & Fricker, 1984b) and polymer interlayers are indicated for glass disc bending tests in DIN 52292 (1984).

9.3 **Theoretical elastic stress analysis**

9.3.1 *Square bar bending tests*

Figure 9.1 shows a uniform prismatic square bar loaded in 3-point bending through fixed loading and support rods. Since the bar is prismatic, its axial position with respect to the loading and support lines is irrelevant so long as there is adequate overhang (typically 2 mm) of the bar ends beyond the support rods. When the bar bends, the small angulations of the sections at the supports induce the tendency of the lower 'tensile' face to move outwards relative to the supports; the angulations with alumina bars are however sufficiently small for the forces $F/2$ and $\mu F/2$ still to be considered to

Table 9.2 *Some previous friction test results for ceramics sliding against metals.*

Ref. no.	Ceramic			Metal			Lubricant	Coefficient of friction (μ)
	Material	Surface finish	Roughness (R_a; μm)	Material	Surface finish	Roughness (R_a; μm)		
1	Alumina	Ground	2.5	Co-Cr-Mo	Ground	0.15	None	0.21
2	Alumina	Ground	0.5	Co-Cr-Mo	Ground	0.75	None	0.21
3	RBSN	As-fired	2.3	Steel	Polished	0.02	None	0.24
4	HPSN	Ground	0.45	Steel	Polished	0.02	None	0.20
5	RBSN	As-fired	2.3	Steel	Polished	0.02	PTFE tape	0.07
6	RBSN	Lapped	0.85	Steel	Polished	0.02	PTFE tape	0.05
7	RBSN	Polished	0.05	Steel	Polished	0.02	PTFE tape	0.05

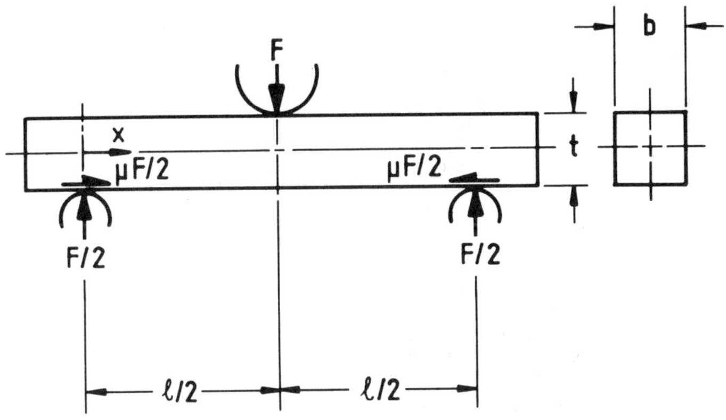

Fig. 9.1 Three-point bending of a prismatic square bar with equal contact friction at the two fixed support rods.

act in the transverse and axial directions, respectively, and the lengthening of the lower face due to tensile strains is also a secondary magnitude effect. Thus the frictional force at each support acts in the direction shown in Figure 9.1 and, for static equilibrium of forces and moments, they must be equal. There is no tendency for relative movement at the loading line and hence no frictional force acts there.

Elementary pure bending theory neglecting friction gives the nominal maximum tensile bending stress $\sigma_{nom} = 3F\ell/2bt^2$, located across the tensile face of the bar beneath the loading line at $x = \ell/2$, due to the bending moment $M = F\ell/4$ there. Away from the supports, the contact friction forces $\mu F/2$ are equivalent to an additional constant bending moment $M = -\mu Ft/4$ at all sections plus a force $\mu F/2$ acting at the longitudinal axis, causing a uniform compression stress $\sigma_a = -\mu F/2bt$. If the maximum bending stress with friction is called $\hat{\sigma}$, the bending stress index can be shown to be:

$$\hat{\sigma}/\sigma_{nom} = 1 - \mu t/\ell \tag{9.1}$$

Superimposed on uniform compression and bending loading, there is transverse shear and also localised contact stresses due to the concentrated loading force F. For the important region beneath the load line, Timoshenko & Goodier (1970) quote the exact two-dimensional results which combine both these latter effects. They give shear stresses to be zero on the tensile face of the bar because it is free. Directly beneath the loading line where $\hat{\sigma}$ and σ_{nom} occur, there is an axial compressive stress $-$ 0.266 F/bt. At a distance $t/4$ either side, the contact stress reaches its peak tensile value 0.174 F/bt but for the nominal ISO geometry here $b = t = 4.5$ mm and $\ell = 25$ mm, the maximum combined tensile bending and contact stress still

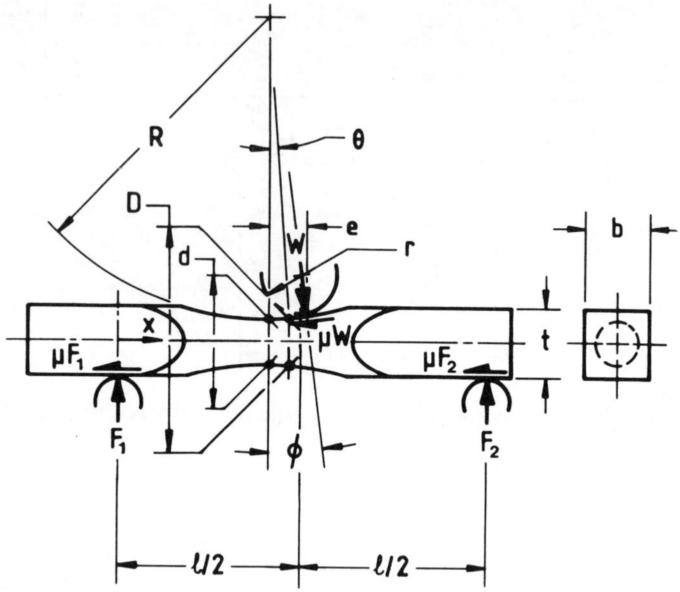

Fig. 9.2 Three-point bending of a bar 'waisted' to a minimum central diameter with contact friction at all three fixed points causing maximum positioning eccentricity (test type A).

occurs across the tensile face directly beneath the loading line. Away from the concentrated load, the parabolic distribution of shear stresses combines with pure bending stresses according to Mohr's circle but for $\ell/t = 5.56$, here, their effect on the flexural strengths of alumina bars is negligible, as indicated by the statistical strength analyses of rectangular section bars in 3-point bending with shear by Chau & Stanley (1985).

9.3.2 *Waisted bar bending tests*

There are two possible effects of contact friction on waisted bars, *viz*. (a) preventing self-centralisation of the specimen, and (b) introducing additional forces and bending moments on sections. The theory below considers the worst, combined situations. Figure 9.2 shows a waisted square bar with an axial positioning eccentricity at its maximum stable value $e = R \sin \phi$ between the minimum diameter d and the plane of loading. In order to establish the maximum eccentricity, it is assumed below that the specimen is initially positioned with the loading rod at the end of the waist radius and then allowed to centralise itself. Thus e has an upper bound $R \sin \phi$ limited by the waist geometry according to $\cos \phi = 1 - [(t - d)/2R]$. Due to asymmetry, the support reactions F_1 and F_2 are different and the load W is applied normal to the local waist surface (at a 'point').

There are so many possible situations with combinations of different friction coefficients at the three contacts, that the analysis below is limited to the more relevant cases of either equal μ or frictionless contact at all three locations.

Test type A

If there are equal μ values at the loading and both support rods, then the forces are as shown in Figure 9.2. All three friction forces oppose self-centralisation. Equilibrium of axial and transverse forces gives two equations from which it can be shown that:

$$\tan \phi = 2\mu/(1 - \mu^2) \tag{9.2}$$

For equilibrium of moments about the loading point where the section diameter equals d + 2R (1 − cos ϕ), it can be shown that F_2 is greater than F_1 in the ratio:

$$\frac{F_2}{F_1} = \frac{\ell + 2r \sin \phi + \mu[t + d + 2R(1 - \cos \phi)]}{\ell - 2r \sin \phi - \mu[t + d + 2R(1 - \cos \phi)]} \tag{9.3}$$

and so both ϕ and F_2/F_1 were easily calculated ($\phi \leq \hat{\phi}$ as described above).

The nominal frictionless maximum bending stress is given by $\sigma_{nom} = 8(F_1 + F_2)\ell/\pi d^3$ since in this particular situation $F_1 = F_2 = W/2$ and $\mu = \phi = \theta_c = 0$. With contact friction and eccentricity e, the maximum bending stress $\hat{\sigma}$ anywhere in the specimen will occur at one particular section defined by θ_c. The value of θ_c will lie between zero (the section of smallest diameter but not subjected to the largest bending moment) and ϕ (the section with maximum bending but not of the smallest diameter). Figure 9.2 shows a general transverse section of diameter $D = d + 2R(1 - \cos \theta)$ with θ in the range $0 \leq \theta \leq \phi$. The total bending moment at this section $M = F_1(x + \mu t/2)$ where $x = \ell/2 + r \sin \phi - e + R \sin \theta$. Hence the maximum elementary bending stress index at this general section can be shown to be:

$$\frac{\sigma}{\sigma_{nom}} = \frac{1 + \dfrac{2R}{\ell} \sin \theta - \dfrac{2(R - r)}{\ell} \sin \phi + \dfrac{\mu t}{\ell}}{\left[1 + \dfrac{2R}{d}(1 - \cos \theta)\right]^3} \times \frac{2}{1 + (F_2/F_1)} \tag{9.4}$$

The critical value of $\theta = \theta_c$ gives the maximum value for this stress index equal to $\hat{\sigma}/\sigma_{nom}$ and it was calculated numerically using the Newton–Raphson iterative method to solve $\partial(\sigma/\sigma_{nom})/\partial\theta = 0$. The contact friction force μF_1 also gives a force acting along the longitudinal axis of the bar causing a uniform tension stress $\sigma_a = 4\mu F_1/\pi D^2$ over this critical section defined by θ_c.

Superimposed on the important bending stresses and uniform tension, there are also local contact stresses. Although there appears to be no exact solution for the three-dimensional contact stresses in a round bar with a concentrated load, it is likely that when $e = 0$ the compressive contact stress at the location beneath the load ($x = \ell/2$) where $\hat{\sigma}$ occurs, approximately equals $-0.266\,F/d^2$. For the general situation $e \neq 0$, $\hat{\sigma}$ does not occur exactly beneath the load and so the absolute magnitude of the relevant contact stress will be lower. These contact stresses also apply in the other test types below.

Test type B

If there are equal μ values at the loading rod and one support rod but the other support has frictionless contact, the maximum bending stress index will depend on which support is frictionless in the situation illustrated in Figure 9.2 with e positive, i.e. the loading rod to the right of the centre of the specimen. Test type B is defined as in Figure 9.2 except for zero friction at F_1, (the other case is test type C). Equilibrium of forces and moments gives the corresponding equations in this case to be:

$$\frac{F_2}{F_1} = \frac{\ell + 2r\sin\phi}{\ell - 2r\sin\phi - \mu[t + d + 2R(1 - \cos\phi)]} \tag{9.5}$$

and,

$$\tan\phi = \mu[2 + (F_1/F_2)]/[1 + (F_1/F_2) - \mu^2] \tag{9.6}$$

Eqn (9.6) for ϕ is transcendental and it was solved numerically using the variable secant method for simplicity. The ratio F_2/F_1 was then easily calculated.

The elementary bending stress index σ/σ_{nom} is given as in eqn (9.4) except with the particular values $\mu = 0$ and F_2/F_1 here inserted. Its maximum value $\hat{\sigma}/\sigma_{nom}$ was calculated by the method used for test type A. The uniform tension stress $\sigma_a = 0$.

Test type C

If there are equal μ values at the loading rod and the left-hand support F_1 as shown in Figure 9.2 but there is zero friction at F_2, equilibrium of forces and moments gives the corresponding equations in this case to be:

$$\frac{F_2}{F_1} = \frac{\ell + 2r\sin\phi + \mu[t + d + 2R(1 - \cos\phi)]}{\ell - 2r\sin\phi} \tag{9.7}$$

and,

$$\tan\phi = \mu[2 + (F_2/F_1)]/[1 + (F_2/F_1) - \mu^2] \tag{9.8}$$

The value for ϕ was calculated by the method used for test type B and then the ratio F_2/F_1 was calculated.

The elementary bending stress index σ/σ_{nom} is given as in eqn (9.4) except with the particular F_2/F_1 value here inserted, and its maximum value $\hat{\sigma}/\sigma_{nom}$ was calculated as above. The uniform tension stress σ_a corresponds to that for test type A, *viz*. $\sigma_a = 4\mu F_1/\pi D^2$ over the section θ_c in this test.

Test type D

If there are equal μ values at both supports as shown in Figure 9.2 but zero friction at the loading rod, equilibrium of forces and moments gives the corresponding equation in this case to be:

$$\tan\phi = \mu \tag{9.9}$$

and the ratio F_2/F_1 is given by eqn (9.3) again. The stress index σ/σ_{nom} is given by eqn (9.4) again and its maximum value was calculated as above. The uniform tension stress σ_a corresponds to that for test type A.

Test type E

If there is friction μW at the loading rod but the supports are frictionless, equilibrium of forces and moments gives $\tan\phi = \mu$ as in eqn (9.9) and:

$$F_2/F_1 = (\ell + 2r\sin\phi)/(\ell - 2r\sin\phi) \tag{9.10}$$

The elementary bending stress index σ/σ_{nom} is given as in eqn (9.4) except with the particular values $\mu = 0$ and F_2/F_1 here inserted. Its maximum value $\hat{\sigma}/\sigma_{nom}$ was calculated as above. The uniform tension stress $\sigma_a = 0$.

Test type F

If there is only friction μF_2 at the right-hand support in Figure 9.2, equilibrium of forces and moments gives the ratio F_2/F_1 as in cqn (9.5) and:

$$\tan\phi = \mu/[1 + (F_1/F_2)] \tag{9.11}$$

Eqn (9.11) is again transcendental and because the derivative $\partial f(\phi)/\partial\phi$ of $f(\phi) = \tan\phi - \mu/[1 + (F_1/F_2)]$ is not too complex, the value of ϕ was calculated using the more efficient Newton-Raphson method.

The elementary bending stress index σ/σ_{nom} is given as in eqn (9.4) except with the particular values $\mu = 0$ and F_2/F_1 here inserted. Its maximum value was calculated as above. The uniform tension stress $\sigma_a = 0$.

Test type G

If there is only friction μF_1 at the left-hand support in Figure 9.2, equilibrium of forces and moments gives the ratio F_2/F_1 as in eqn (9.7) and:

$$\tan\phi = \mu/[1 + (F_2/F_1)] \tag{9.12}$$

so that the value for ϕ was again calculated using the Newton-Raphson method.

The elementary bending stress index σ/σ_{nom} is given as in eqn (9.4) except with the particular F_2/F_1 value here inserted, and its maximum value $\hat{\sigma}/\sigma_{nom}$ was calculated as above. The uniform tension stress corresponds to that for test type A.

9.4 **Predicted friction effects**

Contact friction makes $(\hat{\sigma} + \sigma_a)/\sigma_{nom}$ less than unity so that the flexural strength value calculated neglecting friction is always an overestimate of the tensile strength of the specimen material in every test.

9.4.1 *Square bar bending tests*

Figure 9.3 shows the linear decrease of the square bar bending stress index $\hat{\sigma}/\sigma_{nom}$ with μ for nominal ISO dimensions and Table 9.3 gives σ_a/σ_{nom} and $\hat{\sigma}/\sigma_{nom}$ values and percentage overestimates of strengths for relevant μ values for the alumina specimens to be tested. The values $\mu = 0.21$ and $\mu = 0.05$ apply for unlubricated contacts and PTFE tape interlayers, respectively.

9.4.2 *Waisted bar bending tests*

If there is perfect axial centralisation of waisted square bars ($e = 0$), the maximum bending stress index $\hat{\sigma}/\sigma_{nom}$ is given by eqn (9.1) again and its values as shown in Table 9.3 apply. The uniform compression stress index values σ_a/σ_{nom} are 50% of those shown in Table 9.3 and so they were neglected for all waisted bar bending tests.

At the other extreme, Figure 9.3 shows the largest possible variations of the bending stress indices $\hat{\sigma}/\sigma_{nom}$ with μ for waisted bars of nominal ISO dimensions, for the seven different experimental arrangements considered at their maximum stable positioning eccentricities. Test type A has friction at all three contacts, test types B, C and D at two and test types E, F and G at one contact only, and Figure 9.3 shows corresponding progressively diminishing effects.

In test type A, $\hat{\sigma}/\sigma_{nom}$ decreases approximately linearly with μ up to

Table 9.3 *Predicted effects of friction on the material strength results from three-point bend tests of square bars of nominal ISO geometry*[a].

Coefficient of friction (μ)	Uniform compression stress index (σ_a/σ_{nom})	Bending stress index ($\hat{\sigma}/\sigma_{nom}$)	% overestimate of strength
0.21	− 0.013	0.962	5.3
0.05	− 0.003	0.991	1.2

[a] $b = t = 4.5$ mm and $\ell = 25$ mm.

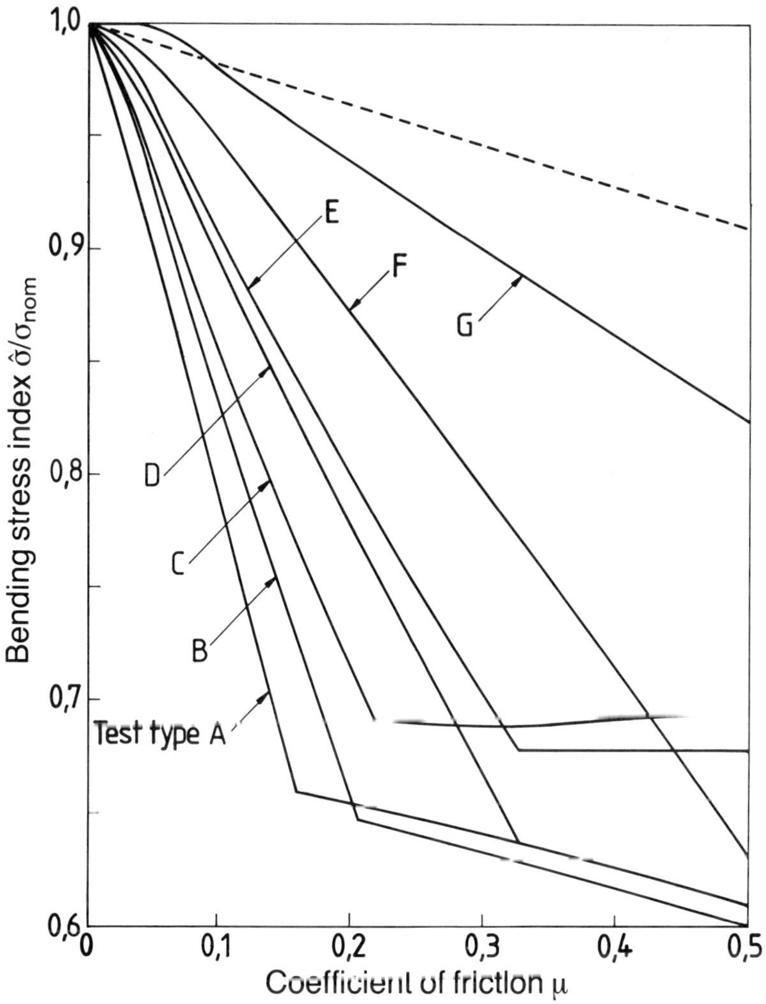

Fig. 9.3 The largest possible effects of the coefficient of friction μ on the maximum elementary bending stress index $\hat{\sigma}/\sigma_{nom}$ for square (broken line) and waisted (solid line) bars with nominal dimensions $b = t = 4.5$ mm, $d = r = 3$ mm, $R = 15$ mm and $\ell = 25$ mm.

$\mu = 0.16$ for which value $\phi = \hat{\phi} \cong 18°$ and there is no longer any self-centralisation from the initial loading position at the end of the waist. Larger μ values do not therefore increase the maximum stable eccentricity *e* but they do increase their negative bending moment; θ_c decreases slightly but F_2/F_1 increases monotonically. Figure 9.3 shows this point of discontinuity and a gradient for $\mu > 0.16$ of similar magnitude as with square bars.

In test type B, $\hat{\sigma}/\sigma_{nom}$ again decreases approximately linearly with μ until $\mu = 0.21$ when $\phi = \hat{\phi}$. Larger μ values do not increase either e or θ_c but F_2/F_1 increases to give a similar magnitude gradient for $\mu > 0.21$ as in test type A.

In test type C, $\hat{\sigma}/\sigma_{nom}$ again decreases approximately linearly with μ until $\mu = 0.22$ when $\phi = \hat{\phi}$. Larger μ values do not increase e but θ_c decreases slightly whilst F_2/F_1 increases. For $\mu > 0.22$, $\hat{\sigma}/\sigma_{nom}$ remains fairly constant as shown in Figure 9.3.

In test type D, $\hat{\sigma}/\sigma_{nom}$ again decreases approximately linearly with μ until $\mu = 0.33$ when $\phi = \hat{\phi}$. Comparing eqns (9.2) and (9.9) shows that these ϕ values are approximately 50% of the corresponding values in test type A (for $\mu < 0.16$) and this gives an initial gradient for $\hat{\sigma}/\sigma_{nom}$ approximately half that in test type A. For $\mu > 0.33$, all variables are the same in types A and D so that their $\hat{\sigma}/\sigma_{nom}$ curves coincide.

In test type E, $\hat{\sigma}/\sigma_{nom}$ decreases until $\mu = 0.33$ when $\phi = \hat{\phi}$ as in test type D. Larger μ values do not change any variables so that $\hat{\sigma}/\sigma_{nom}$ remains constant for $\mu > 0.33$.

In test types F and G, Figure 9.3 shows that $\hat{\sigma}/\sigma_{nom}$ decreases approximately linearly with μ over the range studied. Comparing eqns (9.11) and (9.12) with (9.2), ϕ values in test types F and G are approximately 25% of the corresponding values in test type A. Values of F_2/F_1 are higher in test type F than in G however, and this is the main reason for the larger $\hat{\sigma}/\sigma_{nom}$ gradient in the former test.

Table 9.4 gives the measured dimensions of the three batches of waisted alumina bars to be tested using fixed loading and support rods, i.e. test type A. The importance of perfect axial positioning was studied by calculating the greatest effects friction $\mu = 0.05$ or $\mu = 0.21$ could have on the overestimates of material strength and Table 9.4 shows these results. It can be seen that friction effects can cause significant axial eccentricity of the specimens; for unlubricated contacts, where $\mu = 0.21$, there is no self-centralisation effect for any of the batches. However the critical sections where the highest bending stresses occur are quite close to the minimum diameter section in all batches, as given by $R \sin \theta_c$ values. Table 9.4 shows that there can be large overestimates of material strengths, even when friction is low. Similar overestimates for the batch 1 and 3 specimen geometries are shown, whilst the maximum overestimate is larger for batch 2 specimens owing to their larger waist radius R.

9.5 **Predicted contact stress effects**

In square bar bending tests, the theoretical stress analysis predicts that localised contact stresses will cause an additional 3% overestimate of material strength for the nominal ISO geometry $\ell/t = 5.56$.

Table 9.4 *Greatest predicted effects of friction on waisted bar results from three-point bend tests of type A for different specimen batches* (r = *3.2 mm).*

Batch ref.	Span (ℓ; mm)	Mean ± 1 SD of specimen dimensions (mm)	Coefficient of friction (μ)	Maximum axial eccentricity (ϵ; mm)	Critical section distance ($R \sin \theta_c$; mm)	Minimum bending stress index ($\hat{\sigma}/\sigma_{nom}$)	Maximum % overestimate of strength
1	25.45	$\bar{d}$ = 2.91 ± 0.07 $\bar{t}$ = 4.47 ± 0.01 $\bar{R}$ = 13.7 ± 0.0	0.21 0.05	4.6 1.4	0.7 0.5	0.664 0.910	50.6 9.9
2	25.04	$\bar{d}$ = 3.05 ± 0.04 $\bar{t}$ = 5.02 ± 0.04 $\bar{R}$ = 20.6 ± 0.0	0.21 0.05	6.3 2.1	1.2 0.9	0.558 0.870	79.2 14.9
3	25.04	$\bar{d}$ = 3.07 ± 0.02 $\bar{t}$ = 5.14 ± 0.02 $\bar{R}$ = 14.7 ± 0.0	0.21 0.05	5.4 1.5	0.8 0.6	0.604 0.904	65.6 10.6

In waisted bar bending tests with zero positioning eccentricity, contact stresses are predicted to cause an additional 1% overestimate of material strength for the nominal ISO geometry $\ell/d = 8.33$. When there is eccentricity and $\hat{\sigma}$ is not located directly beneath the loading rod, contact stress effects will be negligible compared with friction effects.

9.6 Experimental result

The small number of alumina specimens available precluded any comparative strength tests to confirm the predicted effects of contact friction. However, before its fracture test, the first specimen from batch 1 waisted bars was lightly loaded in the 3-point bending fixture using PTFE tape interlayers, with sufficient initial positioning eccentricity that it centred itself. The stable eccentricity was measured to be approximately $1\frac{1}{2}$ mm and this value agrees with the Table 9.4 prediction for batch 1 specimens with $\mu = 0.05$.

9.7 Discussion

In this chapter, friction at the loading and support contacts in 3-point bending tests is included by means of statically equivalent forces and bending moments. For prismatic square bars, Table 9.3 shows the slight advantage of using PTFE tape interlayers between the specimen and the fixed support rods. For waisted square bars, Table 9.4 shows that large overestimates of material strength are possible unless precautions are taken in experiments to minimise the axial positioning eccentricity of each specimen. The subsequent fracture tests (Fricker, 1987) did this, and also used PTFE tape interlayers to minimise contact friction effects. The friction at the contacts should be minimised, made consistent, measured and together with contact stresses, included in the strength calculations if accurate values are required. A statistical analysis of strength results is recommended in order to account for the two alternative specimen geometries and their different stress distributions.

The accuracy of the theoretical predictions of the contact stress effect beneath the loading rod is confirmed by the exact agreement of predictions with the effect computed by Stanley *et al.* (1976) using finite element analysis of their geometry of 3-point square bar bending tests.

For square bars, the theoretical stress analysis above has not considered the case where one support is frictionless but there are equal μ values at the other support and the loading rod. A corresponding analysis shows $\hat{\sigma}/\sigma_{nom} = 1$ for all μ values but the distribution of bending moments along the length of the bar is different from that in the frictionless case.

For waisted square bars, elementary bending theory has been used to predict $\hat{\sigma}$ although the waist is not prismatic. The errors in using elementary theory are likely to be small here because: R/d values were large; and the critical angles θ_c were small (θ_c is also the inclination of the local surface where $\hat{\sigma}$ occurs from the axial direction and its value was typically 3°). For comparison, Neuber (1946) gives the stress concentration factor for a waisted round bar with the lowest value $R/d = 4.7$ here, subjected to a constant bending moment (pure bending) to be $K = 1.032$ for Poisson's ratio $\nu = 0.22$ of alumina. The stress concentration effect in 3-point bending with zero eccentricity is probably slightly lower but it does tend to counterbalance strength overestimates due to friction and contact stresses. The low values in Table 9.4 for $R \sin \theta_c$ indicate that the effects of inadvertent high friction are not likely to be noticed from fractography of waisted bar specimens.

The ISO Standard 6474 'recommends' (but not 'requires') that the 3-point bending fixture should have two of the three contacts 'movable'. The usual arrangement for this is that the two supports can roll outwards from their initial position against fixed stops. Such a fixture would provide effectively frictionless loading of square bar specimens. However for waisted bars with initial positioning eccentricity, this fixture gives test types B or C conditions and so considerable strength overestimates could arise. Even with friction only at the loading rod (test type E) or friction at only one support (test types F or G), specimen self-centralisation is inhibited and considerable strength overestimates could arise.

9.8 **Conclusions**

Friction at the fixed support lines in 3-point bending fracture tests of alumina specimens of prismatic square bar geometry loaded according to ISO 6474, causes a 5% flexural strength overestimate for unlubricated contacts ($\mu = 0.21$) but only a 1% overestimate when PTFE interlayers ($\mu = 0.05$) are used. Contact stresses beneath the loading line further increase these overestimates to 9% and 5%, respectively.

For the alternative ISO specimen geometry of a bar waisted to a minimum central diameter, contact friction can cause significant axial eccentricity of the specimen with respect to the loading fixture, even with lubricated contacts, unless precautions are taken in experiments to eliminate this eccentricity. If this is not done, such eccentricity could cause the flexural strengths to be overestimated by as much as 15% with PTFE tape interlayers. When the positioning eccentricity is zero, friction again causes 5% and 1% flexural strength overestimates for unlubricated contacts and PTFE tape interlayers respectively. Contact stress effects further increase these overestimates to 6% and 2%, respectively. Stress concentration effects will however reduce these overestimates.

Acknowledgement

The author thanks Howmedica International Inc. for permission to publish this chapter.

References

Chau, F. S. & Stanley, P. (1985). An assessment of systematic errors in beam tests on brittle materials. *Journal of Materials Science*, **20**, 1782–6.

DIN 52292 Part 1 (1984). Testing of glass and glass ceramics; determination of bending strength; coaxial double ring bending test on flat specimens with small test area. Deutsches Institut für Normung, Berlin.

Fessler, H. & Fricker, D. C. (1984a). Friction at broad and narrow contacts between silicon nitride and hardened steel. *Proceedings of the British Ceramic Society*, **34**, 129–43.

Fessler, H. & Fricker, D. C. (1984b). Multiaxial strength tests for brittle materials. *Journal of Strain Analysis*, **19**, 197–208, 275.

Fessler, H. & Fricker, D. C. (1989). Friction in femoral prosthesis and photoelastic model cone taper joints. *Proceedings of the Institution of Mechanical Engineers Part H, Journal of Engineering in Medicine*, **203**, H1, 1–14.

Fricker, D. C. (1987). Failure probability analysis of flexural strength results for alumina for orthopaedic implants. In: *Ceramics in Clinical Applications*, ed. P. Vincenzini, pp. 139–48. Elsevier, Amsterdam.

ISO 6474 (1981). Implants for surgery – ceramic materials based on alumina, 1st edn. International Organisation for Standardisation, Geneva.

Neuber, H. (1946). *Theory of Notch Stresses*, pp. 89–93. J. W. Edwards, Ann Arbor, Michigan.

Stanley, P., Sivill, A. D. & Fessler, H. (1976). The unit strength concept in the interpretation of beam test results for brittle materials. *Proceedings of the Institution of Mechanical Engineers*, **190**, 49/76, 585–95.

Timoshenko, S. P. & Goodier, J. N. (1970). *Theory of Elasticity*, 3rd Edn, pp. 113–22. McGraw-Hill Kogakusha, Tokyo.

10

Mechanical behaviour of an osteoarticular replacement material

B.F. Shahgaldi, A.A. Amis and F.W. Heatley

10.1 **Introduction**

The problem of painful synovial joints, due to cartilage degeneration, is often dealt with by surgical intervention, including total joint replacement, surgical interposition of a foreign body between the opposing joint surfaces, rotational/translation osteotomy, or removal of the degenerated cartilage and drilling of the underlying bone (shaving of cartilage and drilling of bone).

Many patients, including young people, with early or moderate cartilage degeneration are advised to wait until the defect is large enough to justify total joint replacement. However, many of these people may benefit from treatment of the localised defect, particularly in cases where a non-progressive isolated defect is surrounded by normal cartilage and bone, as in degeneration of cartilage following trauma or chondromalacia patellae (Karlson, 1940).

Shaving the cartilage and drilling the bone is one such treatment. The affected cartilage is shaved to bone to remove the damaged tissue, then the underlying bone is drilled to expose the cancellous bone. This method often provides early good results, but is disappointing in the long run owing to later degeneration of the fibrous tissue that fills the drill hole, as has been demonstrated on rabbits and human patellae (Mitchell & Sheppard, 1976; Ghadially, 1983; Ficat *et al.*, 1979; Coletti *et al.*, 1972; Milgram, 1985).

Another approach has been to replace the localised defect with an implant. For years a large number of viable and non-viable biological and man-made materials have been tried for both humans and experimental animals (Kon & de Visser, 1981; Speer *et al.*, 1979; Minns *et al.*, 1982; Heatley & Revel, 1985; Breck, 1976; Cobey, 1967). Although not many have met with uniform success in restoration of a durable articular cartilage surface, the results have been sufficiently encouraging to warrant systematic testing of other materials.

Matching the mechanical properties of the implant with the host tissue is an important part of the systematic testing which has not received much attention, even though some researchers have expressed the need for it (Hastings, 1980; Black & Frust, 1978). The concern has always been to match the implant to the host biologically, an equally important concept, but unmatched mechanics could put strain on the biology and lead to further degeneration of cartilage (Radin *et al.*, 1975) or total failure; localised pressure concentration has been shown to cause failure of cup hemiarthroplasty of the hip (Enneking & Singsen, 1974).

A successful replacement requires a material that closely matches the *in vivo* mechanical behaviour of the articular cartilage in place on the subchondral bone, particularly its response to loading and unloading during joint movements. Furthermore, the presence of an implant in the joint could alter the congruency of the joint, leading to an increased or reduced contact pressure, either of which could damage the articular cartilage. Wear properties of the implant are also important: it should neither get worn nor wear the opposing cartilage counterface excessively.

The forces acting in a load-bearing joint, during normal activities, are mainly compressive. In the laboratory the response of an implant to compression has been matched, using indentation tests and contact pressure mappings, to that of its surrounding intact articular cartilage in the knee joint of the goat. The laboratory findings were re-examined by an *in vivo* study, in which the implants were used for six months: this also demonstrated the chemical and biological compatibility of the implant with the cartilage and bone of mature goat's knees.

10.2 ***In vitro* studies**

10.2.1 *Materials and methods*

The implants were plugs of 6 mm diameter and 6 mm length, removed from glutaraldehyde-fixed bovine fibrocartilage menisci. They were inserted into osteochondral defects created by drilling through the joint surface in the knee joint of the goat, in the femoral condyles for indentation tests and the patellae for contact pressure measurements. The plugs were inserted into same size or undersized defects at various heights: above, level with, or below the cartilage level.

The experimental apparatus was a modified version of Kempson's indentation rig (Kempson, 1970). A spherical indenter of 3.17 mm diameter and dead load of 9.2 N produced stresses within the physiological range when dropped onto the cartilage surface at a constant, known velocity.

In indentation testing, a punch under constant load is pressed normally into the articular surface (Figure 10.1A). The deformation with time is

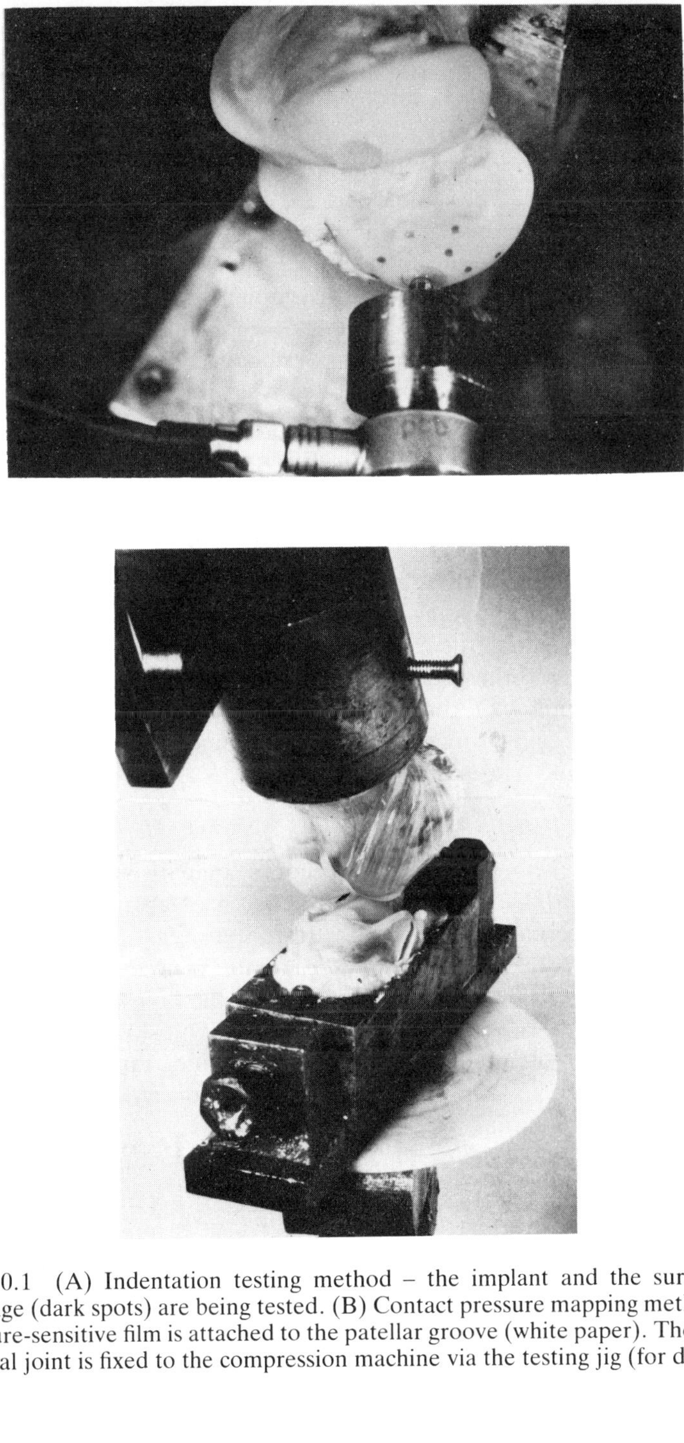

Fig. 10.1 (A) Indentation testing method – the implant and the surrounding cartilage (dark spots) are being tested. (B) Contact pressure mapping method – the pressure-sensitive film is attached to the patellar groove (white paper). The patello-femoral joint is fixed to the compression machine via the testing jig (for details see text).

recorded and properties such as elastic and creep moduli are calculated. With this technique the cartilage is tested *in situ*, thus the experimental method resembles localised physiological loading of cartilage and bone.

The responses of the implant, inserted into a same size osteochondral drill hole in the knee joint of the goat, and that of the surrounding cartilage on application and removal of a compressive load was studied using the Instron 1122 machine. The materials were compressed by pressing a flat-ended, 3.17 mm diameter, punch at a rate of $10\,mm\cdot min^{-1}$ against their flat surfaces until a maximum load of 22.2 N was reached. The load was then taken off at the same rate. Five implants and their surrounding cartilages, maintained in Ringer's solution, were studied.

The contact pressure measurements were carried out using a pressure-sensitive material, Prescale Film, made by Fuji Film Company and a specially designed jig (Figure 10.1B). The pressure-sensitive film consists of two parts which, when sandwiched in the joint with their appropriate surfaces in contact, produce a red colour the density of which depends on the applied pressure; the pressure corresponding to any particular density is obtained from a calibration curve. The jig was fixed to a compression machine and provided flexion angles of 0–90°, for varus and valgus angulation and rotation for testing the knee: the load was chosen to produce contact pressures within the physiological range of $2\text{–}5\,MN\cdot m^{-2}$. The physiological load and the pressures quoted here are those determined *in vitro* for the human knee (Ahmed *et al.*, 1983). The pressures in the knee joint of the goat are not known, but are of the same order of magnitude as those of the human (Alexander, 1980).

10.2.2 *Results*

Figure 10.2 shows the behaviour of the plugs in comparison with that of the cartilage of goat's knee under the compression of an indenter. Curves a and b are the matched indentation with time behaviour of the implant and the cartilage, respectively, both tested on the same subchondral bone. The indentation with time of both materials is the behaviour of a visco-elastic material: an elastic deformation followed by a slow creep.

The responses of the implants and their surrounding cartilage to loading and unloading are shown in Figure 10.3. The deformation and recovery curves of the two materials are similar.

Figure 10.4 compares the cartilage – cartilage contact pressure (normal pressure) with that of implant – cartilage contact pressure. The best position for the implant was when the implant – cartilage contact pressure was similar to a normal pressure distribution. When the implant was proud of the surface the contact pressure was considerably larger than the normal pressure, while a sunk implant did not bear any of the contact pressure. This meant that the cartilage in contact with the proud implant was under larger than normal pressure and the one in contact with the sunk implant

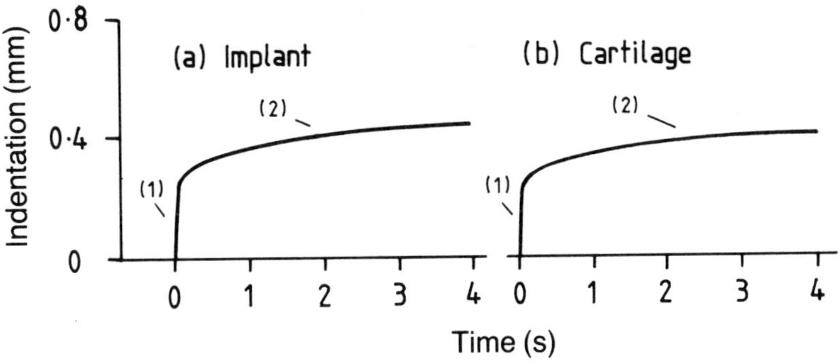

Fig. 10.2 Matched indentation with time curves of the implant (a) and the articular cartilage of goat's knee (b). (1) and (2) are the elastic and creep deformations, respectively.

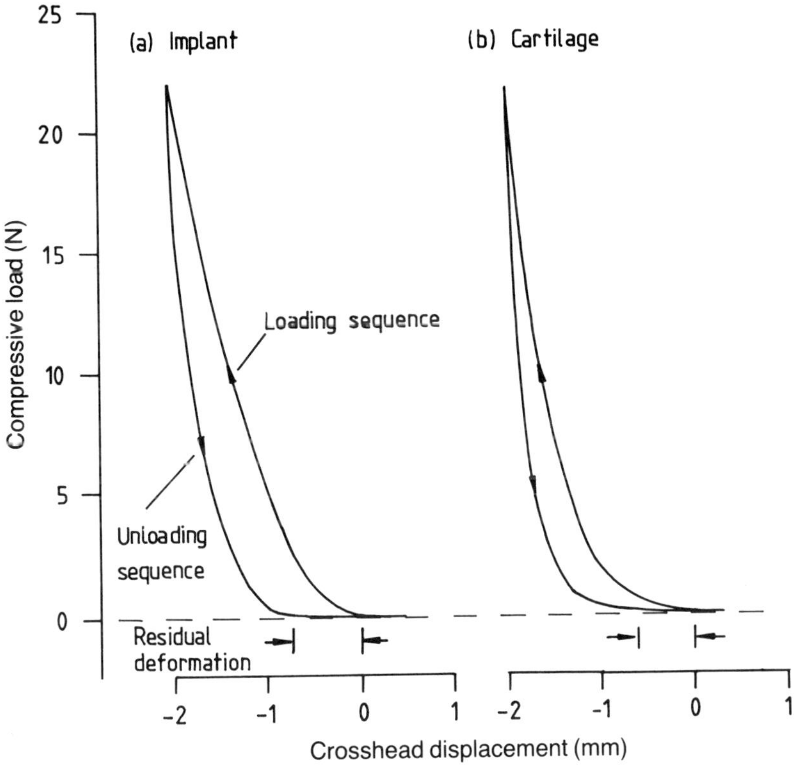

Figure 10.3 Deformation and recovery curves of the implant (a) are compared with those of the goat's knee cartilage (b), both materials were tested on the same subchondral bone.

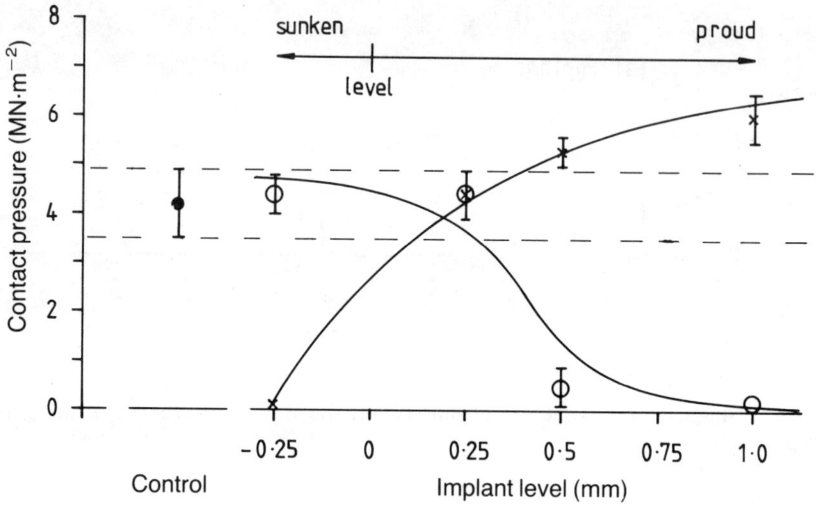

Fig. 10.4 Contact pressures of implants inserted into osteochondral holes in the medial facets of the patellae of four goats at various heights with respect to the surrounding cartilage: (●), cartilage–cartilage contact; (x), implant–cartilage contact; and (○), cartilage surrounding the implant–cartilage contact.

was subjected to no pressure or considerably less pressure than normal. The matching position, about 0.25 mm proud, was obtained when the contact pressure of an area comprising the implant and surrounding cartilage was similar to that of the cartilage – cartilage contact pressure of the same joint tested under the same conditions.

10.3 ***In vivo* studies**

10.3.1 *Materials and methods*

Having optimised the implant in the laboratory, 48 such plugs were prepared and implanted in the left knee joint of 12 goats; in four different anatomical sites, the patella and the patello-femoral groove (low stress area), and the two femoral condyles (high stress area), for six months *in vivo*.

Plugs of 6 mm length were used to ensure good anchorage by the growth of repair tissue from the highly vascular cancellous bone for encapsulation and firm fixation. They were implanted in osteochondral defects which were the same size or undersized in diameter and at various heights, above, level with, and below the cartilage level, so that the laboratory findings could be reassessed. Mature goats were used so that the performance of the implants in a mechanically-demanding joint in a large animal could be studied.

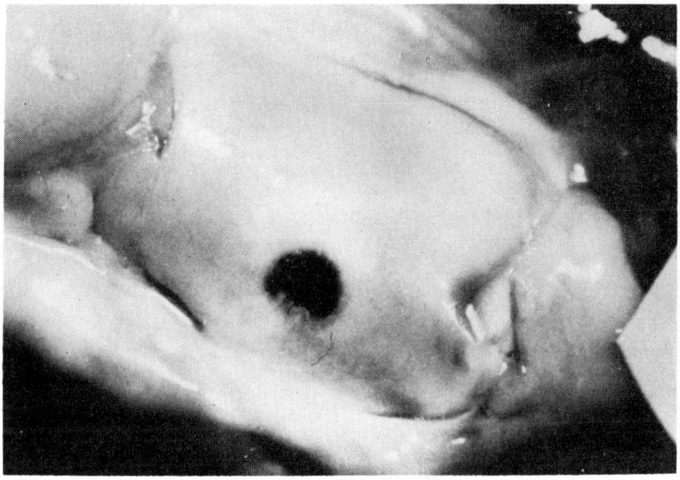

Fig. 10.5 An implant in the patella of the goat after 6 months use *in vivo*, the dark colour of the implant is due to staining with glutaraldehyde. (Scale indicated by the 6 mm implant diameter.)

10.3.2 *Results*

The laboratory findings were confirmed by the *in vivo* studies: the best results were obtained when the implants were placed in the optimal positions, about 0.25 mm proud and in tight-fitting holes. Mechanically, they were compatible with the bone and the cartilage in the joint: there was no evidence of adverse mechanical interaction between the host and the implant, as would have been evident as erosion of cartilage and resorption of bone, or premature breakdown of the implant. Equally, in cases where the implants had been placed sunken (unmatched positions), the collapse or flow of cartilage into the holes was observed. This was anticipated from the contact pressure results of the laboratory studies; it was found that the sunken implant did not make contact with opposing cartilage, leaving the surrounding cartilage subjected to greater than normal pressure. However, the rim of the hole was not supported by the implant and it was found (mainly in the high pressure joints, the condyles) that the rim of the hole, approximately 1 mm wide radially, actually supported less pressure than the area beyond it.

The implants demonstrated satisfactory wear properties; in general their surface remained smooth and that of opposing cartilage remained undamaged (Figure 10.5). In a few cases the surfaces of the implants had become rough, but had not damaged the opposing cartilage.

The implants remained chemically and biologically inert, also there

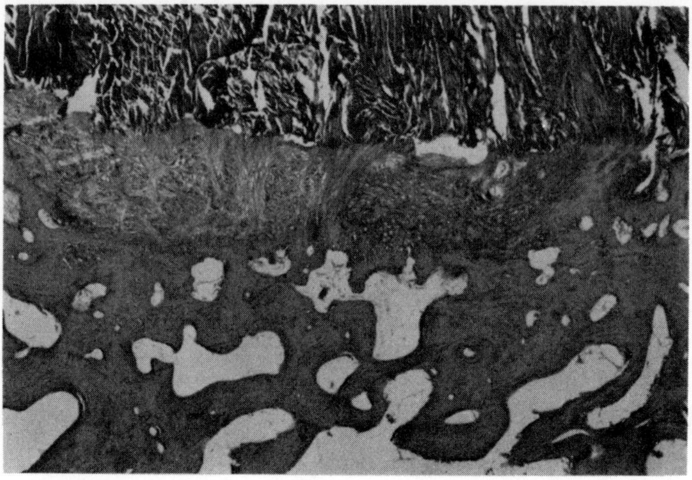

Fig. 10.6 Histological section showing the close adaptation of the host to the implant via a fibrous tissue layer linking the bone trabeculae (bottom) to the implant (top). (× 30; Azan Trichrome stain.)

was no evidence of adverse host tissue reaction. In fact, there was clear evidence that the host had accepted and incorporated them by fibrous tissue encapsulation to the extent that in some cases the junction between the host and the implant could not be discerned (Figure 10.6). There was however, no evidence of host tissue invasion, but the union between the host and the implant was firm, as assessed by mechanical implant pull-out tests (B. F. Shahgaldi, A. A. Amis and F. W. Heatley, unpublished observations).

10.4 **Discussion**

An implant of biological origin, such as fibrocartilage meniscus, is pliable; if implanted in an optimal position it deforms under a compressive load and takes the shape of the surrounding tissue. In so doing it spreads the load, avoiding localised pressure concentration. Its behaviour under compression is similar to that of articular cartilage on the bone, and on removal of the load it recovers to regain its full thickness.

Equally, an implant such as metal is markedly harder and stiffer than the opposing cartilage and would not deform to spread the load, thus leading to localised pressure concentration at the metal-cartilage contact. The results of experimental and clinical studies on the use of metal and plastic prostheses in hemiarthroplasty of the hip and the knee, published during the past decades, clearly show this point. In a series of studies (Harmon,

1943) the relief of pain in patients who had metal, glass and polymethylmethacrylate cup arthroplasties was compared and it was found that the patients with plastic cups had greater relief of pain than those with metal and glass cups. The relief of pain was probably due to the fact that the plastic was less stiff than the metal and glass, so it was a better match to the opposing articular cartilage and bone. It has been shown, in a series of experimental arthroplasties of the canine hip (Akeson *et al.*, 1969), that erosion of cartilage had occurred at the point of the most immediate contact between the cup and the acetabulum as early as two weeks post-operatively. The high failure rate of clinical cup arthroplasty was attributed to localised pressure concentration caused by hard metal in contact with articular cartilage (Radin *et al.*, 1975; Enneking & Singsen, 1974). Prosthetic resurfacing of the patella using cobalt-chromium alloy was anticipated (Worrell, 1979) to cause degeneration of the cartilage in the patellar groove of the femur after a few years of contact. These authors pointed out that a desirable material is one that has similar mechanical properties to articular cartilage.

A material softer than the articular cartilage in contact with it may reduce the contact pressure below what is neccessary for the physiological functioning of the cartilage. Articular cartilage is avascular and takes its nutrition from synovial fluid by diffusion. The diffusion of nutrients into the deeper layer of the cartilage is only possible if the fluid is squeezed in and out of this layer under pressure (Mow & Torzilli, 1975).

In the normal knee, the fibrocartilage menisci rub against the articular cartilage surfaces without being damaged by, or damaging them. The present study showed that the glutaraldehyde-fixed meniscal plugs did not wear off the cartilage during six months use *in vivo*, thus it was concluded that glutaraldehyde fixation does not alter the wear properties of bovine menisci significantly.

A laboratory study to match the wear properties of an implant material to those of articular cartilage is likely to be less useful than *in vivo* studies, such as that reported above, because of the rapid autolytic changes which would occur in the cartilage counterface whilst running a wear-testing machine.

10.5 **Conclusion**

Isolated non-progressive osteochondral defects in the knee joints of goats were satisfactorily replaced by meniscal fibrocartilage implants. The mechanical behaviour of the implants matched that of the normal articular cartilage on bone. It is believed that matching mechanical properties would enhance both the biological repair, and the long-term performance of the natural articular cartilage counterfaces.

Acknowledgements

The authors gratefully acknowledge the support of the Arthritis and Rheumatism Council of Great Britain in funding this project. We are also grateful to Professor S. A. V. Swanson, head of the Biomechanics Section of Imperial College, London, for his helpful suggestions for improvement of this manuscript.

References

Ahmed, A. M., Burke, D. l. & Yu, A. (1983). *In vitro* measurement of static pressure distribution in synovial joints- Part 11: retropatellar surface. *Journal of Biomechanical Engineering*, **105**, 226–36.

Akeson, W. H., Miyashita, C., Taylor, T. K. F., Laviolette, D. & Amiel, D. (1969). Experimental arthroplasty of the canine hip. *Journal of Bone and Joint Surgery*, **51A**, 149–64.

Alexander, R. M. (1980). Forces in animal joints, *Engineering in Medicine*, **9**, 93–7.

Black, J. & Frust, L. (1978). The response of cartilage to artificial articular surfaces. In: *The Human Joint in Health and Disease*, ed. by W. H. Simon. University of Pennsylvania Press, Philadelphia.

Breck, L. (1976). The use of certain plastic materials in reconstructive surgery of the knee joint. *Clinical Orthopaedics and Related Research*, **54**, 133–9.

Cobey, M. C. (1967). Arthroplasties using compressed ivalon sponge (intra-medic sponge). *Clinical Orthopaedics and Related Research*, **54**, 139–46.

Coletti, J. M., Akeson, W. H. & Woo, S. L-Y., (1972). A Comparison of the physical behaviour of normal articular cartilage and the arthroplasty surface. *Journal of Bone and Joint Surgery*, **54A**, 147–60.

Enneking, W. F. & Singsen, E. T. (1974). Pathological Changes in the idiopathic painful cup arthroplasty, *Clinical Orthopaedics and Related Research*, **101**, 236–41.

Ficat, P. P., Gedeon, P. & Toussaint, J. B (1979). Spongialisation: a new treatment for diseased patella. *Clinical Orthopaedics and Related Research*, **144**, 74–83.

Ghadially, F. N. (1983). Deep (full thickness) defects in articular cartilage. In: *Fine Structure of Synovial Joints*, ed. F. N. Ghadially, pp. 280–305. Butterworth, London.

Harmon, P. H. (1943). Arthroplasty of the hip for osteoarthritis utilzing foreign body cups of plastic. Cited by Black *et al*. In: *The Human Joints in Health and Disease*, ed. by W. H. Simon (1978), University of Pennsylvania Press, Philadelphia.

Hastings, G. W. (1980). Biomedical engineering and materials for orthopaedic implants. *Journal of Physics. E, Scientific* Instruments, **13**, 599–607.

Heatley, F. W. & Revel, W. J. (1985). The use of meniscal fibrocartilage as a surface arthroplasty to effect the repair of osteochondral defect. *Biomaterials*, **6**, 161–8.

Karlson, S. (1940). Chondromalacia patellae. *Acta Chirurgica Scandinavica*, **83**, 347–81.

Kempson, G. E. (1970). Mechanical properties of human articular cartilage, PhD Thesis, London University.

Kon, M. & de Visser, A. C. (1981). A poly (HEMA) sponge for restoration of articular cartilage defect, *Plastic and Reconstructive Surgery*, **67**, 289–93.

Milgram, J. W. (1985). Injury to articular cartilage joint surfaces. 1. chondral injury produced by patellar shaving: a histopathologic study of human tissue specimens. *Clinical Orthopaedics and Related Research*, **192**, 168–73.

Minns, R. J., Muckle, D. S. & Donken, J. E. (1982). The repair of osteochondral defects in osteoarthritic rabbit knees by the use of carbon fibre. *Biomaterials*, **3**, 81–6.

Mitchell, N. & Sheppard, N. (1976). The resurfacing of adult rabbit articular cartilage by multiple perforations through the subchondral bone. *Journal of Bone and Joint Surgery*, **58A**, 230–5.

Mow, V. C. & Torzilli, P. A. (1975). Fundamental fluid transport mechanisms through articular cartilage. *Annals of Rheumatic Disease*, **34** (Suppl.), 82–93.

Radin, E. L., Maquet, P. & Parker, H. (1975). Rationale and indications for the "hanging hip procedure": A clinical and experimental study. *Clinical Orthopaedics and Related Research*, **112**, 221–30.

Speer, D. P., Chvapil, M., Volz, R. & Holmes, M. D. (1979). Enhancement of healing in osteochondral defects by collagen sponge implants. *Clinical Orthopaedics and Related Research*, **144**, 326–34.

Worrell, R. V. (1979). Prosthetic resurfacing of the patella. *Clinical Orthopaedics and Related Research*, **144**, 91–7.

11

The structure and mechanical properties of a non-woven vascular prosthesis

T.V. How and D. Annis

11.1 Introduction

Prosthetic arteries made from synthetic polymers are now widely used for the replacement or bypass of diseased arteries. It was Voorhees *et al.* (1952) who, over 35 years ago, first demonstrated that a tube fashioned from Vinyon 'N' fabric could function satisfactorily as a blood conduit. Since then a number of different polymers have been tried and tested and the two materials which have stood the test of time are polyethylene terephathalate (Dacron) and polytetrafluoroethylene (Teflon). Moreover, developments in textile technology have made it possible to produce a wide range of seamless tubes and bifurcation prostheses based on the basic woven and knitted constructions. The wide range of fabric prostheses available today indicates that the ideal prosthesis suitbale for all situations has not yet been developed.

The performance of prosthetic arteries is dependent on the anatomical site; the results being best when the central large blood vessels with a high rate of flow are bypassed and worst when small peripheral vessels with low flow blood are involved. The high failure rate of small diameter arterial grafts (internal diameter less than 5–6 mm) are due either to early thrombosis, within a few days of implantation, or later occlusion due to the development of neo-intimal hyperplasia at or near the union between the synthetic graft and the host artery.

The patient's own long saphenous veins are firmly established as the best bypass grafts for peripheral arterial construction. However, in about 30% of the patients they are either not available because they have been removed previously, or they are not suitable because of their small size, or they are diseased. For these patients the only artificial arterial grafts available at present are made from Dacron or Teflon, both of which are relatively stiff polymers. The unsatisfactory performance of these grafts in peripheral arterial bypass has been attributed partly to their stiffness compared to the natural arteries (Walden *et al.*, 1980). The importance of the elastic properties of a blood vessel for the efficient propagation of pulsatile blood flow, without excessive energy losses or disturbances to blood flow, has been discussed by Baird & Abbott (1976). They argued from a theoretical hydrodynamic standpoint that the ideal graft should have mechanical properties closely matched to those of the natural arteries. Although it has been reported that long-term performance of small diameter grafts is related to elastic properties (Kidson & Abbott, 1978; Seifert *et al.*, 1979 Abbott *et al.*, 1987) the evidence is far from conclusive (Fisher *et al.*, 1985).

The elastic properties of blood vessels can be measured by recording the dynamic changes in diameter in response to pulse pressures. An important parameter describing vessel wall properties is compliance and it is defined by:

$$C = \Delta D_o / D_o \Delta P \qquad (11.1)$$

were ΔD_o is the change in external diameter D_o, of a vessel subjected to a change in pressure ΔP. Arteries and grafts are anisotropic and exhibit non-linear behaviour. The compliance is therefore not a constant but is a function of strains or stresses. Thus C should be measured over a range of mean pressures and longitudinal strains and comparison with other data can only be made at the same mean pressures and longitudinal strains.

Most studies of compliance of arterial grafts have been based on *in vitro* measurements under static or quasi-static loading. However, because of the viscoelastic properties of arteries and grafts their response is dependent not only on the level of stress but also the rate at which the stress is applied. Thus, *in vivo*, where rapid changes in blood pressure occur with each heartbeat, the compliance may be different from the static values. It is important, therefore, that mechanical properties of prosthetic arteries be studied under dynamic loading.

A non-woven microfibrous vascular graft produced by electrostatically spinning Biomer, a proprietary poly (ether-urethane-urea) [PEUU], has been developed. The spinning technique has been described in detail elsewhere (Annis *et al*, 1978). Briefly, the PEUU contained in a syringe is fed at a controlled rate through a series of stainless steel needles into a DC electrostatic field surrounding a charged mandrel. The syringe is supported on a traversing mechanism which travels to and fro, at a constant linear speed, across the mandrel. The latter is driven by an electric motor. When droplets of the polymer are injected from the needles into the electrostatic field, they elongate into jets. From the ends of the jets fine fibres, 1–2 μm in diameter, are formed and are attracted onto the rotating mandrel. Layers of fibres gradually build up and, upon removal from the mandrel, a microfibrous tube is formed. This manufacturing technique has been shown to be stable and controllable and grafts with a range of anisotropic properties can be produced reliably (How & Clarke, 1984). Because of the inherent symmetry of the manufacturing process the grafts have been shown to be orthotropic.

We have characterised the viscoelastic properties of the PEUU grafts by carrying out separate quasi-static and dynamic tests. It is shown that the viscoelastic properties, over the range of normal *in vivo* stresses, can be described by the superposition of the quasi-static and dynamic behaviour. The quasi-static properties are based on finite deformation theory using a strain energy function and the dynamic properties were determined by recording the response to small sinusoidal loads.

11.2 **Materials and methods**

Grafts of internal diameter ranging from 3.4 to 3.8 mm and wall thickness from 0.25 to 0.55 mm were produced from Biomer PEUU by the electrostatic spinning process. Because of the porosity of the grafts, gelatin was used to fill the insterstices, as described previously (How & Clarke, 1984; How & Annis, 1987). Below 25°C, the graft thus treated was able to withstand pressures up to 33 $kN \cdot m^{-2}$ (250 mmHg) with minimal fluid loss. Gelatin was not suitable for tests carried out at higher temperatures since it reverts to the solution state at around 30°C. Instead, a dilute latex rubber solution was used to coat the luminal surface of the graft leaving a thin latex film when dry. Three coats of latex were applied, having a total thickness of approximately 15 μm.

A diagram of the experimental set-up is shown in Figure 11.1. A graft segment about 7 cm long is mounted horizontally in a perspex chamber filled with water. The water temperature was controlled to ±0.5° of the set temperature by circulating water from a temperature-controlled waterbath through a heat exchanger placed in the specimen chamber. The sides of the chamber were fitted with glass optical windows through which the graft external diameter was continuously monitored. A specially designed infrared diameter gauge was used for this purpose (How *et al.*, 1984). This device has a resolution of ±2 μm at the rated output of 1.0 $V \cdot mm^{-1}$ and a linear range (to 0.35% full scale) from 1.6 to 8 mm. Calibration of the device was performed prior to each measurement using a rod of known diameter (measured by means of a digital micrometer). This one-point calibration was sufficient since the error due to non-linearity, over the range 4–6 mm, was only 5 μm.

The intraluminal pressure was measured by means of a miniature pressure transducer mounted on a 16-gauge hypodermic needle (Gaeltec Ltd, 12CT). The longitudinal force was measured by means of a strain-gauge

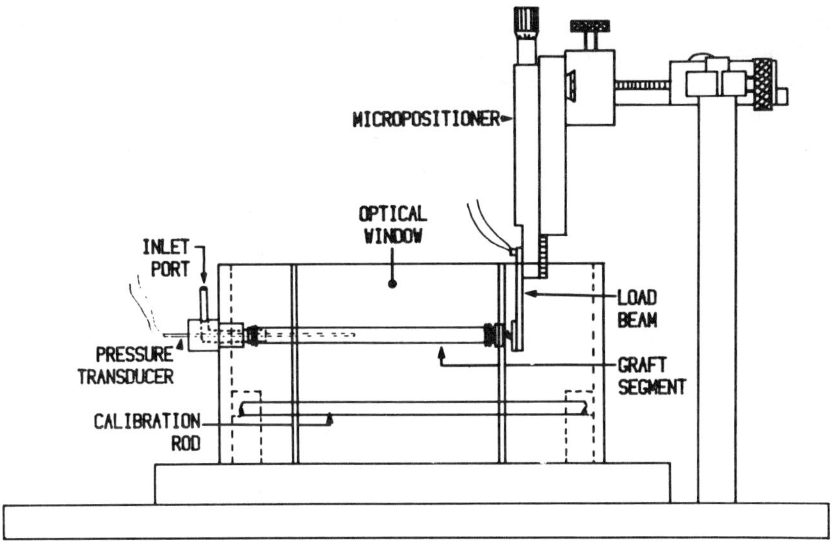

Fig. 11.1 Diagram of the experimental set-up. The pressure generator and diameter gauge are not shown. (Taken from T. V. How & D. Annis (1987); reprinted by permission of the *Journal of Biomedical Materials Research*, **21**, 1093–1108.)

load beam (Strain Measurement Devices Ltd., SMD 100) having a rated load of 5 N. The beam was rigidly attached to a three-axis micropositioner which was used to align the graft horizontally and to apply accurate longitudinal strains.

The quasi-static properties of the graft were determined by applying a slowly increasing pressure from a water-filled syringe pump. The infusion rate was set in order to produce a circumferential strain rate not greater than $0.03\,\text{min}^{-1}$. Pressure, outer diameter and longitudinal force were recorded continuously. Each specimen was tested at four longitudinal strains between 4 and 10%. For the dynamic tests the input pressure consisted of a sinusoidal component, of amplitude $5.3\ \text{kN}\cdot\text{m}^{-2}$ peak-to-peak and fixed frequency in the range 0.02–20 Hz, superimposed on a steady pressure of $13.3\,\text{kN}\cdot\text{m}^{-2}$. Pressure and diameter signals were recorded for each frequency. The phase lag of diameter behind pressure was measured using a storage oscilloscope operated in the X-Y mode.

11.3 **Results and discussions**

11.3.1 *Quasi-static properties*

The static elastic properties of arteries have been extensively studied using strain energy functions. Patel & Vaishnav (1973) developed a two-dimensional model based on a polynomial form of the strain energy function. Although this model was originally used to interpret canine aortic data it has been shown that it can adequately describe the elastic properties of microfibrous prosthetic arteries made from poly (ether-urethane) (How & Clarke, 1984). Whether this model can also be applied to Biomer PEUU grafts is investigated here.

The grafts are assumed to be thin-walled and curvilinearly orthotropic and are made from an homogeneous incompressible and perfectly elastic material. The constitutive equations were derived for this type of material and are given below as a relationship between pressure P, and longitudinal force, T, and mid-wall radius, R, in terms of the graft dimensions and the material constants $A, B, \ldots, G$ (How & Clarke, 1984).

$$P = K_1 \left\{\frac{R}{R_i}\right\}^4 + K_2 \left\{\frac{R}{R_i}\right\}^2 + K_3 \tag{11.2}$$

$$\begin{aligned} T = 2\pi RH(&B(a + 2ab) + C\,(2b + 4b^2) + E\,(a^2 + 2a^2b) \\ &+ F\,(2ab + 4ab^2) + G\,(3b^2 + 6b^3) - PR/2H) \end{aligned} \tag{11.3}$$

where,
$$\begin{aligned} K_1 &= K(3D/4) \\ K_2 &= K(A - 3D/2 + Eb) \\ K_3 &= K(-A + Bb + 3D/4 - Eb + Fb^2) \\ K &= V/2\pi LR_i \end{aligned}$$

V is the volume of the graft wall of length L, initial undeformed mid-wall radius R_i, and wall thickness H; a and b are the Green-St. Venant strains given by $1/2\,[(R/R_i)^2 - 1]$ and $1/2\,[(L/L_i)^2 - 1]$ respectively. For each graft tested a set of material constants was determined. Ideally, all the grafts tested should have the same values of constitutive constants, in which case all grafts having a range of diameters and wall thickness can be described by equations (11.2) and (11.3).

The mean, standard error and coefficient of variation of each of the constants for the Biomer PEUU grafts are given in Table 11.1. The

Table 11.1 *The mean, standard error (SE) and coefficient of variation (CV) of the constitutive constants for electrostatically spun Biomer grafts. The internal diameter of the grafts ranged from 3.4 to 3.8 mm and wall thickness from 0.25 to 0.5 mm.*

	A	B	C	D	E	F	G
Mean ($\times 10^5$ N·m^{-2})	3.2962	1.9843	7.3117	−1.3753	1.2299	0.4318	−11.9818
± SE ($\times 10^5$ N·m^{-2})	0.1379	0.1283	0.1958	0.2974	0.3377	0.3562	0.5781
CV	0.1256	0.1940	0.0803	0.6095	0.8237	2.4745	0.1447

constants A, B and C which have the greatest influence on the constitutive equation have similar absolute values. Although the constants D, E and F have a much greater variation their contribution in the constitutive equations are relatively less important. Substituting the material constants in equation (11.2), the pressure – radius relationship of Biomer grafts of known wall thickness and diameter can be determined and the compliance at a mean pressure of 13.3 kN·m^{-2} and various longitudinal extensions λ_z ($\lambda_z = L/L_i$) calculated. In order to determine the accuracy of this method, the calculated compliance (C_i) and the actual measured compliance (C_m) were compared.

The mean and maximum absolute differences between C_m and C_c, expressed as a percentage of C_m, were 5.38% and 20.5%, respectively. The accuracy of the model in determining the pressure – radius relationship and compliance for prostheses of known dimensions was considered to be satisfactory for the present purposes. The magnitude of the calculated material constants was found to be dependent on small variations in the experimental data. For this reason great care was taken in measuring the length, diameter and thickness of each specimen. In the model, the assumption was made that the prostheses had thin walls, that is, the radius-to-wall thickness ratios were greater than 10. The circumferential and axial stress distribution could then be considered to be uniform and the radial stress negligible compared with the circumferential stress (Fung *et al.*, 1979). Since the grafts tested had radius-to-wall thickness ratios in the range 3.9–9.7 at 13.3 kN·m^{-2} pressure, this assumption was not strictly correct. Although a three-dimensional, thick-wall model could have been used (e.g. Vaishnav *et al.*, 1973) the amount of computation would be greatly increased but with only a marginal improvement in accuracy.

The model also assumes that the prosthesis wall was composed of a perfectly elastic material. Although Biomer is viscoelastic it exhibits preconditioning when subjected to repeated cyclic deformations. As the number of cycles is increased the response reaches a steady state and becomes stable and reproducible. A material thus preconditioned can be regarded as being pseudoelastic (Fung *et al.*, 1979), and either the loading or unloading part of the response curve can be defined uniquely. Biomer prostheses were found to require only three or four preconditioning cycles and only the data obtained during loading were used in calculating the material constants. The cyclic pressure – diameter curves between 0–26 kN·m^{-2} were almost coincident after preconditioning. When the load was removed, recovery to the original undeformed diameter was complete after only a relatively short period of time. Therefore measurements of initial dimensions were carried out about 10 min after preconditioning. Wall thickness was measured at the end of the experiment using a gravimetric method.

Electrostatically spun prostheses, made from the same polymer and under the same spinning conditions, have been shown to have similar microstructure and consequently similar mechanical properties (How *et al.*, 1982). The anisotropic properties can be altered by changing certain process variables, notably the mandrel speed. It is possible, therefore, to produce prosthetic arteries with circumferential elastic modulus greater than or less than the longitudinal elastic modulus depending on the mandrel speed. Differences in the values of material constants can therefore be expected for prostheses made under different spinning conditions. The prostheses tested here were all produced under identical conditions and the variability in the values of the material constants is due to experimental errors and small differences in the structure of the specimens.

When the prosthesis is used to replace a segment of artery, a longitudinal strain is normally applied by the surgeon at implantation. This strain should be sufficiently high for the graft not to undergo compressive stresses at elevated pressures. Compressive loads may result in buckling, leading to obstruction and eventually to complete occlusion of the lumen. There is a limit, however, to the amount of longitudinal strain which can be imposed since excessive loads on the host vessel may lead to anastomotic rupture. The longitudinal tethering force in the canine carotid artery is about 0.5 N (Patel & Fry, 1969; Patel & Janicki, 1970) and it is desirable, therefore, when such an artery is replaced that this force is not exceeded. The longitudinal force, T, of the Biomer prostheses is a function of wall thickness, λ_z, as well as the internal pressure and the properties of the polymer. It can be calculated, after substituting the material constants, for prostheses of known dimensions, using equation (11.3).

11.3.2 *Dynamic properties*

The amplitude of the input sinusoidal presures applied to the test specimens was set to $\pm$ 2.7 kN·m^{-2} in order to reproduce normal *in vivo* values as closely as possible. Moreover, at this amplitude the diameter changes in grafts of wall thickness greater than about 0.3 mm were also sinusoidal, with no observable distortion for frequencies up to 20 Hz. In thinner wall grafts which underwent diameter changes of greater than 5% ($C >$ 0.13% mmHg) marked distortion of the diameter signal was observed at frequencies above 15–20 Hz. Under these conditions, therefore, these grafts exhibited non-linear viscoelastic behaviour and they were excluded from this study.

The dynamic (C_{dyn}) and quasi-static (C_{stat}) compliances were calculated for each graft specimen and plotted as the ratio C_{dyn}/C_{stat} against log frequency. The results for a typical specimen, obtained at four different λ_z between 1.04 and 1.10, are shown in Figure 11.2a. It can be seen that the

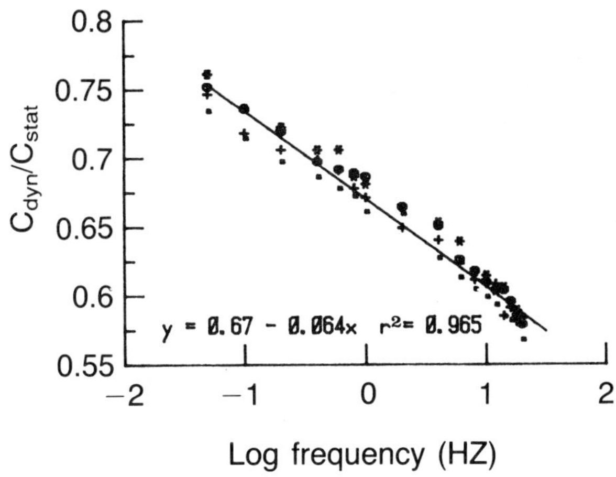

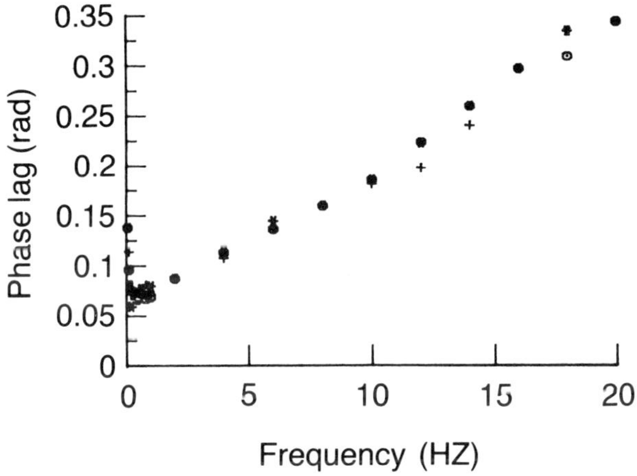

Fig. 11.2 (a) Normalised dynamic compliance of Biomer grafts as a function of the log frequency of the sinusoidal input pressure at four longitudinal extension ratios. (b) Phase lag of diameter behind pressure as a function of frequency and longitudinal extension ratio. For both (a) and (b) λ_z =: (+) 1.04; (■) 1.06; (*) 1.08; (○) 1.10. (Taken from T. V. How & D. Annis (1987); reprinted by permission of the *Journal of Biomedical Materials Research*, **21**, 1093–1108.)

normalised dynamic compliance decreased linearly with log frequency and the regression line through the data points is given by:

$$C_{dyn}/C_{stat} = 0.67 - 0.064 \log f;\ r^2 = 0.965 \qquad (11.4)$$

The phase difference between the diameter and pressure signals are shown in Figure 11.2b, plotted against frequency for various values of λ_z. There is a sharp reduction in phase lag from about 0.14 rad. at 0.05 Hz to about 0.075 rad at 1 Hz, above which it increased linearly with frequency.

Similar results were obtained for grafts of various wall thicknesses, with the data points falling within a narrow band. In this case the normalised dynamic compliance also varied linearly with log frequency. The regression curve, $C_{dyn}/C_{stat} = 0.67 - 0.061 \log f$; $r^2 = 0.963$, has similar coefficients to those calculated from the data of Figure 11.2a. These results clearly show that the prosthesis becomes stiffer as the frequency of the pressure oscillation is increased. Although under quasi-static loading compliance is determined by wall thickness, amongst other factors, the frequency-dependent properties were not significantly altered.

Hasegawa & Azuma (1979) determined the viscoelastic properties of Dacron and Teflon fabric prostheses by carrying out uniaxial stress-relaxation measurements on narrow strip samples. They indicated that the viscoelastic properties of these prostheses depend on the polymeric material and not on their construction or structure. However, preliminary measurements which were carried out on electrostatically-spun Biomer prostheses having different fibrous structure to those tested above, but made from the same polymer, suggest that the fibrous structure has a significant influence on the frequency-dependent behaviour.

The effect of temperature on the viscoelastic properties was examined by testing several graft specimens of the same wall thickness at temperatures of 25°C, 30°C and 37°C at a constant extension ratio of 1.06. These specimens were coated with latex rubber since gelatin could not be used above 30°C. These specimens were more compliant than those treated with gelatin by about 18%. A small increase in quasi-static compliance of about 2.5% was observed when the temperature increased from 25 to 37°C. Similarly, the normalised dynamic compliance appeared to be slightly dependent on temperature especially at frequencies above 8 Hz (Figure 11.3). However, the effect was not significant and may be ignored. There was a significant difference in the frequency-dependent behaviour between the grafts clotted with gelatin and those coated with latex; the decrease in normalised dynamic compliance with frequency being less marked in the latter ($C_{dyn}/C_{stat} = 0.80 - 0.051 \log f$). Gelatin in the clotted graft acts as the ground substance within the microfibrous structure of the graft and it plays a significant role in the quasi-static and dynamic behaviour. The effect of the ground substance on the mechanical behaviour of soft tissues has been studied by Minns *et al.* (1973). They showed that the removal of the ground substance from the human aorta by treatment with EDTA greatly reduced the load relaxation properties.

Preclotting of the Biomer PEUU prostheses with gelatin is necessary only for mechanical testing but is not required prior to implantation. Following the release of the artery clamps after the completion of the

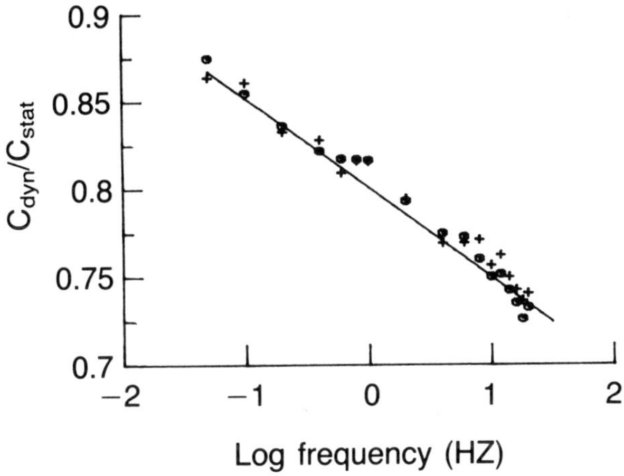

Fig. 11.3 Normalised dynamic compliance *versus* log frequency for Biomer grafts tested at three different temperatures: +, 25°C; ●, 30°C; ○, 37°C. Longitudinal extension ratio = 1.06.

anastomoses, the interstices are filled with platelets, red cells and blood plasma. After slowly oozing out through the wall for a few minutes, the blood plasma coagulates forming an impervious conduit. The clotting procedure using gelatin is therefore similar to that which takes place soon after implantation. With time, however, healing of the prosthesis and its incorporation into the surrounding tissue occurs. Thus the implanted prostheses would be expected to have different viscoelastic properties to those tested here.

11.3.3 *Dynamic modulus: comparison with arteries*

Dynamic mechanical tests, where small sinusoidal strains superimposed on a mean strain level are applied to the material, have been widely used on arterial tissue (e.g. Bergel, 1961; Gow & Taylor, 1968). For a linear viscoelastic material a complex dynamic modulus, E', is defined as:

$$E' = E_{dyn} + j\eta\omega \tag{11.5}$$

where the real part, E_{dyn}, is the dynamic elastic modulus and the imaginary part $\eta\omega$ is the viscous modulus. E_{dyn} and $\eta\omega$ can be expressed in terms of E' and the phase angle φ between the pressure and diameter sinusoids:

$$E_{dyn} = E'\cos\varphi \tag{11.6a}$$
$$\eta\omega = E'\sin\varphi \tag{11.6b}$$

Assuming arterial tissue to be isotropic and incompressible Bergel (1961) derived an expression relating E' to the vessel dimensions and the pressure amplitude as follows:

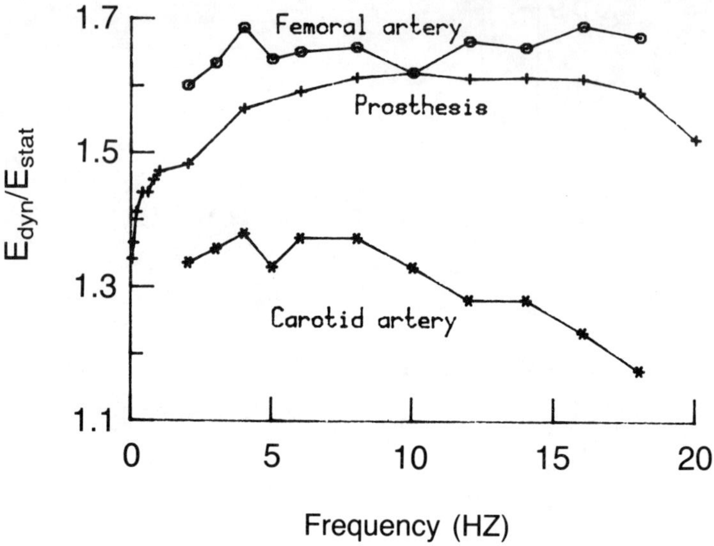

Fig. 11.4 Effect of the frequency of sinusoidal input pressure on the ratio of dynamic to static incremental modulus; Comparison between Biomer prostheses and canine arteries. (Taken from T. V. How & D. Annis (1987); reprinted by permission of the *Journal of Biomedical Materials Research*, **21**, 1093–1108.)

$$E' = \frac{1.5\, R_i^2\, R_o\, \Delta P}{(R_o^2 - R_i^2)\, \Delta R_o} \tag{11.7}$$

where R_i is the inner and R_o the outer radius, ΔR_o is the change in the outer radius in response to an increment of pressure ΔP.

Using equations (11.6) and (11.7) E_{dyn} and $\eta\omega$ were calculated for the Biomer grafts. The dynamic elastic modulus E_{dyn} showed a marked rise in magnitude with frequency up to about 4 Hz after which it assumed a relatively constant value of $1.3 \times 10^6\ N{\cdot}m^{-2}$. The absolute values of E_{dyn} were similar to those of the carotid and femoral arteries of the dog (Bergel, 1961). This was also true for the viscous modulus at frequencies below 2 Hz. At higher frequencies however, $\eta\omega$ was higher in the prostheses than in the arteries.

A plot of the dynamic elastic modulus, normalised with respect to the static modulus, against frequency is shown in Figure 11.4. The data for canine carotid and femoral arteries obtained by Bergel are also plotted on the same figure. The frequency dependence of normalised dynamic elastic modulus show a similar pattern to that of the femoral arteries. Below 1 Hz, E_{dyn}/E_{stat} increased sharply but between 2 and 18 Hz it remained relatively constant. Thus over the physiological range of frequencies the Biomer PEUU prostheses may be assumed to be independent of frequency. Above

2 Hz, the ratio of dynamic to static elastic modulus was 1.6 for the prostheses, compared with 1.66 for the femoral arteries and about 1.3 for the carotid arteries. It may be concluded, therefore, that the frequency dependent behaviour of the Biomer PEUU grafts approximated closely to that of the dog femoral arteries.

References

Abbott W. M., Megerman J., Hasson E., L'Italien G. & Warnock D. F. (1987). Effect of compliance mismatch on vascular graft patency. *Journal of Vascular Surgery*, **5**, 376–82.

Annis D., Bornat A., Edwards R. O., Higham A., Loveday B. & Wilson J. (1978). An elastomeric vascular prosthesis. *Transactions of the American Society for Artificial Internal Organs*, **24**, 209–14.

Baird R. N. & Abbott W. 1976. Low compliance and arterial grafts. *The Lancet*, **2**, 948–9.

Bergel D. H. (1961). The dynamic elastic properties of the arterial wall. *Journal of Physiology*, **156**, 458–69.

Fisher A. C., How T. V., de Cossart L. & Annis D. (1985). The longer term patency of a compliant small diameter arterial prosthesis. The effect of the withdrawing of aspirin and dipyridamole therapy. The effect of reduced compliance. *Transactions of the American Society for Artificial Internal Organs*, **31**, 324–8.

Fung Y. C., Fronek K. & Patitucci P. (1979). Pseudoelasticity of arteries and the choice of its mathematical expression. *American Journal of Physiology*, **237**, H620-H631

Gow B. S. & Taylor M. G. (1968). Measurement of viscoelastic properties of arteries in the living dog. *Circulation Research*, **23**, 111–22.

Hasegawa M. & Azuma T. (1979). Mechanical properties of synthetic arterial grafts. *Journal of Biomechanics*, **12**, 509–17.

How T. V., Clarke R. M. & Annis D. (1982). Uniaxial tensile properties of the Liverpool artificial artery. Proceedings of the Intenational Conference on Biomedical Polymers, University of Durham, 12–15 July, 1982.

How T. V., Bhuvaneshwar G. S. & Annis D. (1984). Infrared diameter gauge for in vitro mechanical testing of vascular grafts. *Journal of Biomedical Engineering*, **6**, 195–9.

How T. V. & Clarke R. M. (1984). The elastic properties of a polyurethane arterial prosthesis. *Journal of Biomechanics*, **17**, 597–608.

How T. V. & Annis D. (1987). Viscoelastic behaviour of polyurethane vascular prostheses. *Journal of Biomedical Materials Research*, **21**, 1093–1108.

Kidson I. G. & Abbott W. M. (1978). Low compliance and arterial graft occlusion. *Cardiovascular Surgery*, **58** (Suppl. 1), I1-I4.

Minns R. J., Soden P. D. & Jackson D. S. (1973). The role of the fibrous components and ground substance in the mechanical properties of biological tissues: A preliminary investigation. *Journal of Biomechanics*, **6**, 153–65.

Patel D. J. & Fry D. L. (1969). The elastic symmetry of arterial segments in dogs. *Circulation Research*, **24**, 1–8.

Patel D. J. & Janicki J. S. (1970). static elastic properties of the left coronary circumflex artery and the common carotid artery in dogs. *Circulation Research*, **27**, 149–58.

Patel D. J. & Vaishnav R. N. (1972). The rheology of large blood vessels. In: *Cardiovascular Fluid Dynamics, Vol. 2*, ed. D. H. Bergel, pp. 2–64. Academic Press, London.

Seifert K. B., Albo D., Knowlton H. & Lyman D. J. (1979). Effect of elasticity of prosthetic wall on patency of small-diameter arterial prostheses. *Surgical Forum*, **30**, 206–8.

Vaishnav R. N., Young J. T. & Patel D. J. (1973). Distribution of stresses and of strain-energy density through the wall thickness in a canine aortic segment. *Circulation Research*, **32**, 577–83.

Voorhees A. B., Jaretski A. & Blakemore A. H. (1952). The use of tubes constucted from Vinyon 'N' cloth in bridging arterial defects. *Annals of Surgery*, **135**, 332–7.

Walden R., L'Italien G. J., Megerman J. & Abbott W. M. (1980). Matched elastic properties and successful arterial grafting. *Archives of Surgery*, **115**, 1166–9.

12

The role of finite element modelling in biomechanics

H. S. Ranu

12.1 Introduction

The human spinal column is subject to many complex loading situations in daily activities and there is enough evidence to suggest that low back syndrome is mechanical in origin. There is, therefore, a need to investigate further the biomechanics of the lumbar spine.

Extensive use has recently been made of the finite element method of stress analysis in biomechanics. The finite element method is ideally suited to study of the modelling of the irregularly-shaped structure and composite nature of the vertebral bone and the disc. It is also able to identify locations of high stress due to surgical procedures. Therefore, a three-dimensional finite element model has been developed which is capable of simulating the *in vivo* situations. In this chapter a number of clinically relevant loading situations of the lumbar spinal unit are modelled and computer graphics of them are presented.

12.2 Anatomy of the human spine

The design of the human spine is such that it provides flexibility, in three planes of motion, and strength, yet it protects the delicate spinal cord. The human spinal column consists of 24 vertebrae, held in place by the ligaments, the intervertebral discs and the facet joints. There are seven cervical (neck region), twelve thoracic (chest region) and five lumbar (lower back) vertebrae; these three sections make up the spine.

The shape of the vertebral body is roughly cylindrical. The posterior section of the vertebra has spinous and transverse processes, the laminae, the pedicles and the facet joints. The spinal cord is protected by the pedicles which form the two sides of the spinal canal and the laminae. The articular facets are between the laminae and the pedicles (Figure 12.1A).

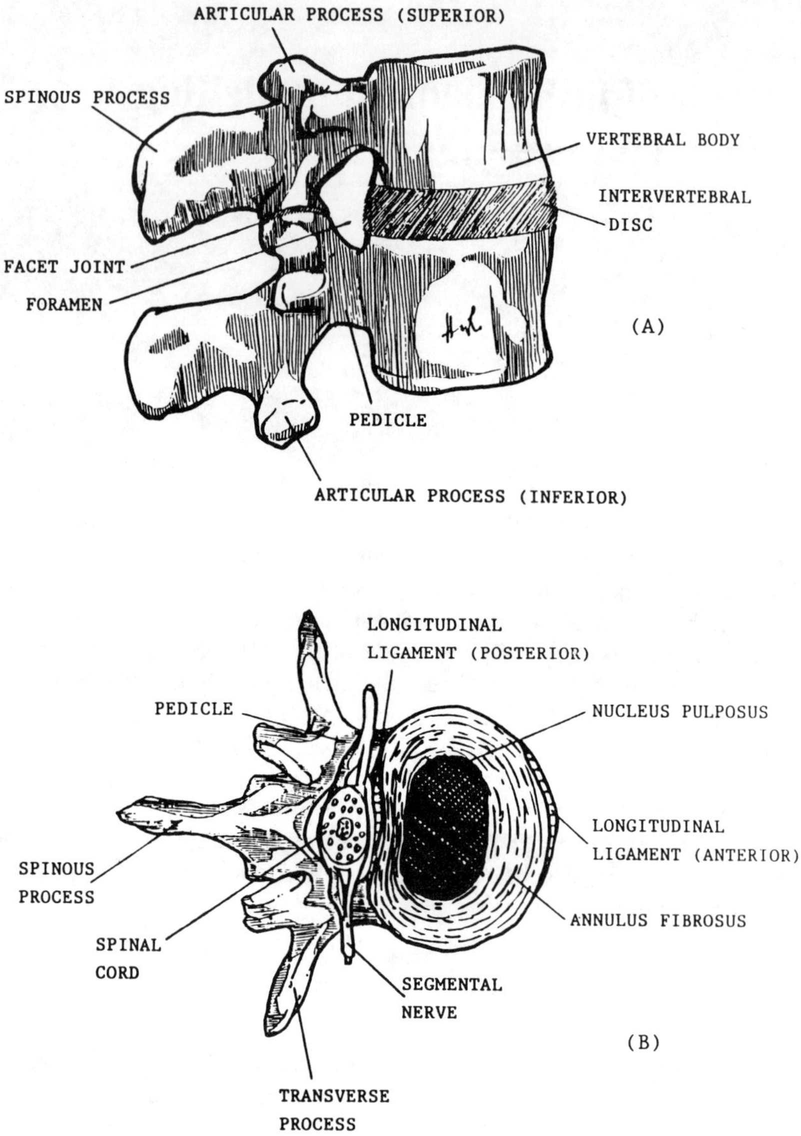

Fig. 12.1 (A) Schematic side view of the lumbar spinal unit. (B) Schematic plan view of the vertebra with a section through the intervertebral disc.

The vertebral bodies are joined together by the intervertebral discs which give stability and flexibility to the spinal column. The discs consist of the nucleus pulposus and the annulus fibrosus. The cartilaginous endplates separate the nucleus and the annulus from the vertebral bodies (Figure 12.1B).

The nucleus pulposus occupies the central portion of the disc. It has a glistening gel-like appearance – this is due to the high content of proteoglycans (PGs) and its behaviour is hydrostatic in nature. The annulus fibrosus forms part of the outer layers of the disc. It consists of lamellae, which have fibres of different orientations enmeshed in mucopolysaccharide matrix.

The behaviour of the disc is very dependent on its degree of degeneration: with advancing age, the clear distinction between the nucleus and the annulus disappears. The entire disc is then transformed into a fibrous cartilage, no longer able to distribute the stresses uniformly in the lumbar section of the spine.

12.3 Description of the model

The three-dimensional model of a single vertebra developed by Ranu (1982) consists of eight-noded hexahedral (brick) elements used to represent the spongy bone of the centrum, the cortical bone of the pedicles, the articular facets and the laminae. The vertebral endplates, the cortex of the vertebral body, the transverse and spinous processes are modelled by thick plate and/or thin shell elements, quadrilateral in shape and having both membrane and bending resistance. Three-dimensional spring elements were used to simulate the intervertebral disc, which resisted axial or shear loads. The disc was not simulated on the superior endplate where concentrated nodal loads were applied. For the centrum, there were a total of 72 brick elements arranged in three transverse layers, each containing 24 such elements. The posterior structure was composed of 22 brick elements, two for the two pedicles, 12 for two lamina and six for articular facets.

The superior facet was modelled by a single hexahedral element, while the inferior facet was represented by two such elements. There was a total of 91 thick/thin shell elements, 86 for the cortex and endplates, three for the spinous process and one each for the transverse process. Eighty-four axial spring elements were used across the inferior endplate, of which 32 provided compressive support. Around the outer periphery of the body, 28 axial springs were located which gave shear support. A total of 24 springs were located on the inferior articular process to give compressive and shear support. The physical properties of vertebral bone and spinal disc were obtained from Ranu (1981, 1985). The NISAOPT code was used to com-

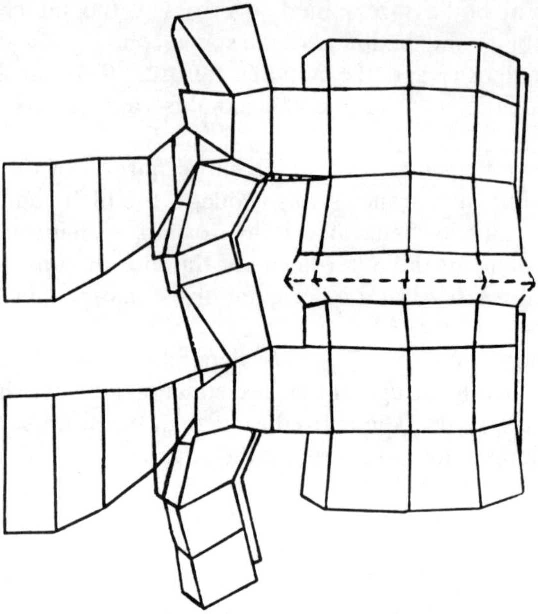

Fig. 12.2 Side view of the spinal unit.

pute the vertebral response to combined static loads of axial 2400 and shear 430 N.

The three-dimensional model was extended to include a spinal unit (Ranu, 1986a) consisting of two vertebrae and an intervertebral disc as shown in Figure 12.2. Figure 12.3 shows the top view of the model and the elements under study are also numbered in this figure. The number of elements and nodes were reduced for computational reasons. The following cases were studied:

(1) bilaterally symmetric vertebra subjected to loads of 2400 N axial and 430 N shear;

(2) asymmetric vertebra with two laminae removed on the right side (elements 133 and 135);

(3) asymmetric vertebra with the top element removed on the right side (element 135);

(4) model predictions of the nucleus and the annulus pressures;

(5) simulation of disc degeneration;

(6) bilaterally symmetric vertebra subjected to concentrated axial loads of 380 N.

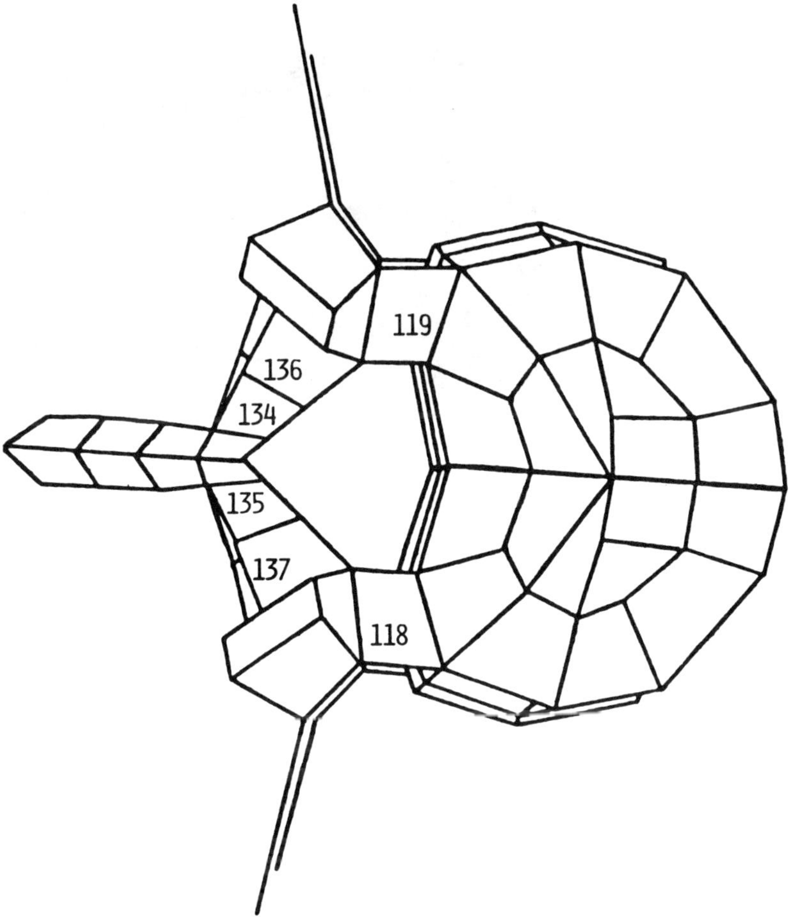

Fig. 12.3 Plan view of the three-dimensional model.

12.4 **Model results**

Case 1

The geometry and loading were bilaterally symmetric. The displacements and stresses were found to be symmetric. All nodal displacements were exaggerated 40 times relative to the vertebral model scale. A lateral view of the displacements for a single vertebra is shown in Figure 12.4. High stress distribution was found in the region of the laminae and at the junction of the pedicle with the lamina.

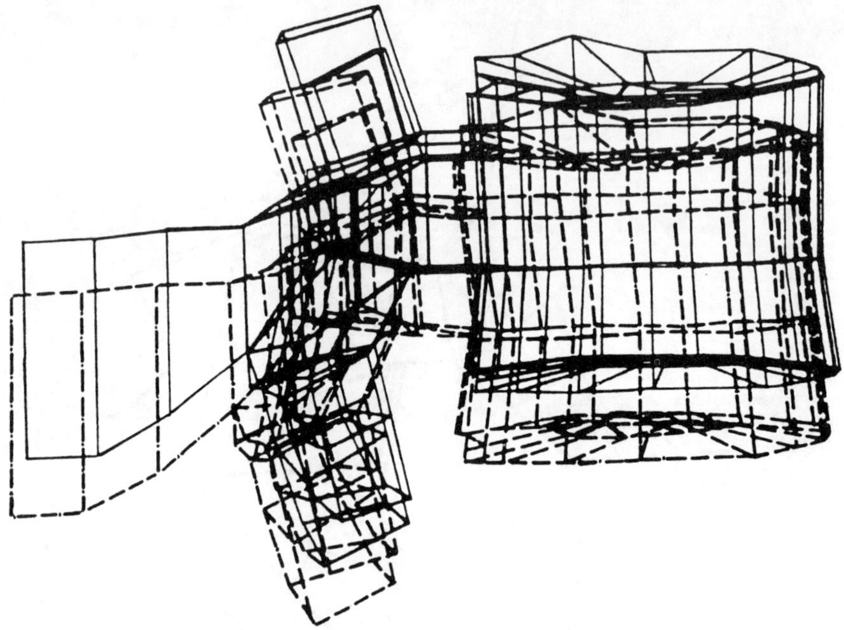

Fig. 12.4 Side view of the displacements for Case 1.

Case 2

Figure 12.5 shows the stress patterns when the entire right lamina is removed (elements 133 and 135). The displacement patterns did not change significantly. However, in the region of the pars, there were increases in tensile stresses. Figure 12.6A shows the tensile stress patterns in the lamina region of Case 2. Stresses for Case 1 are presented in parentheses in Figure 12.6A for the purposes of comparison.

Case 3

When a superior portion of the right lamina (element 135) was removed, the tensile stresses increased by a large amount compared with Case 1. Figure 12.6 B shows the tensile stress patterns in the lamina region of Case 3 and stresses for Case 1 are again presented in parentheses in this figure.

Case 4

The model was used to simulate the distribution of pressures in the nucleus and under the annulus for an axial loading in compression. Table 12.1 shows a comparison of these pressures. Pressures in the nucleus and under the annulus were also obtained by instrumenting the human spinal segments with miniature pressure transducers as described by Ranu & King

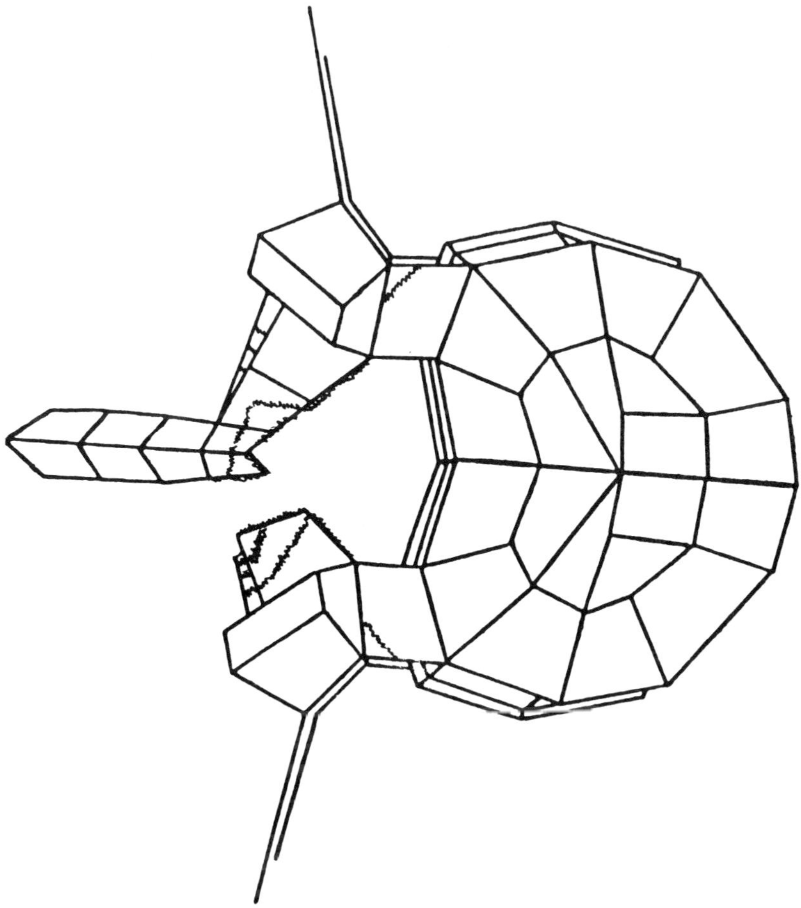

Fig. 12.5 Plan view of Case 2 with predicted stress contour patterns.

(1980) and Ranu (1986b, 1987). These are also compared with the model results as shown in Table 12.1.

Case 5

The model was used to simulate the disc degeneration on the spinal unit by altering the elastic properties of the nucleus pulposus and the annulus fibrosus. This resulted in a decrease in height of the intervertebral disc as shown in Figure 12.2. As a result of this simulation, there was a significant increase in the loading of the posterior elements of the vertebra. Pressures in the nucleus and within the annulus were also increased by a large amount (53%).

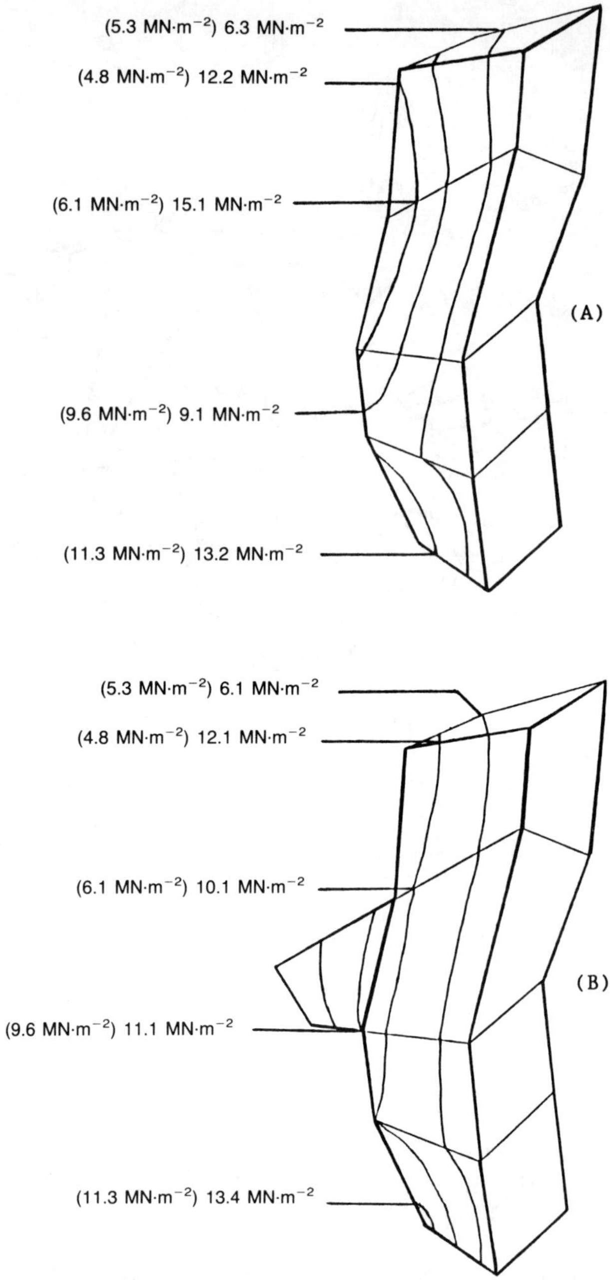

Fig. 12.6 (A) and (B) Side view of the stress patterns in the laminar region of Cases 2 and 3. Stresses for Case 1 are in parentheses.

Table 12.1 *Comparison of nucleus and annulus pressures for the model with experimental values for a normal disc.*

Load (N)	Nucleus pressure ($MN \cdot m^{-2}$)		Annulus pressure ($MN \cdot m^{-2}$)	
	Model	Experimental	Model	Experimental
200	0.28	0.22	0.26	0.24
600	0.76	0.58	0.73	0.65
1000	1.17	0.92	1.14	1.1
1400	1.59	1.18	1.46	1.36

Case 6

Concentrated nodal loads totalling 380 N were applied to the right lamina representing a surgical procedure. The tensile stress patterns due to this procedure are shown in Figure 12.7. High tensile stresses were found at the inferior junction of the pedicle and the lamina and at the inferior tip of the lamina.

12.5 **Conclusions**

A three-dimensional model of a spinal unit has been developed, which includes the simulation of an intervertebral disc in between two vertebrae. The conclusions drawn from this study are:

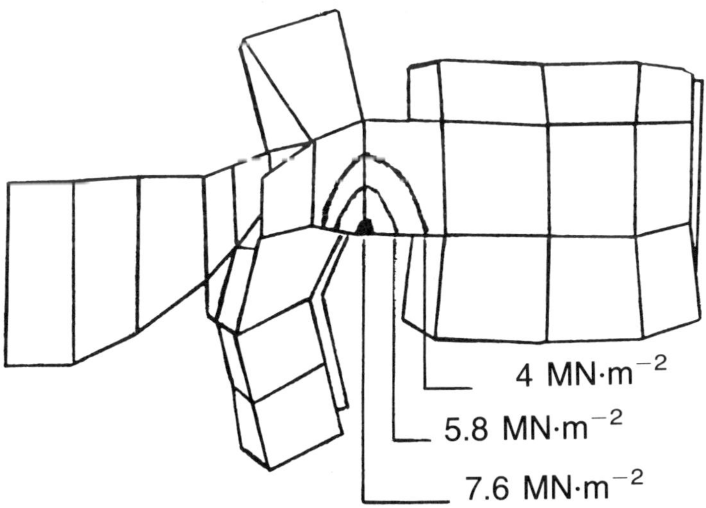

Fig. 12.7 Side view of Case 6 with predicted stress contour patterns.

(1) During the simulation of spinal surgery, changes in stress distribution and location of high tensile stresses were identified. Complete removal of an entire lamina or the removal of the superior part of the lamina undoubtedly increased the stress concentration in the laminar region.

(2) High tensile stresses due to spinal surgery also occurred in the region of the pedicles and the pars interarticularis.

(3) The model was able to predict the nucleus and the annulus pressures fairly accurately.

(4) Simulation of disc degeneration increased the stresses significantly in the posterior region of the vertebra.

(5) Simulation of the surgical procedure on the lamina produced high tensile stresses near the pars interarticularis and at the posterior junction of the lamina and the pedicles.

(6) The model also allows simulation of other clinical situations, e.g. loading due to Dwyer instrumentation, fixation of endplate fractures, etc.

(7) The model can be utilised to simulate the pathological and traumatic lesions of bone and disc, and the response to artificial materials used in their replacement, by altering the material properties of different sections of the model, including simulation of the vertebral osteoporosis.

Acknowledgements

The author wishes to thank Dr A. I. King and Dr Harpal Kaur Ranu for their assistance in this research.

References

Ranu, H. S. & King, A. I. (1980). Correlation of intradiscal pressure with vertebral endplate pressure. In: *Engineering Aspects of the Spine*, pp. 37–42. Institution of Mechanical Engineers, London.

Ranu, H. S. (1981). Viscoelastic behaviour of the human spinal disc and the vertebra. *Biomaterials, Medical Devices and Artificial Organs*, **9**, 365–6.

Ranu, H. S. (1982). Three-dimensional surgical simulations of the spine. *Journal of Biomedical Engineering*, **4**, 285–8.

Ranu, H. S. (1985). Time-dependent response of human intervertebral disc to loading. *Engineering in Medicine*, **14**, 43–5.

Ranu, H. S. (1986a). Relation between the nucleus and the annulus pressures of the lumbar spinal disc. *Scandinavian Journal of Rheumatology*, Suppl. 60, 27.

Ranu, H. S. (1986b). 3-D modelling of the spinal segment. *Scandinavian Journal of Rheumatology*, Suppl. 60, 27.

Ranu, H. S. (1989). Measurement of pressures in the nucleus and within the annulus of the human spinal disc due to extreme loading. *Journal of Engineering in Medicine* (In press).

13

Application of the finite element method to the reconstructed tooth and middle ear

K. R. Williams, J. S. Rees and T. H. J. Lesser

13.1 Introduction

The finite element method (FEM) was originally introduced some 20 years ago as a procedure for solving problems in structural mechanics. The range of applications has become wider and it is now recognised as a general numerical approximation method for any physical problem capable of being modelled by differential equations. Development of the FEM and its applications continues and the present work utilises softeware based on parabolic isoparametric elements and a frontal solution process. These elements, with relative ease, allow the modelling of complex irregular geometries of natural tissue and biomaterials. It is also relatively straightforward to incorporate the non-homogeneous, anisotropic properties of natural materials and combinations of mechanical and thermal loading.

However, it is the mechanical properties and behaviour of the natural tissues which are least well understood, due to difficulties in obtaining vital material, its variable shape and size, and the testing procedure and environment. For example, it is unclear whether or not natural materials behave elastically, plastically or in an elasto-plastic manner. Time is another important consideration and can alter the material's response to stress. Other materials, like ceramics, exhibit brittle behaviour and break at a certain stress level, while still in the elastic region.

The data file input to the numerical analysis program requires material property values. In the simplest linearly elastic analysis, only the elastic modulus (E), the Possion ratio (ν) and specimen geometry are necessary. These two fundamental properties are readily measured and tabulated for common engineering components, but are less well established for natural tissue and some modern prosthetic materials. As finite element stress analyses are attempted in more complex natural situations, where tissue response may depend on stress, strain rate, temperature and the environment, then it becomes increasingly more difficult to set up a meaningful data file. In addition, the exact loading sequence may only be approximately understood, leading to further uncertainty in the output data files.

Nevertheless, because of the efficiency of modern numerical analysis software and the speed of modern computers, many of these variables may be examined within a model of the clinical situation, and the resulting material response compared with observed behaviour. In this way, it is possible, with experience, to accumulate a reasonable assessment of stress and temperature response to a given loading situation. The numerical results also allow the examination of problem areas, where deformation or fracture of the natural tissue or prosthetic material may occur. These factors in turn lead on to alternative design consideration or materials.

In the work reported in this chapter the finite element method has been applied to two problem areas:

(1) reconstruction of the middle ear;

(2) reconstruction of the tooth with modern posterior composites.

13.2 **Reconstruction of the middle ear**

13.2.1 *The surgical problem*

The most common cause of conductive hearing loss in adults is infection, either secondary to cholesteatoma, or chronic tubotympanic disease. The usual reason for conductive hearing loss is necrosis of part or all of the long process of the incus. The disconnection of the sound transmitting system in the middle ear leads to severe conductive hearing loss. This hearing loss may be treated by ossiculoplastic surgery to reconstruct the sound transmission system. The anatomy of this system is altered to provide a more direct route for sound to pass between the tympanic membrance and stapes. This can be achieved using the patient's bone (autograft), donated bone (homograft) or prosthetic materials such as ceramics, porous plastics or combinations of these. The patient's own bone is not always available and recently donated bone has caused concern because of the possible risk of transference of viral diseases. Hence, increasing attention is being paid to the use of prosthetic materials in middle-ear surgery.

13.2.2 *The mechanical problem*

The reconstructed ossicular chain is required to be stable for many years, as well as providing adequate sound transmission. The stability is provided by a fibrous capsule of varying thickness surrounding the implant.

Several candidate ossicular replacement materials have been examined recently (Williams & Blayney, 1986a and b, 1987) ranging from hard brittle ceramics to much softer plastics. Bio-inert ceramics, such as aluminium oxide, Al_2O_3, offer potential particularly in situations where good wear-resistance is a prerequisite. However, ceramic materials, due to their inherent hardness and brittleness, are not generally ideal in load-bearing situations. Carbon implants similarly fail by brittle or fatigue failure in high stress positions. Attempts have been made to reinforce the carbon matrix using carbon fibres, i.e. carbon-reinforced carbon (CFRC). Components such as CFRC bone plates and bone pins have been successfully manufactured and clinically applied.

The problem of the hardness and brittleness of reconstructive materials is probably less severe in middle-ear reconstruction because of the low load and stress amplitudes and the relatively small size and weight of the moving bone or bone replacement. The choice of material is therefore often a case of compatibility and tolerance when inserted alongside living tissue. Prosthetic reconstructive ossicular materials appear to behave quite differently in this respect. For example, plastics introduced into middle-ear surgery in the early 1950s showed that, after initially good results, inflammatory tissue reactions were confirmed following the use of polyethylene in the oval windows of cats (Withers *et al.*, 1963). The use of Teflon (Shea *et al.*, 1962) led to greater compatibility resulting from the encapsulation of the implant in a fibrous sheath. However, these materials have not found acceptance in tympanoplasty due to problems with resorption of the ossicular interfaces and spontaneous rejection from the tympanic membrane (Jobe *et al.*, 1973). More recently, composites made up of PTFE and glassy carbon, such as Proplast and porous polyethylenes such as Plastipore, have been introduced into ear surgery. Connective tissue can grow into these materials leading to good mechanical stability. However, the presence of multinuclear cells has been found in tissue surrounding such material via follow-up otological revision.

In spite of this, and difficulties such as perforation of the tympanic membrane, bone resorption at interfaces and penetration of the ossicle through the stapes footplate, plastic materials remain an attractive possibility due to the ease of manufacture and close matching of mechanical properties to the surrounding tissue.

In the animal implant work described by Williams and Blayney (1986a and b, 1987), fibrous encapsulation occurs at a rate depending on the implant material. The most rapidly occurring encapsulation occurs over glassy carbon implants, with ceramic materials intermediate and plastics the slowest. Encapsulation is thought to be detrimental in the case of load-bearing devices since the gap between fibrous tissue and implant can lead to looseness and ultimately to inflammation and rejection; it is probable that extremely low loads and small displacements are transmitted through the ossicular chain and consequently a direct bony bond may be unnecessary. Indeed, the fibrous encapsulation need only possess the cohesive properties of the tympanic membrane in this respect. Thus a simple fibrous encapsulation of the implant to the surrounding ossicular chain may suffice in middle-ear reconstruction. However, direct bone growth onto the surface of both glassy carbon and bioactive ceramics has been observed from surrounding bone lamellae.

Adhesions of the implant to the surrounding tissue are clearly necessary to stabilize the prosthesis, but may detract from the normal vibration and displacement of the ossicular chain. Vibration patterns of the tympanic membrane have been analysed using FEM, but as yet have not been addressed to the total ossicular chain (Funnell & Laszlo, 1978; Funnell, 1983). In their work Funnell & Laszlo (1978) pointed out the important effects of the mechanical properties of natural soft and hard tissue on vibratory patterns, and it would, therefore, seem reasonable that ossicular replacements should closely match the original material wherever possible. Fibrous adhesions of the implant to the surrounding middle-ear cavity would clearly alter the rotational stiffness value and hence vibratory patterns. Funnell (1983), however, found only small changes in the natural frequency of vibrations when the rotational stiffness was altered by 20%. Thus, fibrous adhesions may have only a modest effect on ossicular transmission efficiency. Indeed, these adhesions are not confined to plastic materials alone, but occur on ceramic and carbon implants and are part of the normal suspensory ligaments connecting the ossicles to the tympanic cavity in a normal healthy ear.

Sound transmission depends upon the positioning of the implant. Whilst this is mainly a consequence of the anatomy of the middle ear, the positioning of the implant along the length of the malleus is at the discretion of the surgeon. It has been shown, both scientifically, using laser doppler interferometry, and clinically in long-term trials, that the most efficient sound transmission is along an imaginary line between the centre of the footplate and the stapes head leading out to the drum and malleus (Vlaming & Feenstra, 1986; Smyth, 1987). However, such a columella effect can cause abnormal stresses at the implant stapes junction and under certain circumstances these stresses may rupture the fibrous capsule surrounding the implant leading to its displacement and eventual rejection. This rejection rate is known to increase with time from 4.5% at 1 year to 6.8% at 3 years (Williams & Blayney, 1987). Many of the abovementioned variables may be investigated using FEM.

The clinical situation was modelled and the discretisation is shown in Figure 13.1. This is a cross-section taken through the middle ear with the tympanic membrane on the left-hand side with the ossicles and stapes projecting to the right. The impedance of the inner-ear fluids was controlled by allowing the stapes footplate limited movement in the x-direction. A physiological load was placed on the tympanic membrane and the stress at

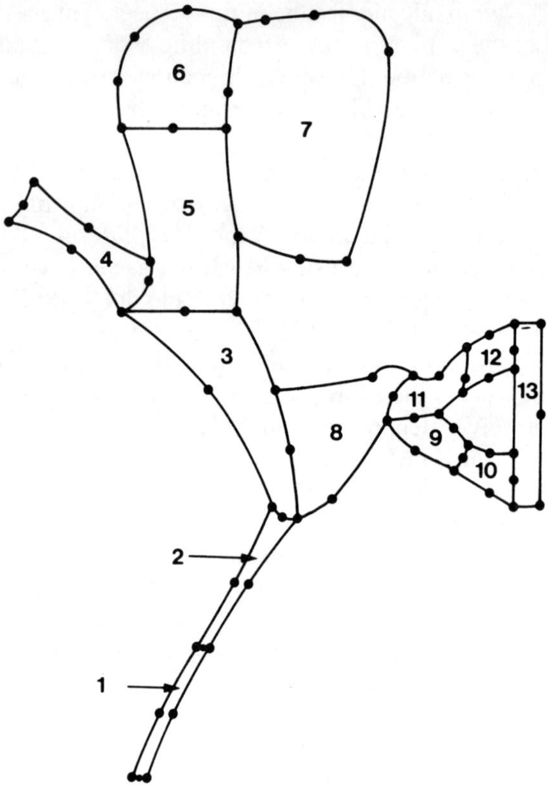

Fig. 13.1 The finite element mesh discretisation of a reconstructed human middle ear. The diseased incus has been removed and replaced (element 8) by a suitable prosthesis.

the incostapedial joint analysed for various implant materials and varying degrees of fibrous encapsulation. The material properties chosen for the natural and artificial materials are shown in Table 13.1. Both dynamic and static finite element stress analyses were examined, but for the purposes of the present contribution only the static analysis results will be described. Because of the relatively rapid loading rates across the tympanic membrane and along the reconstructed ossicular chain, a linearly elastic analysis was carried out.

13.2.3 *Results*

Two examples of the repositioning of the implant are shown in Figure 13.2. The results show that the stresses at the implant stapedial joint increase as the implant is moved lower down the malleus handle.

Table 13.1 *Physical values used in the stress analysis program.*

Material	Elastic modulus ($MN{\cdot}m^{-2}$)	Poisson's ratio
Enamel	46 000	0.3
Dentine	11 760	0.31
Occlusin (composite resin)	12 327	0.3
Calcium hydroxide liner*	367	0.4
Polycarboxylate cement*	4 950	0.4
Chemfil II (liner)*	12 161	0.25
Amalgam	69 000	0.35
Soft tissue	0.5–100	0.30
Hard bone	13 800	0.26
Soft bone	345	0.38

* Materials used as cavity bases.

13.2.4 *Conclusion*

To minimise these stresses and improve the longevity of the implant, its lateral end should be placed as high as possible on the malleus handle. This portion mimics the 'normal' middle ear. It may not, however, maximise sound transmission and a compromise between stability and transmission exists.

13.3 **Reconstruction of the tooth**

In the past 10 years, visible light-activated polymeric filling materials have challenged amalgam as the material of choice to restore posterior teeth. Essentially, the polymeric materials consist of a Bis-Gma resin, together with large amounts of filler particles which are normally quartz or silica-based. During polymerisation of unsaturated monomers, the double bond is broken and short-range covalent bonds form between the monomer units resulting in shrinkage.

13.3.1 *Clinical sequelae of polymerisation shrinkage*

Although composite resins shrink by only a few per cent on setting, this process generates a considerable force which is exerted on the surrounding tooth tissue. Bowen (1963) initially measured these forces as high as 24 $MN{\cdot}m^{-2}$. More recently, values of 2.8–3.9 $MN{\cdot}m^{-2}$ (Davidson & de Gee, 1984) and 6.1–7.3 $MN{\cdot}m^{-2}$ (Bowen *et al.*, 1983) have been reported for conventional and microfilled resins.

The effect of this shrinkage on tooth substance generally determines the longevity of the restoration. Several workers (Asmussen & Jorgensen, 1972; Asmussen, 1985) have demonstrated that polymerisation may be

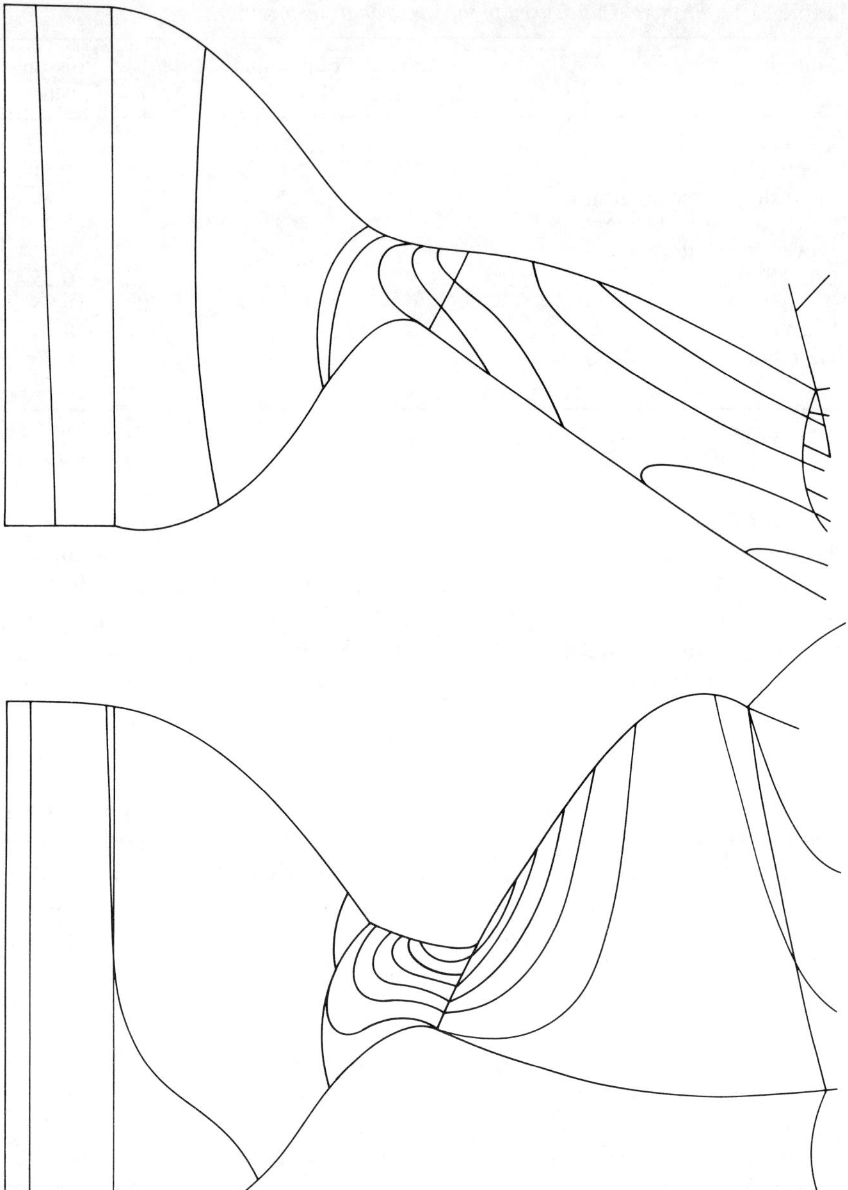

Fig. 13.2 Effective stress distributions at the incudo-stapedial joint for two geometries of middle ear reconstruction. Stress levels increased from 4.7 $MN{\cdot}m^{-2}$ to approximately 8.0 $MN{\cdot}m^{-2}$ as the prosthesis is positioned further down the malleus.

responsible for producing cracks in enamel. The relationship between microleakage and polymerisation shrinkage is also well known (Luescher *et al.*, 1977).

Until recently a gap at the dentine–restorative resin interface was considered inevitable, simply because no chemical bond was possible. However, because of recent improvements in dentine adhesives, this interface is in need of further evaluation.

Davidson & de Gee (1984) examined the forces developed during the first 30 min of curing and the effect of these forces on the integrity of the resin–dentine interface. When shrinkage was restricted to one dimension, the forces where in the region of $2\,MN{\cdot}m^{-2}$ and the interface remained intact. However, when the same material was used to restore a class V cavity, disruption of the interface was observed in numerous cases. This difference was attributed to restriction of the flow during setting in three dimensions. A force estimated to be $20\,MN{\cdot}m^{-2}$ was developed in the *in vitro* cavity, sufficient to break the dentine–resin bond. In an effort to alleviate these high interface stresses, and to improve marginal adaptation, incremental placement of the composite has been attempted.

Another major effect resulting from shrinkage is cuspal movement, particularly in Class II cavities. Providing the interface remains intact, the cusps can move inwards and it has been suggested that this movement may be responsible for post-operative sensitivity and oblique cuspal fracture.

Jensén & Chan (1985) prepared several MOD cavities in extracted premolars and these were restored with a posterior composite and a dentine-bonding agent. The largest cuspal movement was associated with placement of a single increment of composite (45 μm). Incremental placement in an occluso-gingival direction resulted in less movement (37 μm), but least movement was found when the composite was angled to the cavity wall and cured in three increments (33 μm).

13.3.2 *Stress analysis and results*

Most of the shrinkage takes place within the first hour after activation and *in vitro* investigation involving mounting and sectioning of filled teeth is impossible. Recourse was, therefore, made to the finite element method. The more modern, rounded cavity design was modelled using quadrilateral elements. Initially, the general shape of the tooth and restoration were established in an extracted upper first premolar. The tooth was restored with a posterior composite resin in the normal manner, cured overnight and sectioned longitudinally in a plane parallel to the long axis of the tooth, using a diamond saw cooled with a jet of distilled water.

The outline of the tooth, restoration and the amelo-dentinal junction were traced onto transparent sheet. This outline was then transferred to graph paper and enlarged ten times using a $1\,mm^2$ grid. The mesh used

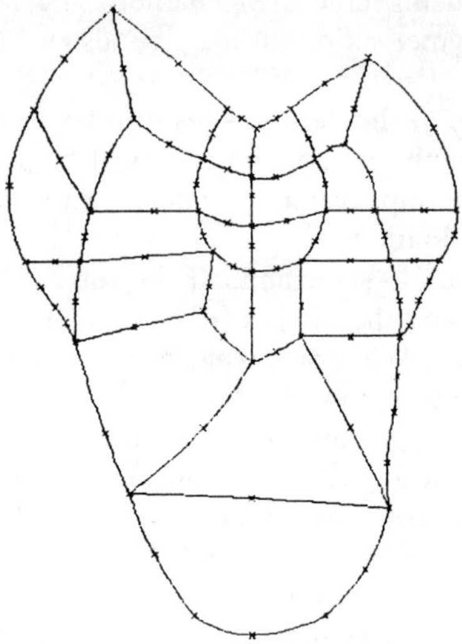

Fig. 13.3 Mesh discretisation for an ovoid cavity design.

for this analysis is shown in Figure 13.3. The force generated by the polymerisation shrinkage was modelled as follows.

A volumetric shrinkage of 1% was adopted as being representative of the shrinkage value of a typical posterior composite resin (Rees & Jacobsen, 1987). This value, however, relates to a three-dimensional situation, whereas the current numerical analysis is a two-dimensional approach. Furthermore, since it has been shown that a volumetric shrinkage of 1% or less is approximately equivalent to three times the linear shrinkage value (Rees & Jacobsen, 1987), it was decided to use a value of one-third of 1% for the change in linear dimension of the shrinking resin.

To simulate a shrinkage stress within the mesh, it was necessary to use the thermal loading option of the program. This was achieved by setting the nodes surrounding the restoration at a temperature of 0°C and the nodes within the restoration at −1°C, since movement was required in an inward direction.

Since the temperature difference across the tooth restoration interface and the original length of the restoration were known, it was possible to calculate an apparent coefficient of thermal expansion which would generate the required shrinkage. The coefficient of thermal expansion A is given by:

$$A = \frac{\mathrm{d}L}{\mathrm{l_o d}T} \qquad (13.1)$$

where $\mathrm{d}L$ = change in length of material; $\mathrm{l_o}$ = original length of material; $\mathrm{d}T$ = changes in temperature.

Table 13.2 *Coefficients of thermal expansion used to simulate the relevant polymerisation shrinkage values.*

Polymerisation shrinkage (volume %)	Coefficient of thermal expansion (per 1°C)
1.00	3.33×10^{-3}
0.75	2.50×10^{-3}
0.50	1.68×10^{-3}
0.25	8.33×10^{-3}
0.10	3.33×10^{-4}

A suitable cavity diameter, measured from the original tracings, was taken as being 2 mm. Therefore, since $L = 2$ mm and $dT = 1$°C,

$$dL = \text{one-third of } 1\% \text{ of } 2\,\text{mm} = 6.66 \times 10^{-3}$$

$$A = (6.66 \times 10^{-3})/(2 \times 1) = 3.33 \times 10^{-3} \text{ per°C}$$

This value for the coefficient of thermal expansion (or contraction in this case) was assigned to all materials used so that a thermal differential of 1°C across the tooth restoration interface would result in shrinkage of the composite resin equivalent to one-third of 1% linear contraction. Furthermore, since the coefficient of thermal expansion is a dimensionless quantity, it is applicable to all cavity sizes regardless of geometry.

The variation in polymerisation shrinkage was initially investigated in order to ascertain the value manufacturers need to achieve to prevent marginal leakage. Values between 1.0 and 0.1% were investigated and the various coefficients of thermal expansion required to achieve these values were input to the data file (Table 13.2).

The movement of the nodes forming the outer surface of the tooth were analysed and the maximum horizontal movement of the cusp is shown graphically as a function of polymerisation shrinkage in Figure 13.4. The plot is linear and shows that the analysis indicates cuspal movements near those found experimentally for shrinkage values near 1%. In addition, it can be seen that the lingual cusp is displaced more than the buccal cusp, presumably as a result of their differences in size.

A further important consideration is the type of cavity base placed beneath the posterior composite. A cavity base is usually used beneath a filling to prevent thermal and chemical irritation from the overlying restorative material. The different cavity base materials, together with amalgam, were modelled and are listed in Table 13.1. To improve the distribution of the moduli the following values were also included: 2500 MN·m^{-2}, 20 000 MN·m^{-2}, 40 000 MN·m^{-2} and 55 000 MN·m^{-2}.

The maximum horizontal movement of the outside of the tooth was

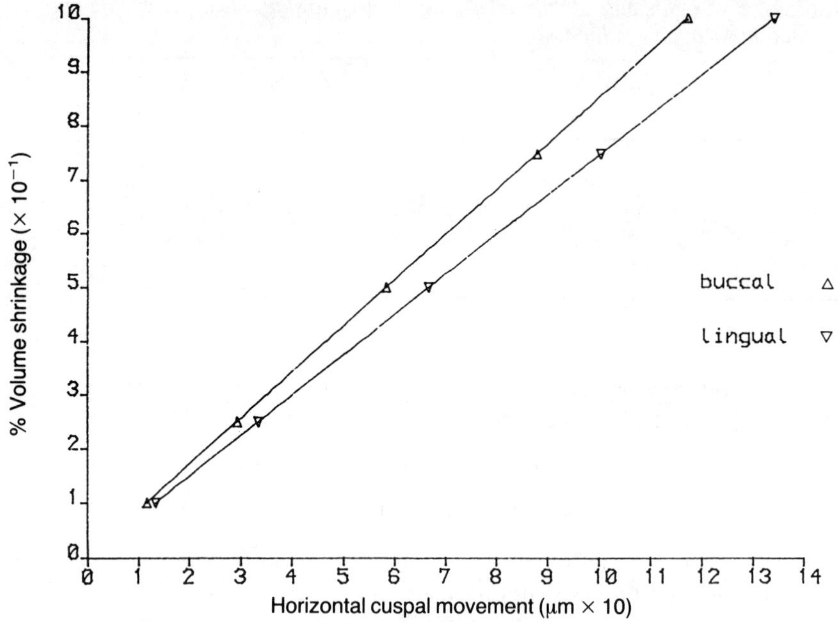

Fig. 13.4 Maximum cuspal movement for various polymerisation shrinkage values: (△) buccal, (▽) lingual.

plotted against cavity base modulus. Again, owing to its smaller size, the lingual cusp shows larger movements. Furthermore, there appears to be a fairly rapid rise in movement with the base material with a modulus of up to 10 000 MN·m^{-2}, which then levels off with further increase in modulus. Translating this to the clinical situation, a lining material with a low modulus, especially below 10 000 MN·m^{-2}, would help to minimise tooth movement and post-operative sensitivity.

As stated earlier, it is possible to bond composite resins chemically to enamel and dentine, producing tensile and/or shear bond strengths of approximately 15 MN·m^{-2} to dentine and 15–20 MN·m^{-2} to enamel. One of the first stress systems that this bond is subjected to is that created by the polymerising resins. However, the magnitude of the forces developed by a shrinking composite resin are not known in detail, since it is difficult to cut a cavity, section and examine the tooth before the majority of shrinkage has taken place. Nevertheless, some workers have attempted to correlate the bond strength of various dentine adhesives with the width of the marginal gap. Extrapolating to the point of zero marginal gap size, these studies have suggested that a bond strength of 20 MN·m^{-2} may be sufficient to prevent marginal gap formation.

A bulk packing example was modelled and this produced very large values for cuspal movement and stresses along the interface (Figure 13.4). Various incremental packing geometries were also modelled, which are clinically more realistic. All the incremental geometries examined reduced cuspal movement, but the stresses along the interface did not reduce in proportion to the size of the increment. In fact, most of the stresses across the interface were of the order of 20 $MN{\cdot}m^{-2}$.

In conclusion, an incremental packing technique must be used clinically to ensure an acceptable degree of polymerisation and to reduce cuspal movement. However, the stresses induced by a shrinking increment are still large enough to rupture the composite resin–dentine bond, suggesting that stronger dentine adhesives are required.

References

Asmussen E. (1985). Clinical relevance of physical, chemical and bonding properties of composite resins. *Operative Dentistry*, **10**, 61–73.

Asmussen E. & Jorgensen K. D. (1972). A microscopic investigation of the adaption of some plastic filling materials to dental cavity walls. *Acta Odontologica Scandinavica*, **30**, 3–21.

Bowen R. L. (1963). Properties of silica reinforced polymer for dental restorations. *Journal of the American Dental Association*, **66**, 57–64.

Bowen R. L., Nemoto K. & Rapson J. E. (1983). Adhesive bonding of various materials to hard tooth tissues. Forces developing in complete materials during hardening. *Journal of the American Dental Association*, **106**, 475–7.

Davidson C. L. & de Gee A. J. (1984). Relaxation of Polymerisation contraction stresses by flow in dental composites. *Journal of Dental Research*, **63**, 146–8.

Funnell W. R. J. (1983). On the undamped natural frequencies and mode shapes of a finite element model of the cat eardrum. *Journal of the Acoustic Society of America*, **73**, 1657–61.

Funnell W. R. J. & Laszlo C. A. (1978). Modelling of the cat eardrum as a thin shell using the finite element method. *Journal of the Acoustic Society of America*, **63**, 1461–7.

Jensen M. E. & Chan D. C. N. (1985). Polymerisation shrinkage and microleakage. In: *Posterior Composite resin Dental Restorative Materials*, eds D. C. Smith & G. Vanherle, pp. 243–62. Peter Szule Publishing Company, The Netherlands.

Jobe R., Ivesan R. & Vistnes L. (1973). Bone deformation beneath alloplastic implants. *Plastics and Reconstructive Surgery*, **51**, 169–73.

Luescher B., Lutz F., Oschsenbein H. & Muhlemann H. R. (1977). Microleakage and marginal adaption in conventional and adhesive class II restorations. *Journal of Prosthetic Dentistry*, **37**, 300–9.

Rees J. S. & Jacobsen P. H. J. (1987). The polymerisation shrinkage of posterior composite resins. BSDR Conference, London, Abstract No. 119.

Shea J. J., Sanabria F. & Smyth G. D. (1962). Teflon piston operation for otosclerosis. *Archives of Otolaryngology*, **76**, 516–21.

Smyth G. D. L. (1987). Development of surgery for conductive deafness during the past 25 years. British Medical Bullation, **43**, 1–10.

Williams K. R. & Blayney A. W. (1986a). An optical and electron microscopy study of materials implanted in the rat middle ear. I. Carbon. *Biomaterials*, **7**, 283–6.

Williams K. R. & Blayney A. W. (1986b). An optical and electron microscopy study of materials implanted in the rat middle ear. II. A bioactive glass ceramic. *Biomaterials*, **7**, 287–91.

Williams K. R. & Blayney A. W. (1987). Tissue response of several polymeric materials implanted in the rat middle ear. *Biomaterials*, **8**, 254–8.

Withers B. T., Hatfield S. E. & Richmond R. W. (1963). Drum and middle ear grafts in cats. *Laryngoscope*, **73**, 1022–43.

Vlaming M. S. M. G. & Feenstra L. (1986). Studies on the mechanics of the reconstructed human middle ear. *Clinical Otolaryngology*, **11**, 411–22.

14

The loading of cemented femoral hip stems: a finite element investigation

M. J. Fagan and A. J. C. Lee

14.1 **Introduction**

In an effort to improve the long-term success of total hip replacement joints, the design of the femoral component has been investigated by researchers using the finite element method with two-dimensional, axi-symmetric and three-dimensional models (Huiskes & Chao, 1983). A total hip joint is difficult to represent because of the complex geometry of the problem, the uncertainty in the material property of the bone, the condition of the stem–cement and cement–bone interfaces and the loading conditions. The last of these factors has to date received little attention, and yet is paramount since, no matter how accurately the system is modelled, the results are invalid if relevant loads are not applied. The work reported here investigates how the femoral component of a hip joint is loaded, in particular in the one-legged stance, examining the contributions from each active muscle. The model described then forms the basis of a detailed design analysis of the prosthesis.

Throughout the work, close attention has been paid to the clinical performance of the stems which, although clearly an important source of information, has thus far been neglected in other design studies. In particular, the mode and location of the fracture in stems that have failed, reveals details of the critical loading mechanism.

Examination of the fracture surfaces of failed stems shows striations characteristic of a fatigue-type failure. The cracks are seen to propagate across the stem from the lateral to the medial face, with the medial edge finally failing in a ductile manner leaving a shear lip. Cases of stem fracture from simple overload leading to a ductile failure of the complete cross-section have not been reported, although examples have been described where the load was sufficient to cause a significant plastic deformation (Carlsson *et al.*, 1977; Ducheyne *et al.*, 1975; Wroblewski, 1979). The origin of stem fractures has been identified to be at the anterior–lateral corner of the stem (Wroblewski, 1979), showing that the main loading on the hip is from the anterior-superior direction. The plastic deformation of failed prostheses also shows this to be the case, with the proximal section of the fractured stems twisted backwards in relation to the distal section.

The clinical evidence therefore clearly indicates the important loading mechanism experienced by the hip. The fatigue-type of failure is undoubtedly due to the walking cycle causing fluctuating stresses on the joint. The loading conditions acting on the femur during walking have been investigated both theoretically and experimentally (English & Kilvington, 1979; Johnston *et al.*, 1979; Paul, 1964). The greatest hip reaction force occurs during the stance phase of the walking cycle, so this would appear to be the most relevant position to consider.

The finite element analysis reported here investigates the design of the Exeter prosthesis under the similar condition of standing on one leg. This ignores the inertia loads experienced by the leg, but allows the muscle forces and hip reactions to be easily calculated by consideration of the equilibrium of the joint. The values and orientations of the calculated loads, together with the results of the design analysis, match favourably with the available clinical and experimental evidence, to show that standing on one leg is a relevant loading condition.

14.2 **The finite element model**

The accuracy and validity of the finite element model used in this research is discussed in another publication (Fagan & Lee, 1986a) which also outlines the complete programme of work that has been carried out into the design and application of cemented hip prostheses. Many models were developed, including one of a natural femur, with up to 1750 isoparametric brick elements defining over 6000 degrees of freedom.

The model of the implanted Exeter femoral component used in the present research is shown in Figure 14.1(a). Part of the model is removed to reveal some of the inner elements. The discretisation of the basic stem is included separately in Figure 14.1(b); it was not necessary to model the head of the prosthesis, with the hip reaction being applied more effectively through a node positioned at its centre. The stem cross-section was defined by eight elements, and the cement and bone both simulated by at least two layers with 12 elements in each.

All the materials were assumed to be linearly elastic and isotropic. The effects of modelling the bone as isotropic was examined thoroughly in the development of the basic model (Fagan & Lee, 1986a). It was found that the only significant error occurred in the radial and hoop stresses of the bone, but the dominant axial stresses were still accurately predicted. A perfect bond was assumed to exist between all material interfaces. It was

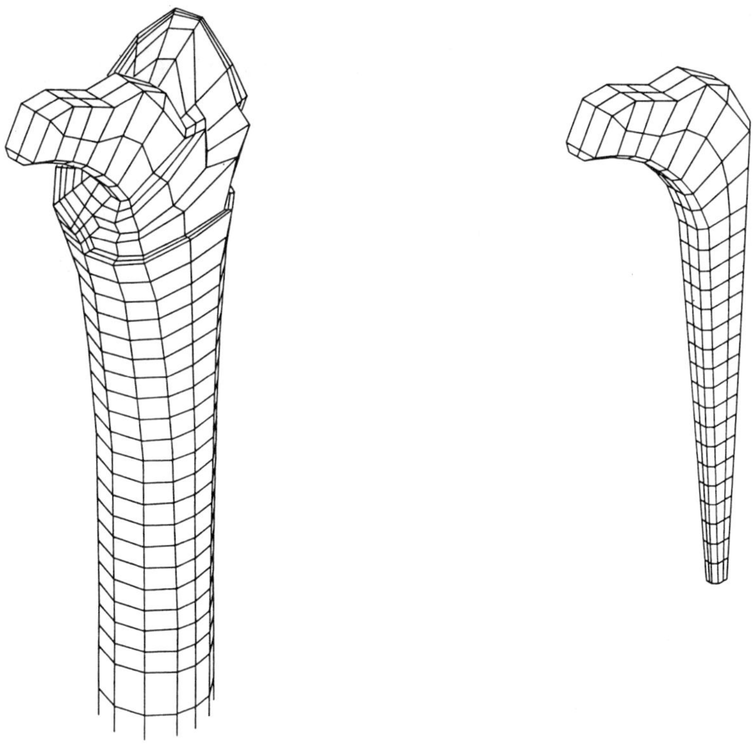

Fig. 14.1 (a) The finite element model of the implanted femoral component. Part of the model is removed to show the internal structure. (b) The representation of the basic stem. (Taken from M. J. Fagan and A. J. C. Lee, 1986. Reprinted with permission from *The Journal of Biomedical Engineering*, **8**, 295–304.)

concluded that this did not significantly affect the stem and bone stresses, but the behaviour of the cement, particularly in the region of the stem-cement interface, was brought into question. The elastic moduli of the cortical and cancellous bones were chosen to be 17.0 and 0.32 $GN{\cdot}m^{-2}$, respectively, while the implant was modelled with a value of 200 $GN{\cdot}m^{-2}$, and the bone cement, 2.80 $GN{\cdot}m^{-2}$.

14.2.1 *Loading modes*

The loading applied to the model of the prosthesis was calculated from a force and moment analysis of the femur and pelvis, when in the one-legged stance (Figure 14.2). To perform the analysis, the actions of the muscles were simplified into three groups: the abductor muscles (Fa), the ilio-tibial band (F_t) and the ilio-psoas muscle (Fp). The muscle forces and hip reaction required to maintain equilibrium were found to be:

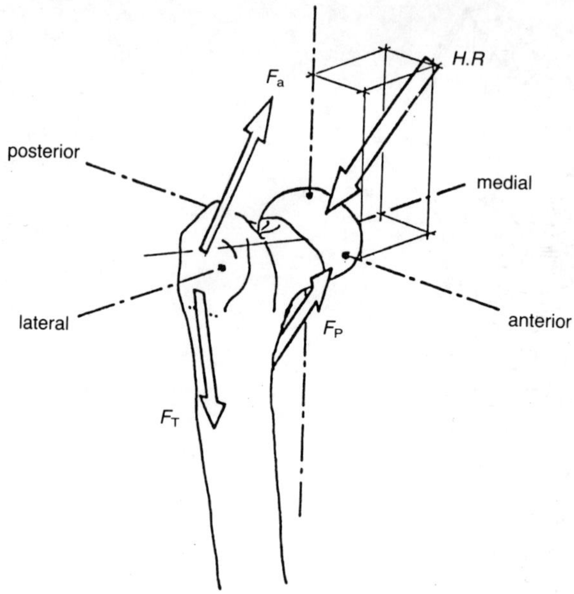

Fig. 14.2 Schematic diagram of the muscles and hip reaction considered to be acting on the upper femur while standing on one leg. (Taken from M. J. Fagan and A. J. C. Lee, 1986. Reprinted with permission from *The Journal of Biomedical Engineering*, **8**, 295–304.)

$$
\begin{aligned}
F_a &= 1.50\ W_B \\
F_t &= 0.36\ W_B \\
F_P &= 1.42\ W_B \\
\\
H.R &= 3.66\ W_B \\
R_X &= 0.58\ W_B \\
R_Y &= -0.60\ W_B \\
R_Z &= -3.56\ W_B
\end{aligned}
$$

where, R_X, R_Y and R_Z are the components of H.R (the hip reaction) in the medial-lateral, anterior-posterior and superior-inferior directions, and W_B is the bodyweight.

It is interesting to note that the resultant hip reaction makes an angle of 9.3° with the vertical in the coronal plane, and the longitudinal axis of the femur is reported to lie at 6° to the vertical (Kapandji, 1970). Therefore, the hip reaction makes an angle of 15.3° with the long axis of the femur. This is the typical inclination of the main load-bearing trabeculae in a natural femur, which transfer the hip reaction from the femoral head into the medial femoral neck.

A comparison of the results of this analysis with other values reported in the literature is not straightforward. Apart from any natural variations in the geometry of the hip joint, the pelvic tilt and general posture will affect the values. The dimensions used in this analysis were defined by the specimen femur used in the definition of the finite element model, and the experimental work of McLeish & Charnley (1970), who determined the position of the body's centre of gravity in the coronal plane, in the one-legged stance.

The most accurate method for determining the magnitude of the hip reaction is to measure the force directly in the body. English & Kilvington (1979) did this with a strain-gauged prosthesis, and found the hip reaction to be 3.59 times the body weight. This compares very well with the value of 3.66 predicted by this present work. However, Rydell (1964), using a similar procedure, found the hip force to be only 2.3 to 2.8 times bodyweight. Rydell used a high offset stem, but this still does not explain such a large discrepancy.

In a three-dimensional approach similar to that carried out here, Williams & Svensson (1968) calculated that the hip reaction would be 6.0 W_B, with an abductor muscle force of 1.27 W_B. This large value for the hip force arises from the choice of dimensions and orientation of the model used.

The work of McLeish & Charnley has already been mentioned. They determined experimentally the geometry of the muscles and the hip joint, then applied two-dimensional equilibrium conditions, to find a hip reaction of 1.8 to 2.7 times bodyweight, and an abductor muscle force of 1.0 to 1.8 times bodyweight, depending on the tilt of the pelvis. When the results of the present work are referred to two-dimensions, (by ignoring the anti-torsion angle of the femur), the hip reaction is found to be 2.32 W_B, with the total abduction force remaining the same, at 1.50 W_B. Inman (1947) also used a two-dimensional approach, and showed the hip reaction to equal 2.4 to 2.6 times bodyweight.

This summary of theoretical and, more importantly, experimental work shows that the force analysis carried out in the present work is reasonable, and is a representative loading condition for applying to the finite element model.

The model was also loaded with a two-dimensional loading scheme, in which the anti-torsion angle of the femur was ignored, implying that the greater trochanter, and therefore the attachment of the abductors to the femur, lay on the medial – lateral axis. With the centre of mass of the body also over this axis, the ilio-psoas was no longer required to maintain equilibrium in the sagittal plane. The muscle forces and hip reactions required to maintain equilibrium in two-dimensions were then found to be:

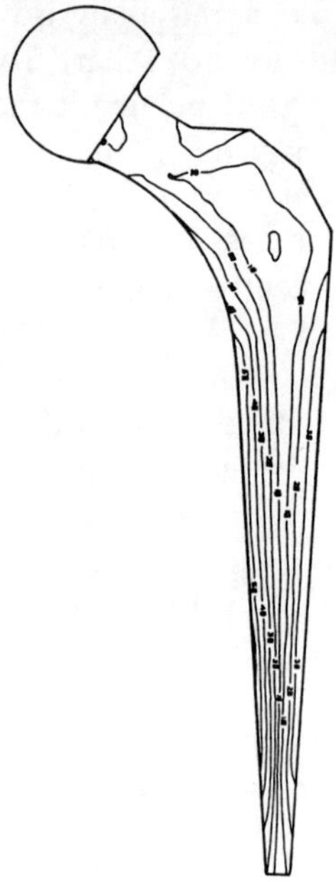

Fig. 14.3 Variation of the octahedral shear stress in the central (coronal) cross-section of the stem. (Contour step of 20 $MN \cdot m^{-2}$)

$$F_a = 1.50\ W_B$$
$$F_t = 0.36\ W_B$$

$$H.R = 2.32\ W_B$$
$$R_X = 0.31\ W_B$$
$$R_Z = -2.30\ W_B$$

The individual stress distributions produced by each group of muscles were also examined, revealing how they combine to give the overall stress characteristics of the basic replacement joint. The action of the ilio-tibial band was investigated in particular detail, since its action between the greater trochanter and epicondyles does not affect the equilibrium of the joint, and the body can use it as a tension band on the femur to limit the stresses in the bone. The force in the ilio-tibial band was varied between zero and three times bodyweight, while the remaining loading conditions were unaltered.

14.3 **Results**

The distribution of the octahedral shear stress in the central (XZ) section of the stem is presented in Figure 14.3. The peak stresses occur in the

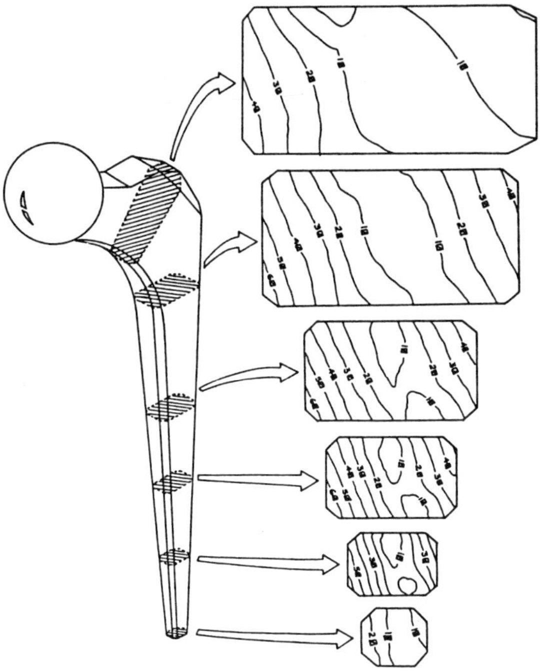

Fig. 14.4 Variation of the octahedral shear stress at different transverse sections of the stem: full three-dimensional loading. (Contour step of 10 MN·m^{-2})

central and distal thirds of the implant. The transverse distribution of the octahedral shear stress at various cross-sections of the stem is shown in Figure 14.4. The greatest stresses are recorded at the anterior – lateral corner of the stem.

When a two-dimensional loading scheme is applied to the model, the contours of octahedral shear stress in the transverse planes of the stem are regularly distributed, and parallel to the medial and lateral faces of the stem. The distribution of the octahedral shear stress down the centre of the medial and lateral faces of the stem for two- and three-dimensional load cases are compared in Figure 14.5. This shows that the distal stresses are increased but the proximal stresses are reduced, when simple two-dimensional loads are considered. In fact it was shown in Figure 14.4 that, for the three-dimensional system, the corner stresses are typically 20% higher than those at the centre of the medial and lateral faces of the stem. Therefore in fact for the implant, the two-dimensional loading underestimates the maximum (lateral) proximal stresses by up to 30%, but only underestimates the maximum distal (lateral) stresses by 8% (on the medial face the proximal stresses are underestimated by up to 40%).

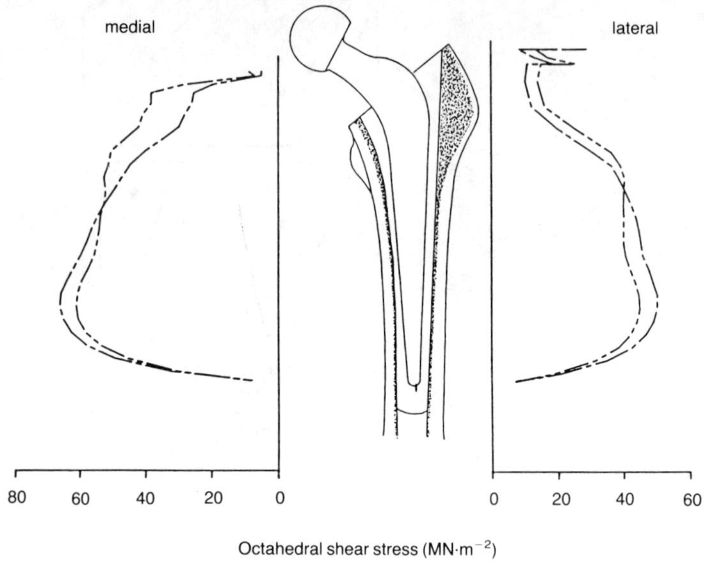

Fig. 14.5 Comparison of the octahedral shear stress distributions produced in the stem by two- and three-dimensional loading regimes: (– - –), two-dimensional; (– - - –) three-dimensional.

The octahedral shear stress produced on the outer medial and lateral surfaces of the cement and bone for the two- and three-dimensional load cases again show that the proximal stresses are underestimated by the two-dimensional loads, while the distal stresses are overestimated. The difference is particularly noticeable in the proximal cement, where the stress is over 30% greater when the more accurate loading regime is applied to the model.

The axial stress distributions developed on the faces of the stem by the individual muscle groups are shown in Figure 14.6. The hip reaction is responsible for the majority of the proximal loading in the implant. On the medial face, the magnitude of the axial stress from the hip reaction decreases from over 100 MN·m^{-2} at the femoral neck to zero at its distal end. Laterally, the axial stress from the hip reaction drops to zero one quarter of the way up the stem. The effect of the abductor group of muscles increases approximately linearly down the length of the stem, reflecting their bending action. The ilio-psoas muscle produces relatively insignificant stresses, except at the distal lateral position of the stem, where values of over 30 MN·m^{-2} are recorded.

The ilio-tibial band leads to a small tensile and compressive stress on the medial and lateral faces of the stem, respectively. Its possible effect is examined in more detail later.

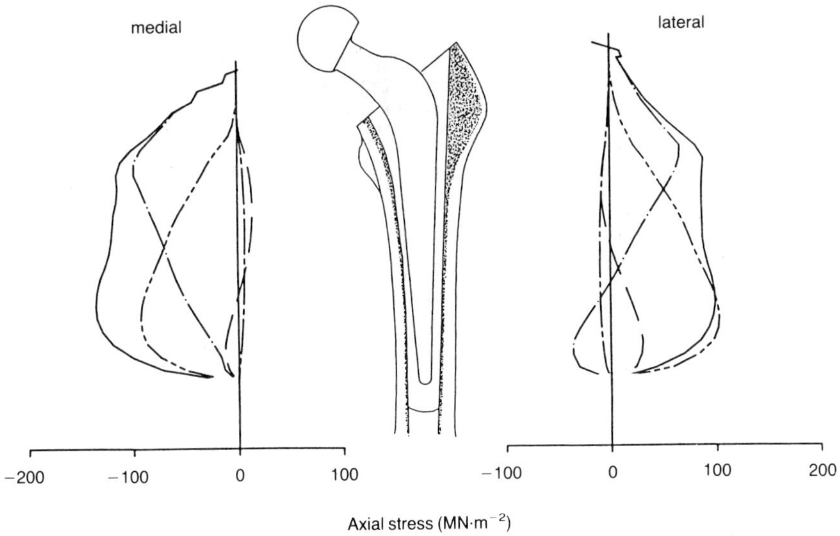

Fig. 14.6 Variation of the axial stress in the stem produced by the hip reaction and the individual muscle loads: (—) full three-dimensional loading; (–·–) hip reaction; ---- abductor muscles; (– - –) ilio-tibial band; (——) ilio-psoas.

The distribution of the axial stress in the cement produced by the separate muscle actions is presented in Figure 14.7. The general trends are similar to those recorded for the femoral stem. The effect of the hip reaction, however, decreases more rapidly, and the stresses from the abductor muscles are more prominent, particularly on the lateral aspect. The ilio-psoas muscle also contributes more to the overall stress level, increasing linearly down the femur.

The effect of varying the force in the ilio-tibial band (while the other applied forces remain unaltered) is shown in Figures 14.8 and 14.9 for the stem and cement. It is obvious that by developing a different force in the muscle, the body can significantly alter the stresses developed in the system. Increasing the force in the ilio-tibial band gives the greatest reductions in stress levels on the lateral side of each material, with the middle and distal thirds being most significantly affected. This has the effect of moving the critical region proximally so that, for large values of the force, the more proximal regions of the stem (for example) appear more prone to failure, rather than the previously susceptible distal parts.

14.4 **Discussion**

The major complication of cemented hip replacement joints is loosening of the femoral stem. The causes are biological and/or mechanical and

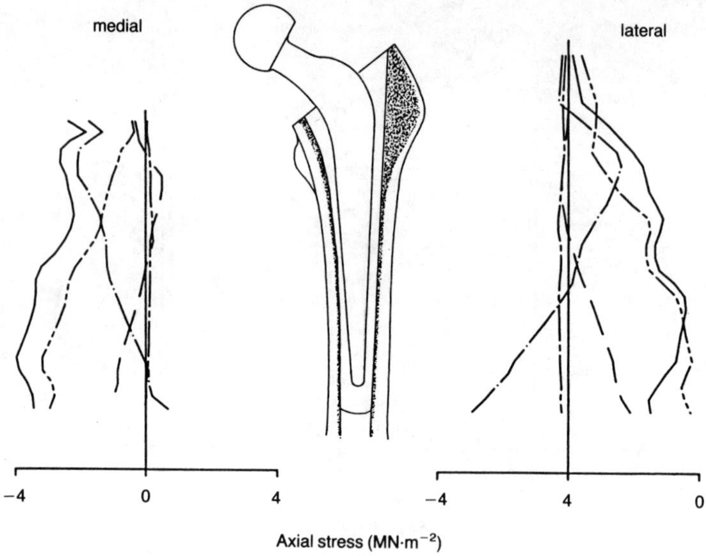

Fig. 14.7 Variation of the axial stress in the cement produced by the hip reaction and the individual muscle loads: (—) full three-dimensional loading; (-·-) hip reaction; (- - - -) abductor muscles; (- - -) ilio-tibial band; (——) ilio-psoas.

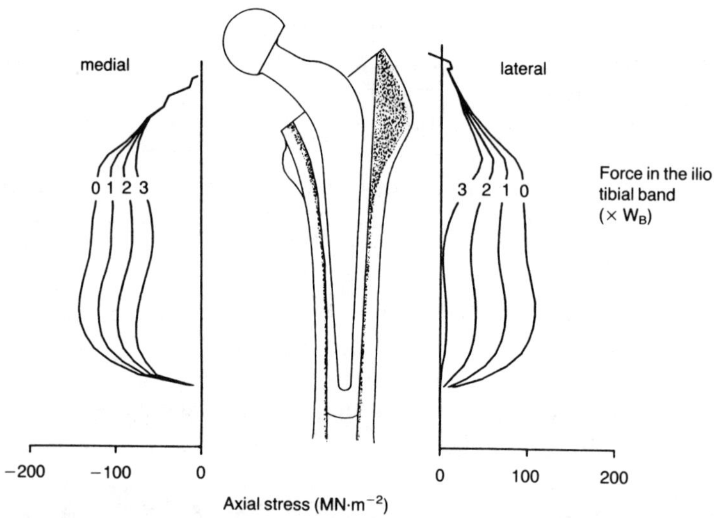

Fig. 14.8 Variation of the axial stress in the stem as a function of the force in the ilio-tibial band. W_B = bodyweight.

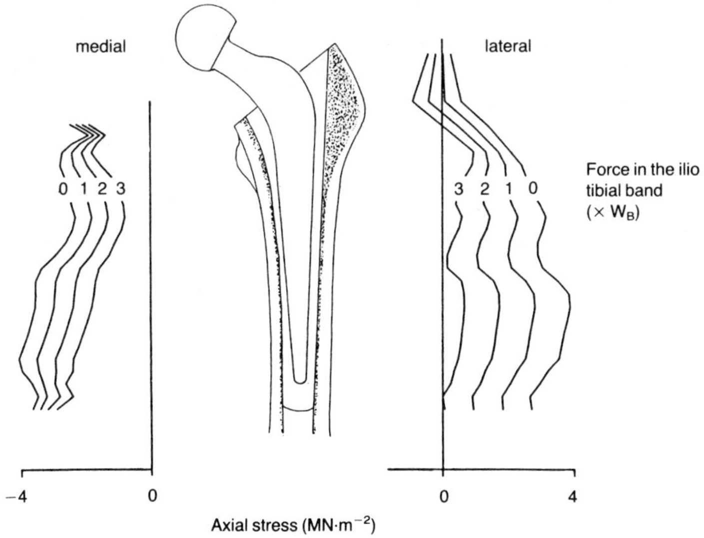

Fig. 14.9 Variation of the axial stress in the cement as a function of the force in the ilio-tibial band. W_B = bodyweight.

may, with time, lead to stem fracture through the increased stress levels developed in the prosthesis. Failure of the cement bone interface and loss of proximal support are the most likely indications of loosening (Carlsson *et al.*, 1977; Charnley, 1979; Beckenbaugh & Ilstrup, 1978; Gruen *et al.*, 1979), and should therefore be avoided with all stem designs. The design analysis of any hip replacement components is particularly awkward because of the large range of possible positions and resultant loading mechanisms, but to perform a realistic evaluation of different prostheses, a relevant physiological loading condition must be considered.

By considering static equilibrium of the proximal femur, the magnitudes of the muscle forces and hip reactions have been derived. The value of these forces clearly depends on the geometry of the chosen femur, the orientation of the muscles and the position of the centre of gravity of the body. Thus, the natural variation of the human femur and the infinite range of possible postures means that by a different choice of parameters, a different set of muscle and hip reaction forces could have been produced. McLeish & Charnley (1970) have investigated the effects of pelvis angle and trunk position on the abductor forces and hip reaction, the analysis was two-dimensional, but it does allow an assessment of the possible range of values.

With three-dimensional loads applied to the finite element model, the area of the stem most at risk is its anterior-lateral corner, so it is from here that a fatigue fracture should originate. To determine the fatigue resistance of the femoral stem, the results of the analysis must be extrapolated to estimate the stresses produced during walking, the cyclic loading responsible for the fatigue fractures. The hip reaction developed during walking depends on the stride length and speed of the patient, but has been estimated to reach five to six times bodyweight (Johnston *et al.*, 1979; Paul, 1964; Seireg & Arvikar, 1975). From the results of this model, the implant might typically be subjected to a cyclic stress of at least zero to 200 $MN \cdot m^{-2}$. The values quoted for the endurance limit of the materials used for femoral stems can vary from 400 $MN \cdot m^{-2}$ for stainless steel to over 650 $MN \cdot m^{-2}$ for the high strength vitalium alloys (Lorenz *et al.*, 1978; Miller *et al.*, 1982). However, these values are the endurance limits for highly polished rotating beam specimens tested in air. In practice, the implant will be in a potentially corrosive environment, and may have microscopic or macroscopic stress concentrations so the endurance limit can be expected to be less than this ideal figure. If factors are applied to the rotating beam values to compensate for the surface finish, temperature, stress concentrations, reliability and size effects, (Shigley, 1977), the endurance limit of the stems may be reduced by up to 40%. The results of the present finite element analysis implies that the Exeter stem modelled here, has satisfactory fatigue characteristics, although the exact safety margin is difficult to determine.

The model assumes the implant is well fixed in the cement and bone, with intact stem-cement and cement-bone interfaces. Clinical evidence shows that fracture of such a prosthesis is rare, and usually occurs in conjunction with a poor surgical technique, inadequate cement support or a loose stem. For example, finite element models of the implanted femoral component with resorption of the medial femoral neck (Fagan & Lee, 1986b), have shown that the stresses may easily be increased by 15% with small degrees of bone loss. Other factors that are consistently implicated in stem-fracture cases are varus placement of the stem and the weight of the patient, which may all combine cumulatively so that even the strongest material can fail. There has already been a report of a failed prosthesis, manufactured from forged Vitallium (Miller *et al.*, 1982).

Although the results of the three-dimensional loading mode agree with the clinical evidence, and may reflect the true loading experienced by the implant, it is important to recognise that the bending stresses in the coronal plane are still dominant. When only two-dimensional loads are applied to the model, the abductor force remains constant, but the ilio-psoas muscle is not considered, leading to a sympathetic decrease in the hip reaction force. Since the hip reaction is responsible for most of the proximal loading in the system, this causes a drop in the proximal stresses while the abductor muscles keep the distal stresses at the same level, resulting in the distal stresses becoming more prominant.

The force in the ilio-tibial band between the greater trochanter and the epicondyles cannot be determined by simple consideration of the equilibrium conditions of the femur. The effect of varying its magnitude was therefore investigated for values of zero to three times bodyweight, while the other loads were kept constant. If the body operates efficiently, one might expect the ratio of the maximum compressive to tensile stress in the femoral cortex to equal the ratio of ultimate strengths. The strength of bone in tension and compression are typically reported to be 130 and 190 $MN \cdot m^{-2}$, respectively (Kraus, 1968; Reilly & Burstein, 1975). These values can vary considerably depending (amongst other things) on the strain rate, position in the bone and method of storage of the femur prior to testing, but the ratio of ultimate strengths might typically be 1.5:1. Such a ratio of maximum compressive and tensile stresses occurs when the force in the ilio-tibial band equals nearly twice the bodyweight. When the effect of the ilio-tibial band was examined in a similar manner with an intact femur (Fagan & Lee, 1986b), the optimum value was found to be 1.2 times bodyweight.

Few references could be found on the action of the ilio-tibial band. Simple beam theory has been used to analyse the stresses in a femur loaded by the hip reaction, abductor muscles and varying values of the tensor fasciae latae (Rybicki *et al.*, 1972). The ratio of the maximum tensile and compressive stresses was reported to be optimised with a muscle force of 1.5 times bodyweight, and the elastic strain energy stored in the femur minimised with a similar value.

The force in the ilio-tibial band has been measured experimentally during the walking cycle in three subjects (Jacob *et al.*, 1982). The results varied considerably between the different tests, but the highest value recorded during heel-strike was approximately one times bodyweight. Such a variation in the muscle force, when applied to the finite element models used in this research, would lead to a reduction of nearly 25% in the cortex stresses.

The finite element method has been applied to the analysis of total hip stems for over 12 years. A thorough review of the literature has been performed by Huiskes & Chao (1983). The results of these analyses (together with the work in this chapter) are in general agreement, taking into account the wide variation in model specifications, material properties and applied loading conditions. As the available computing power has increased over the years, so both the validity and accuracy of the models has and will continue to improve. The models become more valid as their geometries reflect the true three-dimensional nature of the human femur and use realistic material properties, possibly with non-linear material interface conditions. The accuracy of the analyses improves as finer meshes and/or high order interpolation functions are used in the discretisation of the problems.

Rohlmann and colleagues have developed sophisticated finite element models, and verified their performance by comparison with experimental tests (Rohlmann *et al.*, 1983, 1987). Part of their excellent work was concerned with different loading conditions. Again the results that they produce agree with those reported elsewhere in the literature. Also in common with other analyses, the muscle forces and hip reactions are all applied in the frontal plane, as a result of which the maximum stem stresses are predicted to occur at the posterior – lateral corner of the implant. In practice, as mentioned above, failure of the stem is seen to originate from the anterior – lateral corner.

The work presented in this paper shows how the stress patterns in the stem and cement are built up by the muscle actions and hip reactions that may occur at the joint. A three-dimensional loading scheme is introduced which does predict failure at the anterior – lateral corner, indicating that it may well be a representative and relevant loading condition to consider.

The three-dimensional scheme is also compared to the equivalent two-dimensional loading condition, where all the loads are confined to the frontal plane. The two-dimensional analysis suggests the magnitude of the hip reaction will be 40% lower than that predicted by a full three-dimensional analysis and, when applied to a finite element model of the femur with prosthesis, the stress distribution through the structure proves to be different for the two cases.

The analysis of the different muscle actions on the proximal femur should also be of use when examining the effects of changes in the geometry of the hip joint following hip-replacement surgery. For example, reducing the offset of the stem has been assumed, simplistically, to reduce the maximum bending moment, and consequently the stresses developed in the prosthesis by the hip reaction. However, altering the offset of the stem will also move the position and line of action of the abductors, which will then need to produce a larger contraction force to develop the same moment in order to maintain equilibrium in the coronal plane, thus leading to an increase in the hip reaction.

In the field of biomechanics, where prostheses are designed for implantation into the human body, the work of the engineer is greatly hampered since the boundary conditions, loading conditions and even material properties of the body can be unknown. It is hoped that this research will go some way to bridging that gap in our knowledge.

Acknowledgements

The work described in this paper was funded by a grant from the UK Science and Engineering Research Council (GR/B/97893) under the auspices of the Specially Promoted Programmes in Medical Engineering.

References

Beckenbaugh, R. S. & Ilstrup, D. M. (1978). Total hip arthroplasty. *Journal of Bone and Joint Surgery*, **60A**, 306–13.

Carlsson, A. S., Gentz, C. F. & Stenport, J. (1977). Fracture of the femoral prosthesis in total hip replacement according to Charnley. *Acta Orthopaedica Scandinavica*, **48**, 650–5.

Charnley J. (1979). *Low Friction Arthroplasty of the Hip*. Springer-Verlag, Berlin.

Ducheyne, P., de Meester, P., Aernoudt, E., Martens, M. & Mulier, J. C. (1975). Fatigue fractures of the femoral component of Charnley and Charnley-Muller type total hip prosthesis. *Journal of Biomedical Materials Research Symposium*, **6**, 199–219.

English, T. A. & Kilvington M. (1979). *In vivo* records of hip loads using a femoral implant with telemetric output. *Journal of Biomedical Engineering*, **1**, 111–15.

Fagan, M. J. & Lee, A. J. C. (1986a). Material selection in the design of the femoral component of cemented total hip replacements. *Clinical Materials*, **1**, 151–67.

Fagan, M. J. & Lee, A. J. C. (1986b). The role of the collar on the femoral stem of cemented total hip replacements. *Journal of Biomedical Engineering*, **8**, 295–304.

Gruen, T. A., McNeice, G. M. & Amstutz, H. C. (1979). Modes of failure of cemented femoral components – A radiological analysis of loosening. *Clinical Orthopaedics and Related Research*, **141**, 17–27.

Huiskes. R., & Chao, E. Y. S. (1983). A survey of finite element analysis in orthopaedic biomechanics: the first decade. *Journal of Biomechanics*, **16**, 385–409.

Inman, V. T. (1947). Functional aspects of the abductor muscles of the hip *Journal of Bone and Joint Surgery*, **29**, 607–19.

Jacob, H. A. C., Huggler, A. H. & Ruttimann, B. (1982). *In vivo* investigations on the mechanical function of the tractus iliotibialis. In: *Biomechanics – Principles and Applications*, ed. R. Huiskes, D. Van Campden and J. De Wign. Martinus Nijhoff, The Hague.

Johnston, R. C., Brand, R. A. & Crowninshield, R. D. (1979). Reconstruction of the hip. A mathematical approach to determine optimum geometric relationships. *Journal of Bone and Joint Surgery*, **61A**, 639–52.

Kapandji, I. A. (1970). *The Physiology of the Joints*, Volume 2. Livingstone Press, Edinburgh.

Kraus, H. (1968). On the mechanical properties and behaviour of human compact bone. In: *Advances in Biomedical Engineering and Medical Physics*, ed. S. N. Levine. Interscience Publishers, Chichester.

Lorenz, M., Semlitsch, M., Panic, B., Weber, H. & Willert, H. G. (1978). Fatigue strength of cobalt-base alloys with high corrosion resistance for artificial hip joints. *Engineering in Medicine*, **7**, 241–50.

McLeish, R. D. & Charnley, J. (1970). Abduction forces in the one-legged stance. *Journal of Biomechanics*, **3**, 191–209.

Miller, E. H., Shastri, R. & Shih, C. (1982). Fracture failure of a forged Vitallium prosthesis. *Journal of Bone and Joint Surgery*, **64A**, 1359–62.

Paul, J. P. (1964). Bioengineering studies of the forces transmitted by joints. II. Engineering analysis. In: *Biomechanics and Related Topics*, ed. R. M. Kenedi. Pergamon Press, Glasgow.

Reilly, D. T. & Burstein, A. H. (1975). The elastic and ultimate properties of compact bone tissue. *Journal of Biomechanics*, **8**, 393–405.

Rohlmann, A., Mössner, U., Bergmann, G. & Kölbel, R. (1983). Finite element analysis and experimental investigation in a femur with hip endoprosthesis. *Journal of Biomechanics*, **16**, 727–42.

Rohlmann, A., Mössner, U., Bergmann, G., Hees, G. & Kölbel,R. (1987). Effects of stem design and material properties on stresses in endoprostheses. *Journal of Biomedical Engineering*, **9**, 77–83.

Rybicki, E. F., Simonen, F. A. & Weiss, E. B. (1972). On the mathematical analysis of stress in the human femur. *Journal of Biomechanics*, **5**, 203–15.

Rydell, N. (1964). Forces in the hip joint. In: *Biomechanics and Related Topics*, ed. R. M. Kenedi. Pergamon Press, Glasgow.

Seireg, A. & Arvikar, R. J. (1975). The prediction of muscular load sharing and joint forces in the lower extremities during walking. *Journal of Biomechanics*, **8**, 89–102.

Shigley, J. E. (1977). *Mechanical Engineering Design*. McGraw-Hill, New York.

Williams, J. F. & Svensson, N. L. (1968). A force analysis of the hip joint. *Biomedical Engineering*, **3**, 365–75.

Wroblewski, B. M. (1979). The mechanism of fracture of the prosthesis in total hip replacement. *International Orthopaedics* (SICOT), **3**, 137–9.

15

A non-linear finite element model for interface stresses in total hip replacements

A. Rohlmann, E. J. Cheal, W. C. Hayes and G. Bergmann

15.1 **Introduction**

Long-term fixation of total hip replacement is still a problem in younger, more active patients. Loosening rates of as high as 54% five years postoperatively (Chandler & Wixson, 1979) have been reported for patients less than 30 years of age. Currently, most total hip replacements are mechanically fixed in the femur using either bone cement or through cementless, porous ingrowth systems. In order to improve implant fixation and to reduce mechanically caused implant loosening, an improved understanding of force transfer from endoprosthesis to bone is required.

15.1.1 *Stress analyses in hip implants*

Three-dimensional finite element stress analyses of the proximal femur after joint replacement have been performed by several groups (Röhrle *et al.*, 1977; Huiskes, 1980; Crowninshield *et al.*, 1980; Rohlmann *et al.*, 1983, and others). An excellent survey can be found in Huiskes & Chao (1983). In most of these models rigid bonds across the interfaces between different materials were assumed. However, the mesh representations, prosthetic types and loading cases differed widely. In general, tensile stresses were predicted on the lateral surface of the stem and compressive stresses on the medial side. Because of the imposed bending loads, the greatest stress component was generally in the direction of the femoral axis. High shear stresses were predicted at the interfaces between prosthetic stem and bone cement (Crowninshield *et al.*, 1980; Huiskes, 1980; Rohlmann *et al.*, 1983).

Several studies have also explored the effect of interface boundary conditions and loosening in total hip replacements. Svensson *et al.*, (1977) used a two-dimensional model of a cemented hip stem and found stresses in the cement changed significantly when slip-and-no-tension was allowed at the stem–cement interface. However, the stresses in the bone and prosthetic stem changed only slightly. Röhrle *et al.*, (1977), analysed a three-dimensional model of the proximal femur containing a hip endoprosthesis. They studied several interface conditions, including allowing tangential displacement between prosthetic stem and bone cement. Without collar–calcar contact, they found significantly higher radial stresses in the medial cortex in comparison to the rigidly bonded case. Huiskes & Schouten (1980) also investigated stresses associated with loosening of intramedullary fixed implants using two-dimensional and axisymmetric finite element models. Their two-dimensional model also employed a "spanning-element" layer to account for out-of-plane bone continuity. They compared their results with those achived using beams-on-elastic-foundation theory (Huiskes, 1980) and concluded that interface loosening had a great influence on the stresses in cement but only a modest effect on the most significant stem and bone stresses.

Hampton & Andriacchi (1980) modelled interface failure between bone-cement and prosthetic stem by assuming that post-failure the interface can support only compressive normal forces and coulomb shear. These non-linear interface conditions were implemented in a two-dimensional finite element model using incremental linearisation, a modified Newton–Raphson iteration method and an equilibrium-based convergence criterion. For a geometrically simplified model they compared theoretical results with an experimental study and found reasonable agreement when the interface characteristics were properly selected.

Chao *et al.* (1984) used an axisymmetric model to analyse stresses in tumor prostheses which utilised a tapered conical press-fit for fixation. Their models allowed relative motion between implant and bone only in the longitudinal direction. In the radial and circumferential directions a rigid bond was assumed. Two layers on the outer surface of their model represented a porous layer and biological tissue ingrowth. With tissue ingrowth into the porous layer they predicted significantly reduced stresses in the conical coupling when compared to the non-ingrowth state.

Mössner *et al.* (1985) studied a geometrically simplified three-dimensional finite element model of the femur with a cemented hip endoprosthesis by modifying a 'contact' element. Laterally, where compressive stress normal to the interface between cement and implant occurred they found an increase of the dominant principal stress by a factor of approximately two in comparison to the rigidly bonded case. An advantage of their contact element was that the computer time for six iteration steps increased only by a factor of approximately 1.5. On the other hand, the core memory required was much higher than for the rigidly bonded case since the contact element causes the stiffness matrix to become asymmetric.

Himeno & Tsumura (1985) calculated the stresses at the margins of a simplified two-dimensional model of total hip replacement. They used 16 rigid elements which were combined with normal and shear springs along their contact borders. At the initial step all springs were assumed intact. After the first iteration, springs loaded in tension were removed and the stiffness of the shear springs were modified to adjust their stress according to the local maximum friction. The iteration process was repeated until the solution converged. Based on a stance-phase force resultant applied to the femoral head and assuming collar–calcar contact, they found that most of the load was transmitted to the cortex through the calcar region. Without collar–calcar contact, severe stress concentrations appeared proximally and medially around the tip of the stem.

Hipp *et al.* (1985) developed two-dimensional and axisymmetric models of experimental dental implants and screw threads in bone. They assumed both rigidly bonded and frictionless interface conditions. In the frictionless case they calculated greater maximum stress values and stress ranges than for the rigidly bonded case. The stress distribution was qualitatively different for the frictionless and rigidly bonded cases.

Lewis *et al.* (1985) studied the failure processes experimentally at the interface between cancellous bone and bone-cement under cyclic tensile loading. Using a stereomicroscope and cinematography, they observed crack length and motion at the interface. Often they found a second tensile crack in the cement layer and observable deformations away from the main crack. They hypothesised that tensile cyclic loads at the interface initiate and propagate a primary tensile crack. This results in relative motion between bone and cement and increases fluid flow. Relative motion stimulates fibrous tissue formation and, if it is large enough, leads to bone resorption.

15.1.2 *Objectives*

The objectives of this study were:

(1) to model a non-linear interface between different material components;
(2) to compare two different iteration schemes;
(3) to study the changes in stresses and relative motion at the interface during iteration;
(4) to compare the results for two iteration processes.

The method should be applicable for standard finite element programmes without 'contact element'.

15.2 **Materials and methods**

15.2.1 *Finite element model*

A geometrically simplified three-dimensional finite element model consisting of 1110 eight-node hexahedral elements and 1582 nodal points was

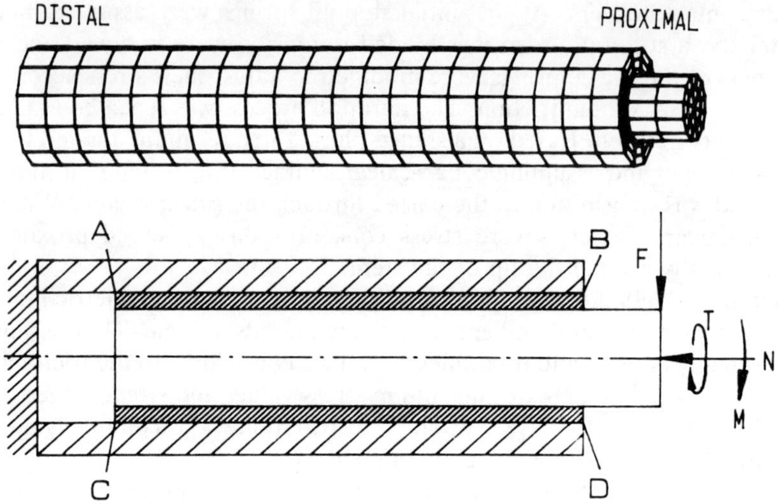

Fig. 15.1 Element mesh and diagrammatic view of the finite element model.

employed (Figure 15.1). The bone was assumed to be a linear, homogeneous and isotropic tube and was represented by two element layers. The prosthetic stem was also assumed to be of constant diameter. In the model the cross-sectional shape of the implant was a polygon with 12 edges. The thickness of the intermediate layer was 1.5 mm. The stem was represented by two element layers and the intermediate layer by one element. The intermediate layer was assumed to be a part of the prosthesis (as in the case of a porous coating). Therefore a rigid bond between intermediate layer and solid prosthetic stem was always maintained during the iteration process.

The elastic moduli chosen were:

(1) 18 000 $MN{\cdot}m^{-2}$ for cortical bone;
(2) 200 000 $MN{\cdot}m^{-2}$ for the prosthetic stem (Co–Cr–Mo alloy);
(3) 5000 $MN{\cdot}m^{-2}$ for the intermediate layer.

The latter value is approximately twice that of bone cement. The model was clamped distally near the points A and C and loaded proximally. Stress analyses were performed for a pure bending moment $M = 1$ Nm.

15.2.2 *Interface conditions*

An uncemented smooth prosthetic stem cannot transfer tensile stresses across the interface between implant and bone. The same is true for a porous coated prosthesis immediately after implantation. For studying these two cases, a new iterative technique was developed for representing non-linear bone/implant interface conditions. The underlying assumption was that the interface could not support tensile stress normal to the interface. In order to study the influence of the interface conditions, all nodal points were duplicated at the interface between bone and prosthesis. Initially, these additional nodal points were fixed and not allocated to any element. The other nodal points with the same co-ordinates were allocated to the elements on both sides of the interface, thus simulating a rigid bond between all elements at the interface. After the first finite element calculation, the stress component normal to the interface was determined for each element at the interface. If the sign of this stress on the prosthetic side was positive, i.e. tensile stress normal to the interface, the connection across the interface between the prosthetic and bone elements was eliminated. In the data set the node number on the prosthetic side of the interface was replaced by the node number with the same co-ordinates and the former restrictions of this nodal point were removed. No changes were made on the bone side of the interface. A new finite element calculation was then initiated. This procedure, however, led in some cases to a physically unrealistic overlap of elements in the model.

15.2.3 *Refixing methods*

In order to avoid element overlap, two different refixing strategies were used.

Refixing strategy A Displacements were determined for all nodal points at the interface. When an overlap occurred, the nodal point on the prosthetic side was fixed and replaced by the nodal point with the same co-ordinates on the bone side. A new finite element calculation was then started using the new data set. This iteration process was repeated until the data set satisfied a convergence criterion based on changes in the stress distribution. Checking for overlap at the interface was straightforward for the simplified model but would be complicated in the general case where the nodal points of an element side are not necessarily in a plane. Therefore a second, more general, refixing strategy was developed which could be used with more complex element geometries.

Refixing strategy B As in the previous case, the stress component normal to the interface between implant and bone was first calculated. For all the prosthetic elements where this stress component was negative, i.e. compressive, it was determined whether there were tensile stresses normal to the interface in the adjacent prosthetic elements. In this case, the connections across the interface between these elements were fixed again. Since this procedure resulted in an excess of elements for which there is rigid bond across the interface, this refixing strategy was only used in the

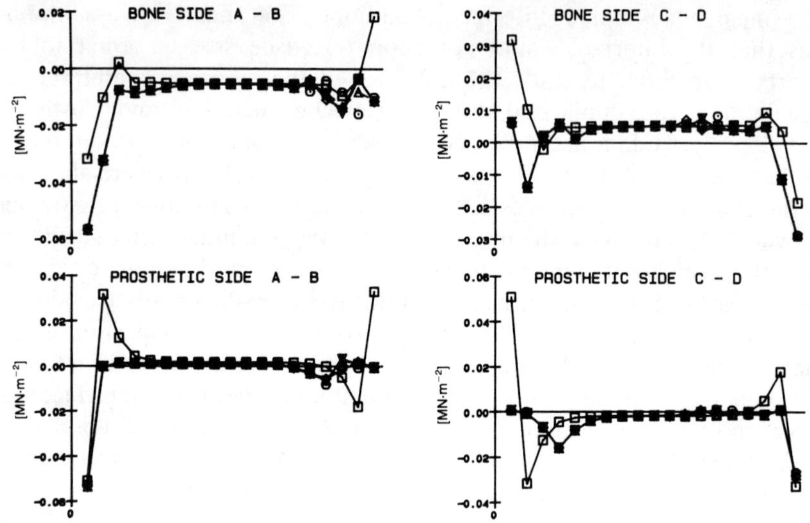

Fig. 15.2 Stresses normal to the bone–intermediate layer interface for the iteration steps: 0 (□), 2 (○), 4 (△), 6 (▽), 8 (◇). The prosthetic stem was proximally loaded with a bending moment M = 1 Nm.

first three iteration steps when the greatest changes in load transmission occurred. After three iterations, the subsequent changes were small and the refixing had only a small effect on the results. In the subsequent iteration steps, only the connection between those elements at the interface were eliminated for which tensile stresses occurred normal to the interface, but no refixing was allowed.

15.3 **Results**

15.3.1 *Changes during the iteration process*

Figure 15.2 shows the stresses normal to the interface between bone and intermediate layer along the lines A–B and C–D for iteration steps 0, 2, 4, 6 and 8. The results are for the pure bending moment loading case and refixing strategy A. Iteration step 0 (□ symbols) represents a rigid bond across the interface. For the rigidly bonded case, the highest stresses occur at points A to D. In the middle region the stresses are constant. During the iteration process, the stresses increase markedly on the bone side at points A and D and decrease significantly at points B and C. The changes are small after two iteration steps.

Relative motion at the interface for lines A–B and C–D and iteration steps, 1, 3, 5, 7 and 9 is shown in Figure 15.3. The greatest values are calculated for points B and C. Relative motion is small in the middle

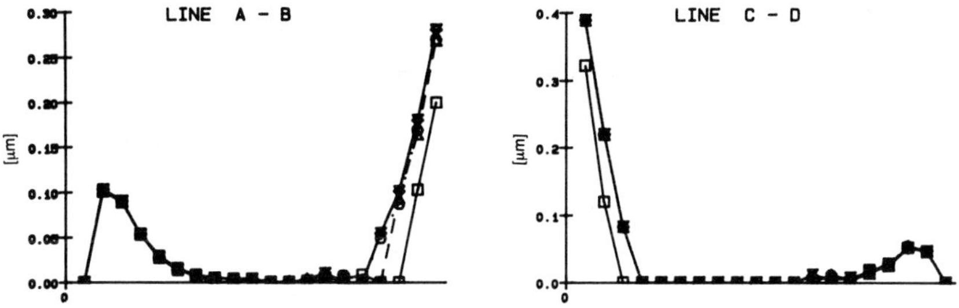

Fig. 15.3 Relative motion at the interface along the lines A–B and C–D for the iteration steps: 1 (□), 3 (○), 5 (△), 7 (▽) and 9 (◇).

region. In the rigidly bonded case (step 0) there is no relative motion at the interface. The changes after three iteration steps are again small.

15.3.2 *Comparision of the two refixing strategies*

Figure 15.4 shows the stresses normal to the interface along lines A–B and C–D for the rigidly bonded case and for refixing strategies A and B after ten iteration steps. Again there are significant differences between the rigidly bonded and the non-linear cases after ten iteration steps, especially near the proximal and distal ends of the interface where the stresses are

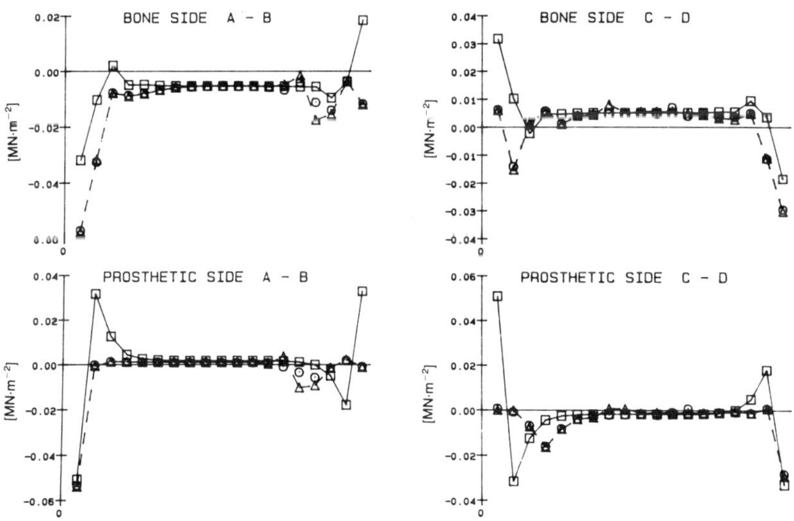

Fig. 15.4 Comparison of the stresses normal to the interface for the rigidly bonded case (□), the refixing strategy A after the tenth iteration step (○), and refixing strategy B after the tenth iteration step (△).

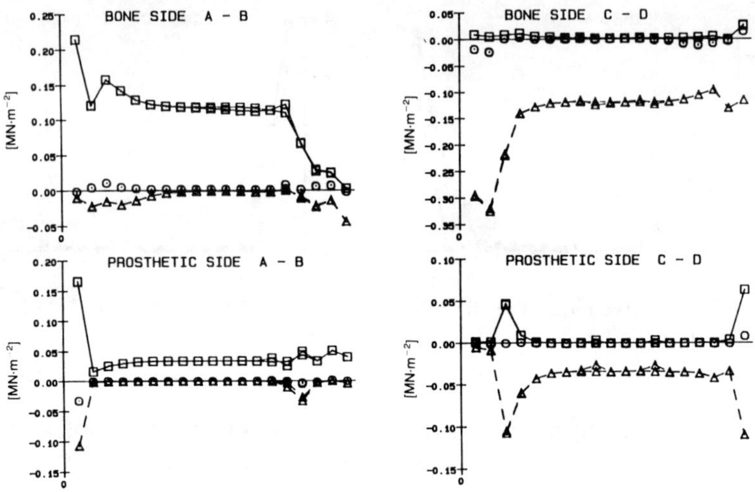

Fig. 15.5 Comparison of the maximum (□) medium (○) and minimum (△) principal stresses for refixing strategies A and B after ten iteration steps.

largest. The differences in the final results for the two refixing strategies are small and occur mainly near point B.

The principal stresses at the interface between bone and intermediate layer for both refixing strategies are shown in Figure 15.5. The maximum (□ symbol) principal stress dominates along the line A–B and the minimum (△ symbols) principal stress dominates along the line C–D. The intermediate principal stress (○ symbol) is small along both lines. The highest stresses are at points A and C on the bone side, and at A and D and near point C on the prosthetic side. Only small differences occur between the final results for the two refixing strategies, mainly in the middle region of the interface.

Relative motions at the interface are shown in Figure 15.6. Maximum values are calculated for points B and C. Again only small differences are found between the two refixing strategies. For the rigidly bonded case (□ symbols) the relative motion is again zero.

15.4 **Discussion and conclusions**

15.4.1 *Stresses normal to the interface*

The results for the rigidly bonded case are similar to those of Huiskes (1980) for a geometrically simplified model of a cemented endoprosthesis. Using beams-on-elastic-foundation theory and the finite element method Huiskes also predicted peak stresses normal to the interface at the proximal and distal ends of the stem. As in the present model, near the ends of the

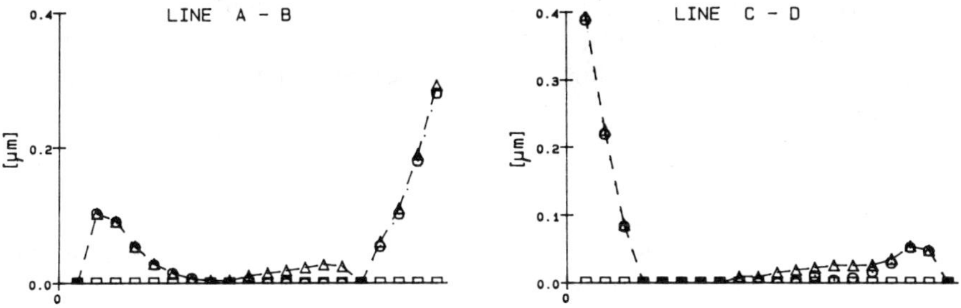

Fig. 15.6 Comparison of the relative motion at the bone–implant interface for the rigidly bonded case (□) and the refixing strategies A (○) and B (△) after ten iteration steps.

interface the stresses changed sign and in the middle region zero stresses were predicted.

The discontinuity in the elastic moduli results in a jump in the stiffness and in some stress components at the interface between bone and femoral stem (Huiskes, 1980; Rohlmann *et al.*, 1983). However, the stress components normal to and the shear stresses parallel to the interface should be equal. In the finite element calculations the stresses normal to the interface on both sides of the interface were often different. Due to the bending load the main stress component was in the direction of the longitudinal axis. The component perpendicular to the interface was only a small fraction of this (often less than 5% of the maximum component). The finite element method is an approximative method. The overall results are mainly in excellent agreement with those expected, e.g. as from beam theory. For instance, in a homogeneous model the principal stresses on both sides of an interface are the same when a rigid bond is assumed (Rohlmann *et al.*, 1983). Some small stress components, however, are often calculated inexactly. For example the stress component normal to a free surface, which theoretically should be zero, is never equal to zero.

In the iteration process the stresses normal to the interface on the prosthetic side were chosen as a criterion for eliminating the connections in the interface between elements representing different materials.

15.4.2 *Non-linear interface conditions*

For loading the prosthesis with a force couple, iteration strategy A converged to two solutions, which alternated. However, the differences in the stresses for these two solutions were very small. The reasons are the same as for the differences in the stress components perpendicular to the interface on both sides of the bone-implant interface.

Separation of the elements at the interface between bone and implant had only a small influence on the global stress distributions. However, the influence on the interface stresses themselves was very significant. These findings agree with those of the two-dimensional model of Huiskes & Schouten (1980).

Since there were only small differences between the two refixing strategies, strategy B is to be preferred, especially for an irregular elementation where checking for element overlap is difficult.

In conclusion, a method has been developed for determing stress distribution at material interfaces in the presence of non-linear interface conditions. The method can be applied to interface problems where no tensile stress should be transmitted across the interface. In the model, convergence occurred after only a few iteration steps. With this method it is possible to determine the stresses and relative motion at the interface between different materials, thus providing more accurate results about the state of stress and relative motion at the interface.

References

Chandler, H. P. & Wixson, R. L. (1979). A five-year review of total hip replacements in patients under the age of thirty with an emphasis on loosening. *Annual Meeting AAOS*, San Francisco.

Chao, E. Y. S., Kwak, B. M. & Kasman, R. (1984). Stress analysis of conical coupling joint in modular prosthetic system design. *Transactions Orthopaedic Research Society*, p. 103.

Crowninshield, R. D., Brand, R. A., Johnston, R. C. & Milroy, J. C. (1980). An analysis of femoral component stem design in total hip arthroplasty. *Journal of Bone and Joint Surgery*, **62A**, 68–78.

Hampton, S. J. & Andriacchi, T. P. (1980). An analytical representation of the non-linear interface condition in a bone-cement-prosthesis system. In: *International Conference Proceedings on Finite Elements in Biomechanics*, ed. B. R. Simon, pp. 193–206. University of Arizona Press, Tucson, AZ.

Himeno, S. & Tsumura, H. (1985). Marginal stress distribution of the femoral component of THR and its 'mode of failure'. *Transactions Orthopaedic Research Society*, p. 143.

Hipp, J. A., Brunski, J. B., Shephard, M. S. & Cochran, G. V. B. (1985). Finite element models of implants in bone: Interfacial assumptions. In: *Biomechanics: Current Interdisciplinary Research*, eds. S. M. Perren & E. Schneider, pp. 447–52. Martinus Nijhoff Publishers, Dordrecht.

Huiskes, R. (1980). Some fundamental aspects of human joint replacement. *Acta Orthopaedica Scandinavica,* Suppl. 185.

Huiskes, R. & Chao, E. Y. S. (1983). A survey of finite element analysis in orthopaedic biomechanics: the first decade. *Journal of Biomechanics* **16**, 385–409.

Huiskes, R. & Schouten, R. Y. (1980). The effect of interface loosening of the stress distribution in intramedullary fixated artificial joints. In: *Advances in Bioengineering*, ed. C. Van Mow, pp. 213–6. The American Society of Mechanical Engineers, New York.

Lewis, J. L., Nicola, T., Keer, L. M., Clech, J. P., Steege, J. W. & Wixson, R. L. (1985). Failure processes at the cancellous bone–PMMA interface. *Transactions Orthopaedic Research Society*, p. 144.

Mössner, U., Rohlmann, A. & Bergmann, G. (1985). Influence of the interface conditions on the force transfer. *Transactions of the 5th European Conference on Biomaterials*, Paris, p. 119.

Röhrle, H., Scholten, R., Sollbach, W., Ritter, G. & Grünert, A. (1977). Der Kraftfluss bei Hüftendoprothesen. *Archiv für Orthopädische Unfall-Chirurgie*, **89**, 49–60.

Rohlmann, A., Mössner, U., Bergmann, G. & Kölbel, R. (1983). Finite element analysis and experimental investigation in a femur with hip endoprosthesis. *Journal of Biomechanics*, **16**, 727–42.

Svensson, N. L., Valliappan, S. & Wood, R. D. (1977). Stress analysis of human femur with implanted Charnley prosthesis. *Journal of Biomechanics*, **10**, 581–8.

16

A loading and stress analysis of the patella

R. J. Minns

16.1 **Introduction**

Numerous explanations have been proposed to account for the destruction of human articular cartilage. In the patello-femoral joint, such lesions have been particularly related to the mechanics of articulation, to contact stresses and to the contact surface topography. Wiberg (1941) proposed one of the earliest explanations for the pathogenesis of articular cartilage change occurring on the medial patellar facet. He described certain positions of high joint loading in which a convex surface of the facet opposes a convex surface on the underlying femoral condyle, producing high contact stresses. This applied particularly to those patellae in which there was a distinct ridge between the medial and 'odd' facets. One argument against this hypothesis is that fewer cartilage changes occur on the medial femoral condyle, even though it is subjected to the same surface stresses. Wiberg's explanation was that the articular cartilage was thicker on the patellar facet than on the condyle and as a result probably receives less nutriment and consequently becomes more prone to degenerative disease. A further argument is that there are patellae which have concave surfaces and yet still undergo degeneration of the articular cartilage on the medial and 'odd' facets, even though these do not contact the condylar surface except in certain positions of extreme flexion.

Outerbridge (1961) described a rim of the upper border of the medial femoral condyle which abraded the medial patellar facet leading to chondromalacic changes on this facet. However, patients without this ridge may also have similar lesions on the medial patellar facet (Emery & Meachim, 1973). Bandi & Brenwald (1974) noted that both extremely high contact pressures and surface friction may lead to patello-femoral arthrosis. They advocated anterior displacement of the tibial tuberosity to reduce the contact pressure. Ficat & Maroudas (1975) cited high contact pressures, combined with lower glycosaminoglycan content of the articular cartilage, as important factors in the frequency of observed destructive changes of the patellar cartilage compared with the hip joint.

There appears to be no relationship between patellar height (Blackburne & Peel, 1977) and chrondromalacia. Insall *et al.* (1976) support the relationship between patella alta and chondromalacia whilst Marks & Bentley (1978) disclaim such a relationship. Although trauma has been implicated as a possible cause of cartilage destruction, some feel it is of more importance as an aggravating factor than as a primary cause (Wiles *et al.*, 1956).

Structural changes in the underlying bone and its effects on the loading of the trabeculae have been suggested to result in a concentration of cartilaginous shear stresses on the central medial facet (Townsend *et al.*, 1977). If stiff or weakened bone is adjacent to normal bone, there exists a difference in resistance to deformation, producing high local shear strains within the cartilage which may be responsible for cartilage damage (Abernethy *et al.*, 1978).

Variations in the location of cartilage lesions and their association with certain clinical syndromes has been described by Ficat & Hungerford (1977). Three main types of lesion have been established. The first type, termed open chondromalacia, is usually confined to the medial and odd facets and involves the secondary ridge between these two facets. Goodfellow *et al.*, (1976b) claim that these changes can be almost limited to the odd facet. The pathological features of this lesion comprise a proliferation of cartilaginous material sprouting above the surface and consisting of massive bundles of collagen fibres arranged in tufts and whorls and associated with an abundance of hyperactive cells.

A second lesion, usually developing on the central medial aspect of the patellar surface and particularly associated with the median ridge, is termed closed chondromalacia. This consists of a localised softening of the cartilage. In this instance, electron microscopy reveals an area of abnormality which is localised to the intermediate zone of the uncalcified cartilage and in which the collagen fibres undergo abrupt changes of direction and disintegration.

A third lesion, termed chondrosclerosis, is usually associated with an extreme lateral pressure syndrome involving the lateral facet. Its gross pathological appearance comprises areas of hard and yellowish cartilage. The collagen fibres within this lesion are usually dense and compact throughout all the zones of the cartilage.

With increasing age, particularly in women, the patellar cartilage invariably becomes thinner and is associated with narrowng of the joint space (Meachim *et al.*, 1977). This leads to a secondary stage of patello-femoral arthrosis, accompanied by the usual underlying bony changes, and in which the three cartilage lesions previously described cannot be isolated (Ficat, 1974).

Bony changes in chrondromalacia have been described by Darracott & Vernon-Roberts (1971). Large areas of oesteoporotic bone, predominantly deep to the medial facet, were associated with discrete nodular aggregates of woven bone attached to the trabeculae. They reasoned that these nodules arose as a result of some form of disordered new bone formation producing a buttressing effect at sites of high trabecular stress. Stougard (1975) observed that the bone was less dense medially and that bone sclerosis did not occur even in the presence of medial cartilage damage. Bony traceculae are modified when subjected to high compression forces, as observed radiologically in the extreme lateral pressure syndrome (Ficat & Hungerford 1977). Haasters (1974) wrote that the spongiosa of the patella, as observed on lateral X-ray of the patella, responds to loads applied through the quadriceps mechanism. Clearly , the localisation of the lesions appears to support the hypothesis that stress and the evolution of chrondromalacia are associated (Wiles *et al.*, 1956).

There are many conflicting views as to the etiology of retropatellar pain (Insall, 1982). These have included misalignment of the patella, tightness of the patellar retinacular mechanism (Merchant & Mercer, 1974), subluxation and dislocation of the patella (Crosby & Insall, 1976), patellar shape (Wiberg, 1941), patellar position (Marks & Bentley, 1978), and congruity during articulation of the patella (Merchant & Mercer, 1974; Aglietti *et al.*, 1983).

Many papers have been written on the misalignment and mechanical derangement of the patella (Reider *et al.*, 1982; Veress *et al.*, 1979; Blackburne & Peel, 1977; Delgardo-Martins, 1979; Hvid *et al.*, 1983; Insall & Salvati, 1971), confirming that retropatellar pain is a multifactorial condition which has a large mechanical component.

In a theoretical study using a simple stress analysis Minns *et al.* (1979) suggested that under certain mechanical conditions such as misalignment or large lateral pressure and with patellae of different shape, the surface stresses could be quite high in tension and shear especially in areas where lesions might be seen (Figure 16.1a).

Goodfellow *et al.* (1976a) implicated the 'odd' facet as a feature of degeneration because of stresses of the medial and odd facet surfaces, and this may be associated with the stresses within the bone and cartilage (Minns & Braiden, 1980; Minns *et al.*, 1979). Bentley (1985) conducted a histomorphological study of the articular cartilage associated with chondromalacia patella and showed distinct changes in the composition and organisation of the matrix compared with normal patellar articular cartilage.

A study on the articulation and position of the patella and standardisation of X-rays so that precise bone structure could be observed were

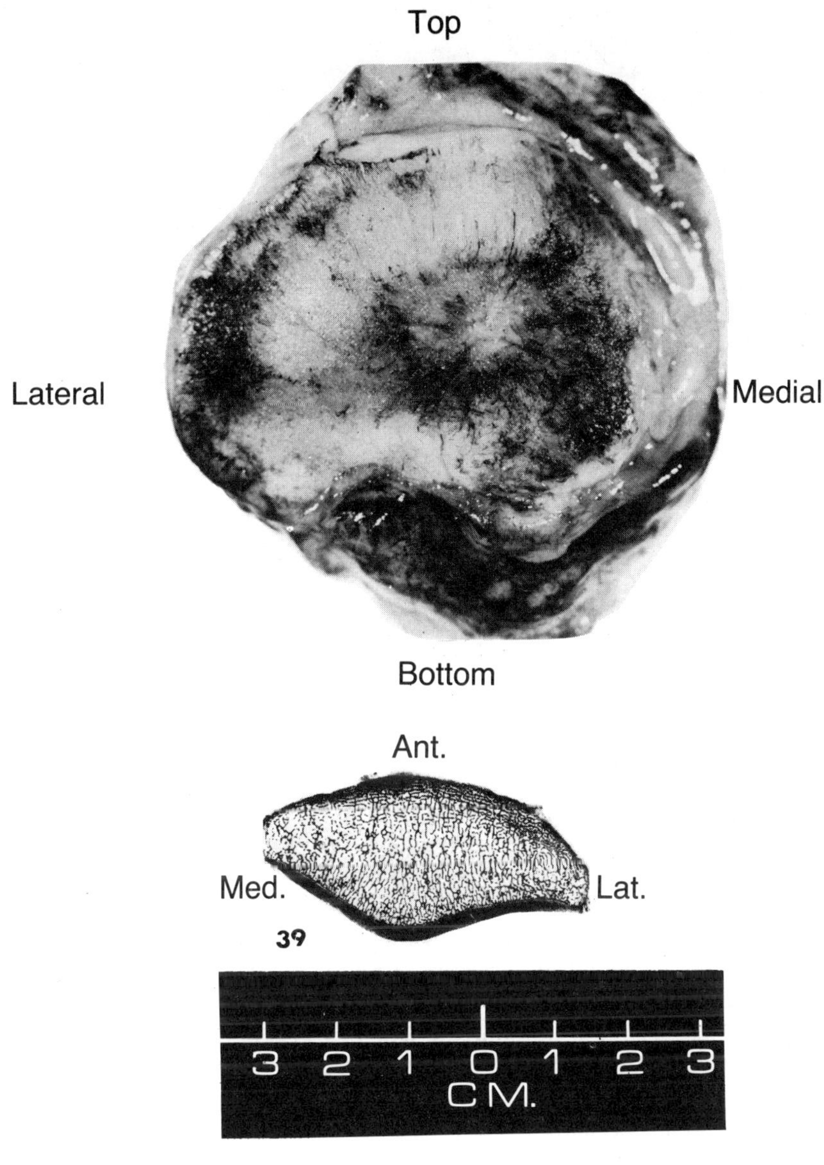

Fig. 16.1 (a) *En face* view of a patellar articular surface stained with Indian ink to show the extent of fissures within the articular cartilage due to wear and stressing of the cartilage matrix. (b) Microradiograph of horizontal mid-section of a patella.

related to the mechanics of the patella (Minns *et al.*, 1989). These data were related to clinical findings and subjective comments from patients presenting with retropatellar pain.

16.2 **Experimental studies**

Fifty-six normal patellae were examined. These comprised:

(1) 16 post-mortem horizontal mid-sections, prepared as described by Abernethy *et al.* (1978); the negatives of these patellar sections included grids marked in cm for scaling purposes (Figure 15.1b);
(2) 12 patellae obtained post-mortem sawn horizontally through the mid-section, laid on a photographic plate with a scale and X-rayed;
(3) skyline radiographs from 28 patients, of average age 20 yrs, who on clinical examination were free of patello-femoral symptoms.

Skyline radiographs of 11 patellae were also studied from 11 patients who underwent surgery for the relief of patello-femoral pain. There were two men and nine women in this group with a mean age of 20 yrs. Of these, nine were considered to have the extreme lateral pressure syndrome (ELPS), of whom two had Wiberg type I patellae and seven had Wiberg type II patellae. Of the two patients who were considered to have medial chondromalacia patellae, one had a Wiberg type III patella and the other a Baumgartl (1964) type patella. One patient with Wiberg type II patella had bilateral fractures of the lateral facets. In this group of patients undergoing surgery, physical dimensions were taken from the patient for scaling purposes and only true skyline views were analysed. The X-ray cassette was positioned at the same distance from the mid-patellar axis in all cases as was the X-ray tube at the distal end of the leg (Figure 16.2).

In order to obtain the angular position of the patella in space, a goniometer was built which allowed sagittal plane movement during knee flexion to be measured utilising the fact that the patellar tendon extends very little. Using a thermosetting plastic sheath which was securely fastened to the subject's calf, an adjustable portion on its proximal end allowed the experimenter to position a small hinged runner centrally on the patella which followed its sagittal movement during knee flexion. Mounted on this runner was a transducer that was used in the angle measuring system. The angle measuring system uses a polarised light goniometer with three angles measured: those of the thigh, leg and patella in the sagittal plane. The variation of the angles during the four tested activities was recorded by an ultraviolet light recorder. The accuracy of this patellar-movement recording was checked by lateral X-rays at three angles of knee flexion on two subjects.

The force between the ground and the foot was measured by using four

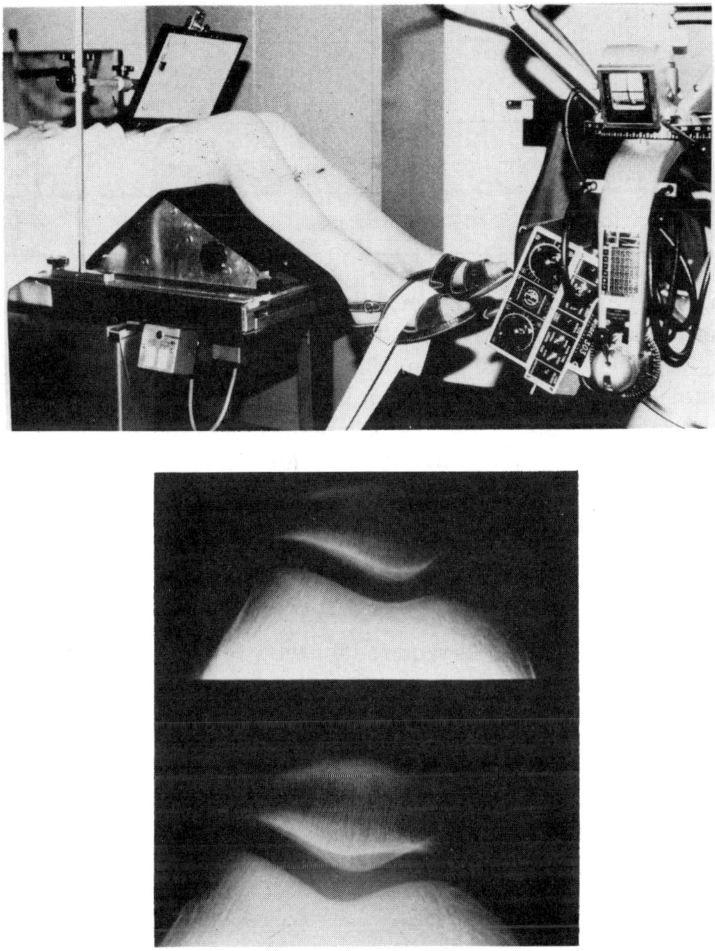

Fig. 16.2 (a) Position for the casette and X-ray camera to produce the standardised skyline X-rays. (b) Example of skyline views of the patella taken by this technique.

miniature compression load cells positioned under the heel, 1st, 3rd and 5th MT head and the big toe in four different sized sandals. The position of each load cell relative to the ankle axis was checked by lateral X-rays and their horizontal position by vertical X-rays. The outputs from each load cell were amplified and the individual loads and total load were recorded on the same UV strip chart recorder as the angular movements. A time marker on the recorder allowed data on cadence and step time to be calculated.

Anatomical data of each subject were recorded: in particular, tibia

Table 16.1

Material	Young's modulus (E; $N{\cdot}m^{-2} \times 10^6$)	Poisson's ratio (ν)
Articular cartilage	7.87	0.473
Subchondral bone	700	0.20
Cancellous bone	410	0.20
Cortical bone	24 100	0.28

length, patellar tendon lever arm, hamstring/gastrocnemius level arm and patellar tendon/tibia angle. These data are necessary to calculate the loads passing through the patellar tendon, the quadriceps mechanism, the patellofemoral joint force and the tibio-femoral joint reaction. These forces were calculated by considering equilibrium of the patella, and shank and foot as free body diagrams. Assumptions were made as to the loads passing through the hamstrings and gastrocnemius and their effect on the patellar loads. Data provided for the stress analysis assumed equilibrium provided by the quadriceps muscle group, and although the weight of the leg and foot were taken into account (6% bodyweight), inertial forces were not.

The mechanical properties required for the stress analyses were Young's modulus, E, and Poisson's ratio, ν; these are recorded in Table 16.1. The materials were assumed to be linearly isotropic, so only one value of E and ν for each material was used. These values for the four materials were obtained (from past research) for articular cartilage (Hori & Mockros, 1976), subchondral bone (McElhaney *et al.*, 1970), cancellous bone (Townsend *et al.*, 1975) and cortical bone (Reilly & Burstein, 1974).The distribution of the four materials in the model was obtained from a previous study (Minns & Braiden, 1980).

16.2.1 *Contact loading areas*

After obtaining the values and direction of the patello-femoral joint force it is necessary to know where this force acts on the patellar surface. It is generally agreed that for the lower values of knee flexion the contact areas occur centrally nearer the inferior pole and progress proximally during an increase in knee flexion (Goodfellow *et al.*, 1976a; Townsend *et al.*, 1977; Matthews *et al.*, 1977). In the model used here, the load positions on the patellar articulating surface nodes acting at the four horizontal levels were assumed from published data at the knee flexions calculated.

16.2.2 *Distribution of patellar material*

The stress analysis model required data on the distribution of the four materials used. These were obtained by cutting post-mortem patellae at four horizontal sections, obtaining radiographs of the surface and estimating the proportions of the materials by thickness, distribution and length. Seven vertical sections were also taken from further post-mortem patellae. The distributions at horizontal mid-section obtained by this method compare favourably with the horizontal sections prepared by Abernethy *et al.* (1978). A mid-vertical section used in the model which shows the distributions and loadings calculated is shown in Figure 16.3.

16.3 **Theoretical studies**

The patella can be considered to be an oblate disc with is subjected to longitudinal tensile loads through the action of ligaments, and to lateral compressive loads on the rear bearing surface. Earlier work has related lesions on the articular surface of the patella to different types of loading and stresses (Minns & Braiden, 1980). Stresses within the patella arise from contact, muscle, ligament and capsule loads and these were estimated using data obtained from the experimental studies. Thirteen cases of loading corresponding to four different physiological conditions: walking on the level, upstairs and downstairs and a 'knees bend', were applied. There were four forces acting on the block, P_J, P_Q, P_T, and either L_C or M_T, balanced in all directions. P_Q, and P_T are the muscle forces applied obliquely by the 'Q' and 'Φ' angles effect (Insall *et al.*, 1976). P_J is the contact force applied to the articular cartilage surface; L_C and M_T are the medial and lateral 'Q' angle component balancing forces, either compressive (L_C) or tensile (M_T) (Figures 16.4 and 16.5)

Each force was uniformly distributed over the relevant face. The uniform load distribution could only be approximated and, over these surfaces, there were a limited number of nodes on which loads could be applied. For each area banded by four nodes, the force on that area was calculated and hence the equivalent force on each of the four nodes. This procedure was repeated for all the sub areas and the forces at each node summed to give the total force on each node. Those forces which were applied at an angle to a face (P_Q and P_T) had to be resolved into vectors in the *xyz* directions.

As was stated earlier the patella loading cases were derived from four activities: a 'knees bend', walking on the level, upstairs, and downstairs. The results were analysed in that order of activity. There were seven loading cases for the 'knees bend', where $\Phi = 60°$, the 'Q' angle varied between 0° and 30°, and there were values each for the lateral loads, L_C and M_T. (see Table 16.2).

Only rigorous X-ray photogrammetry has been undertaken to describe patella movement (Veress *et al.*, 1979) which is costly, damaging to the patient and involves static measurments. Using the goniometer described,

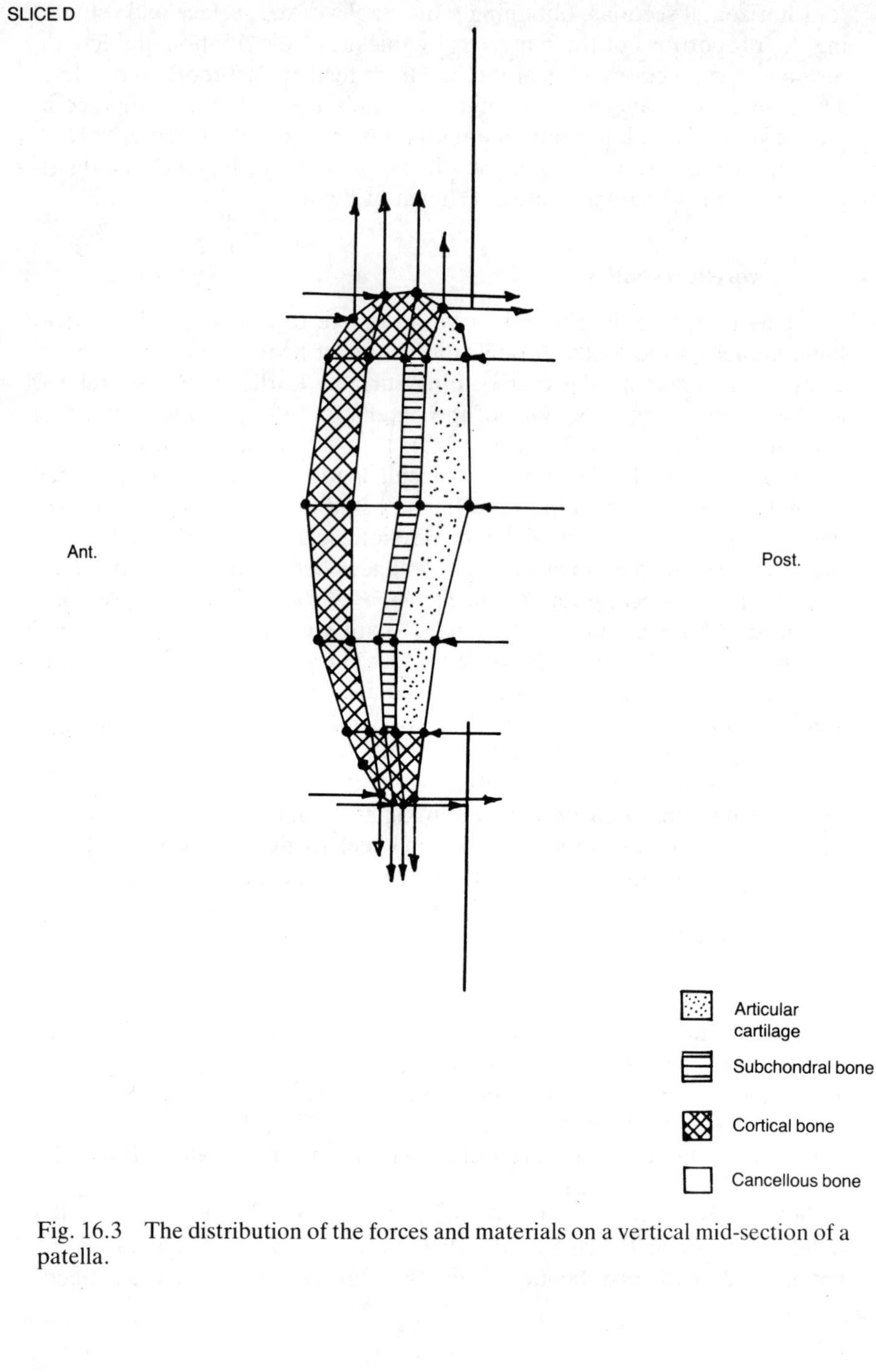

Fig. 16.3 The distribution of the forces and materials on a vertical mid-section of a patella.

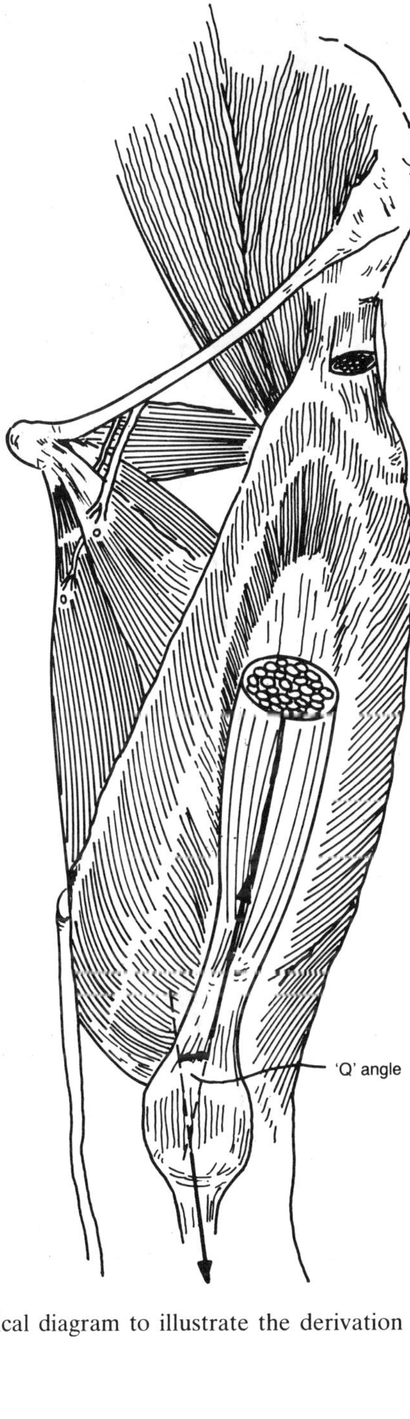

Fig. 16.4 Anatomical diagram to illustrate the derivation of the quadriceps 'Q' angle.

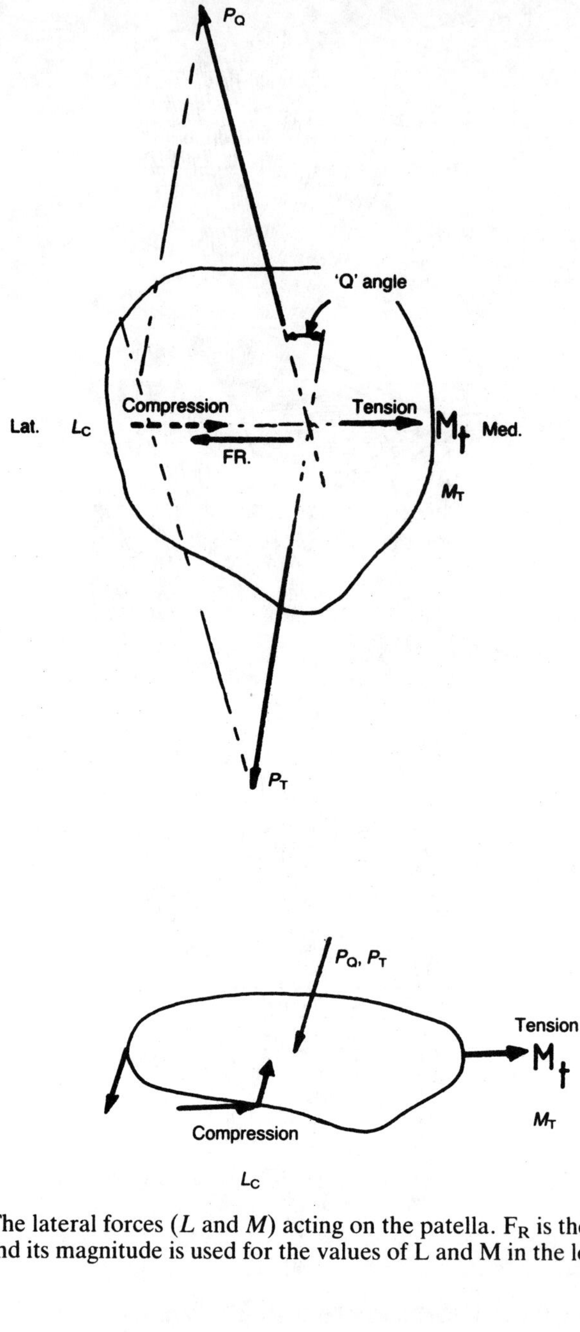

Fig. 16.5 The lateral forces (*L* and *M*) acting on the patella. F_R is the resultant of P_Q and P_T and its magnitude is used for the values of L and M in the loading cases.

Table 16.2

Case no.	Flexion angle (degress)	P_Q, P_T	'Q' Angle	Φ *Angle*	P_J	L_C	M_T
1	45	2433	0	60	2433	0	0
2	45	2433	10	60	2433	420	0
3	45	2433	20	60	2433	840	0
4	45	2433	30	60	2433	1260	0
5	45	2433	10	60	2433	0	420
6	45	2433	20	60	2433	0	840
7	45	2433	30	60	2433	0	1260
8	15 (walking)	647	10	38	422	108	0
9	45 (upstairs)	1923	10	54	1746	335	0
10	62 (downstairs)	1687	10	62	1738	294	0
11	15 (walking)	647	10	38	422	0	108
12	45 (upstairs)	1923	10	54	1746	0	335
13	62 (downstairs)	1687	10	62	1738	0	294

'Q' angle = angle in degrees between P_Q and P_T viewed from front; Φ angle = angle in degrees between P_Q and P_T viewed laterally; P_Q (= P_T) = quadriceps tension (patellar ligament tension) in Newtons uniformly distributed over top and bottom surfaces, respectively; P_J, = patello-femoral joint force in Newtons distributed uniformly over cartilage surface; L_C and M_T = lateral compressive force or medial tensile force in Newtons distributed uniformly over sides.

it was possible to track the sagittal rotation of the patella dynamically by non-invasive means. A device is currently being developed to track lateral displacement during knee flexion, thus providing data on the position of the patella relative to the femur. To determine the forces acting on the patella it is necessary to know the forces acting through the patellar ligament (P_T) and through the quadriceps mechanism (P_Q). Knowing the angles of the tibia to the vertical (θ), the patellar ligament to the tibia (Φ), and axis of the patella to the vertical (ψ) and the angle of the quadriceps action (ε) it is possible, knowing the displacements of the forces and the magnitude of the ground to foot reaction force (R) to calculate the forces on the patella (P_J, P_Q and P_T) by considering equilibrium of the upper tibia and of the patella (Figure 16.6).

To obtain P_T, the equilibrium of the upper tibia as a free body is considered by taking moments about A in Figure 16.6, to give $P_T \cos\theta\, t = R\, d$.

By geometry $d = \ell_1 \sin\theta - (y_c - z)$ where $(y_c - z)$ is the position of ground to foot reaction force relative to the ankle and ℓ_1 is the length of the tibia. If R is the subject weight during static loading ignoring dynamic or inertial forces, then:

$$P_T = [\ell_1 \sin\theta - (y_c - z)]/(t \cos\Phi) \qquad (16.1)$$

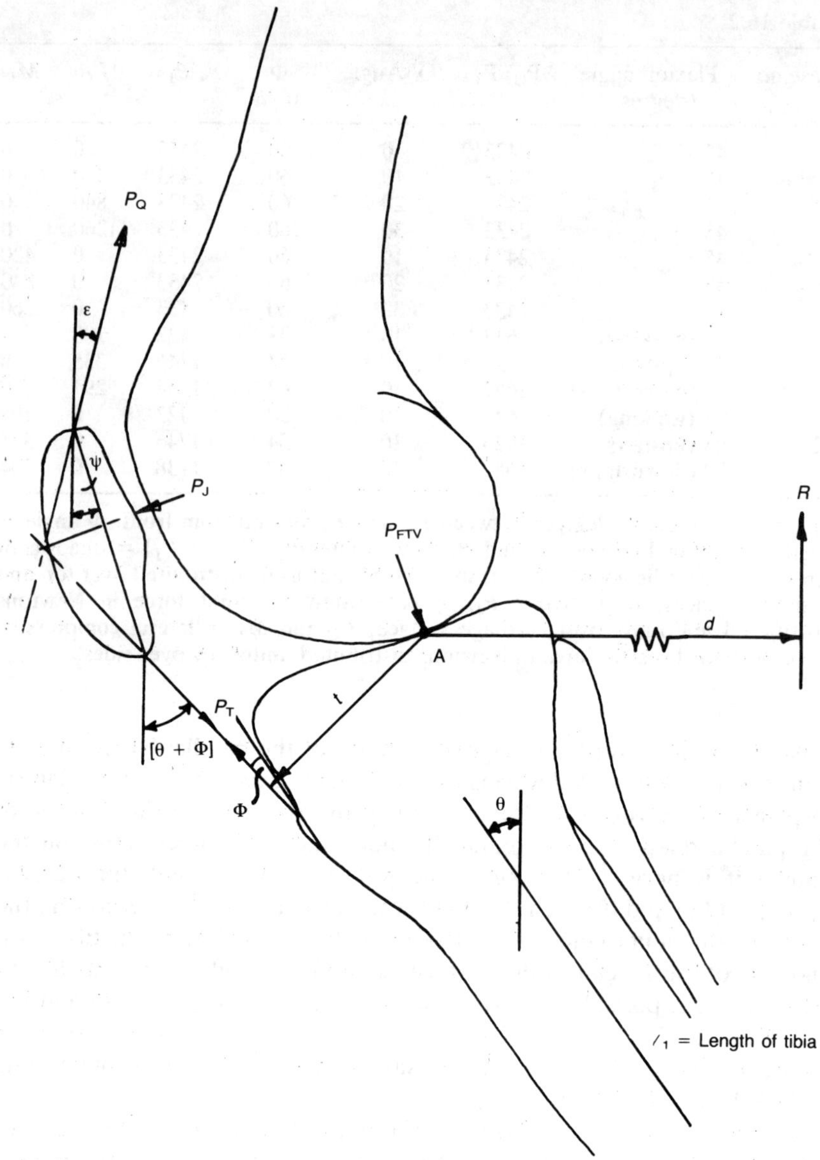

Fig. 16.6 The forces at the knee joint.

Now considering equilibrium of the patella by resolving the forces vertically:

$$P_Q \cdot \cos\varepsilon = P_J \cdot \sin\psi + P_T \cdot \cos(\theta + \Phi) \qquad (16.2)$$

and resolving the forces horizontally,

$$P_J \cdot \cos\Phi = P_Q \sin\varepsilon + P_T \sin(\theta + \Phi) \qquad (16.3)$$

By solving eqns (16.2) and (16.3) for values for P_J and P_Q as a function of P_T one can obtain all the forces on the patella in terms of the patella position in space:

$$P_Q = \{P_T [\cos(\theta + \Phi) + \tan\psi \cdot \sin(\theta + \Phi)]\} / (\cos\varepsilon - \sin\varepsilon \tan\psi) \qquad (16.4)$$

and,

$$P_J = \{P_T[\sin(\theta + \Phi) + \tan\varepsilon \cdot \cos(\theta + \Phi)]\}/(\cos\psi - \sin\psi \tan\varepsilon) \qquad (16.5)$$

By similar methods the vertical resolution of forces on the tibia (P_{FT}) can also be determined.

$$P_{FT} = P_T (\cos\Phi + \cos\theta) \qquad (16.6)$$

Knowing the angles by goniometry, the forces on the patella in certain activities can now be determined for the stress analysis. An example of the variations of these forces during a knees bend is shown in Figure 16.7.

The initial results are shown in Figure 16.8. If the patella followed a course midway between the femur and tibia during a squat, the patella would stay nearly vertical. However, because of the articulation of the patella in the femoral groove and the condylar shape, the patella follows a movement closer to the tibia. If the angle of the patellar tendon is added to that of the tibia it follows closely a rotation similar to the femur but in the other direction. This would suggest that for a squatting activity the loading vectors are similar to those reported by Matthews *et al.* (1977), however it was found that this is not the case for the other activities analysed (Table 16.2). The loadings calculated were all very much dependent on the lever arm of the patellar tendon, and a family of curves can be obtained. As the lever arm is increased the corresponding values of all the loads decrease, related almost to the amount of increase of that lever arm. If the lever arm is doubled, all the loads calculated were approximately halved, this must also produce corresponding increases in other forces occurring within the knee (Denham & Bishop, 1978). As a consequence of changes in loading with lever arm, the contact stresses will also decrease with an increase in patellar tendon elevation, this has been shown experimentally by Ferguson *et al.* (1979) For an average lever arm of 3.7 cm, the variation of loads associated with patella/knee loading are shown in Figure 16.7 for a 'knees

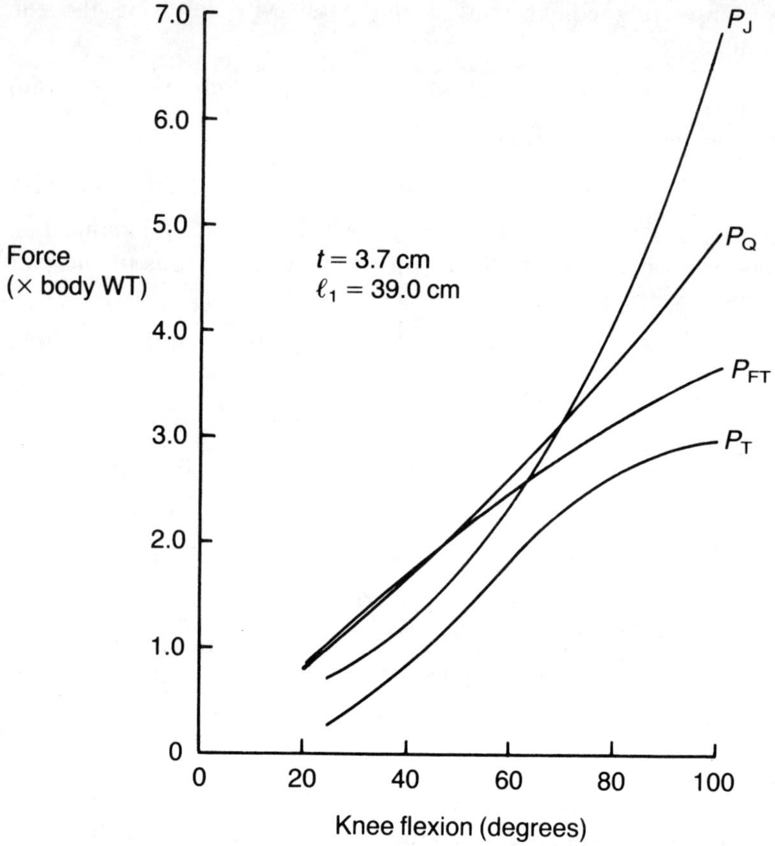

Fig. 16.7 Variation of the forces with knee flexion on the patella during a 'knees bend' squat from a standing position: $t = 3.7$ cm; $\ell_1 = 39.0$ cm.

bend'. The patello-femoral joint force and the quadriceps mechanism tension rise sharply above 45° of knee flexion, whilst the patellar tendon tension and, more interestingly, the tibio-femoral force level off above 70° of knee flexion.

16.4 **Two-dimensional stress analysis**

The PAFEC finite element program was used which gave data on the three principal stresses occurring within all the elements of the model together with maximum shear stresses and principal stress directions. Output was in both numerical and graphical form, this permits actual stress values to be obtained for each element and at any node in the field.

The finite element technique is a numerical method of analysis and

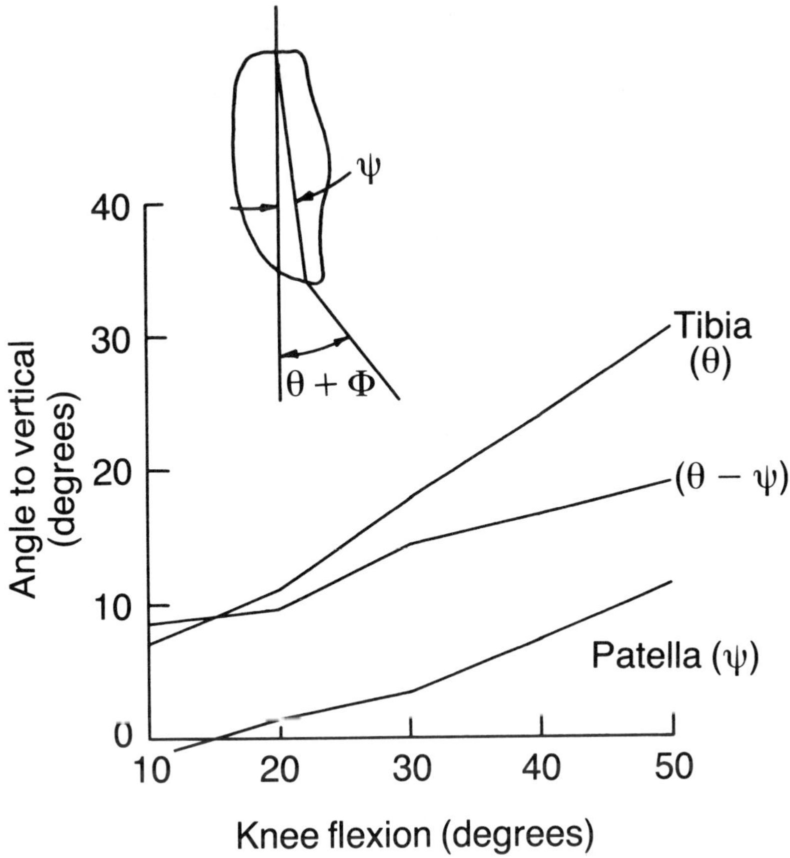

Fig. 16.8 Angular movement of the patella during knee flexion.

Table 16.3

Activity	Vertical ground reaction force	Max P_T	Max P_Q	Max P_J	Max P_{FT}	P_T/P_Q
Level walking (at 85% stance phase)	1.01	2.23	1.66	1.56	2.94	1.34
Climbing stairs	1.30	2.10	3.19	3.65	3.22	0.66
Descending stairs	1.0	5.22	6.94	9.00	5.40	0.75

Values are multiples of bodyweight (approx. 700 N).

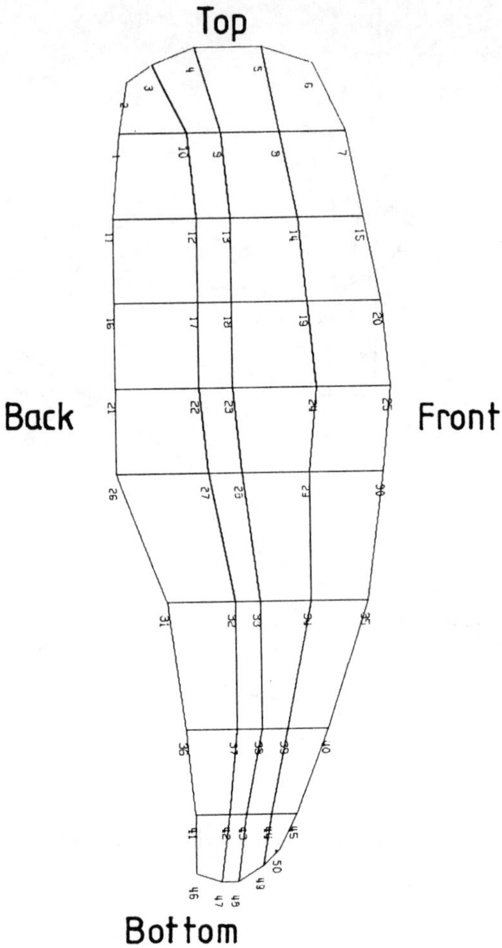

Fig. 16.9 Node numbering scheme used in the PAFEC stress analysis.

modelling, in which the region of interest is divided up into a number of discrete elements (Figure 16.9). In this project the finite element program, PAFEC 75, was used to perform the plane analyses; graphical output was run on the Durham routine, *DURPLOT. Both these packages were available under the Michigan Terminal System (MTS).

A preliminary stress analysis suggested that there exist areas of cartilage subjected to large values of tensile, compressive or shear stress under certain loading conditions (Minns *et al.*, 1979) that could result in lesions similar in appearance to those seen in articular cartilage (Minns *et al.*, 1977), as well as the bony changes predominantly seen within the medial cortex of the patella (Darracott & Vernon-Roberts, 1971). The results from the two-dimensional stress analysis indicate similar values of stress occurring in similar regions with the additional fact that stresses occurring within the patella, when subjected to the loadings calculated previously, can now be visualised. The distribution of the tensile and compressive stresses and their localisation, the patellar displacement, and the Von Mises equivalent stresses for walking and squatting are shown in Figures 16.10 and 16.11. High tensile stresses occur anteriorly and distally and compressive stresses on the articulating surface. From the numerical results the maximum principal tensile stresses are shown in Figure 16.12.

16.5 **Three-dimensional stress analysis**

In the three-dimensional stress analysis, eight-noded isoparametric bricks were used. Other necessary data included the positions, directions and values of the applied loads, the material properties, and any restraint conditions. A technique was used to generate the optimum mesh sizes so reducing the number of input lines.

The computer initially summarised the input data before issuing the displacement and stress results, numerically. The latter provided the following information: the stresses in the global axes (SIGX, SIGY, SIGZ), the principal stresses (SIG1, SIG2, SIG3), and the angles of the first and second principal directions to the global axes (AX to BZ), for every node within each element, for each load case. A rectangular block, 15 × 48 × 48 mm, was used to model the patella (Figure 16.13), 175 nodes in 96 eight-noded isoparametric bricks were used to define the mesh. The model comprised four materials with the properties defined in Table 16.1. In all cases, the maximum principal tensile stresses in the articular cartilage occurred when the medial edge of the patella was restrained by tension, the lower values always occurring when lateral forces were restrained by compression on the lateral facet (Figure 16.14). Activity 2 (climbing upstairs) also produced the maximum values of principal tensile stress in the articular cartilage (Figure 16.15). Compressive stresses within the bone at a 'knees bend' flexion of 45° were largest adjacent to the articular cartilage at the cortex–cartilage interface and areas within the subchondral bone and the front cortex became tensile but lower in value. All values of principal stress rose with the rise in the 'Q' angle and climbing upstairs produced the maximum values in all areas (Figure 16.14).

Clinically, the results suggest that compressive stresses could be reduced if the high lateral retinacular loads were reduced, possibly by a surgical procedure similar to that described by Merchant & Mercer (1974), and that the medially placed tensile stresses that arise could be reduced by correct alignment of the patellar tendon/quadriceps mechanism. However, the

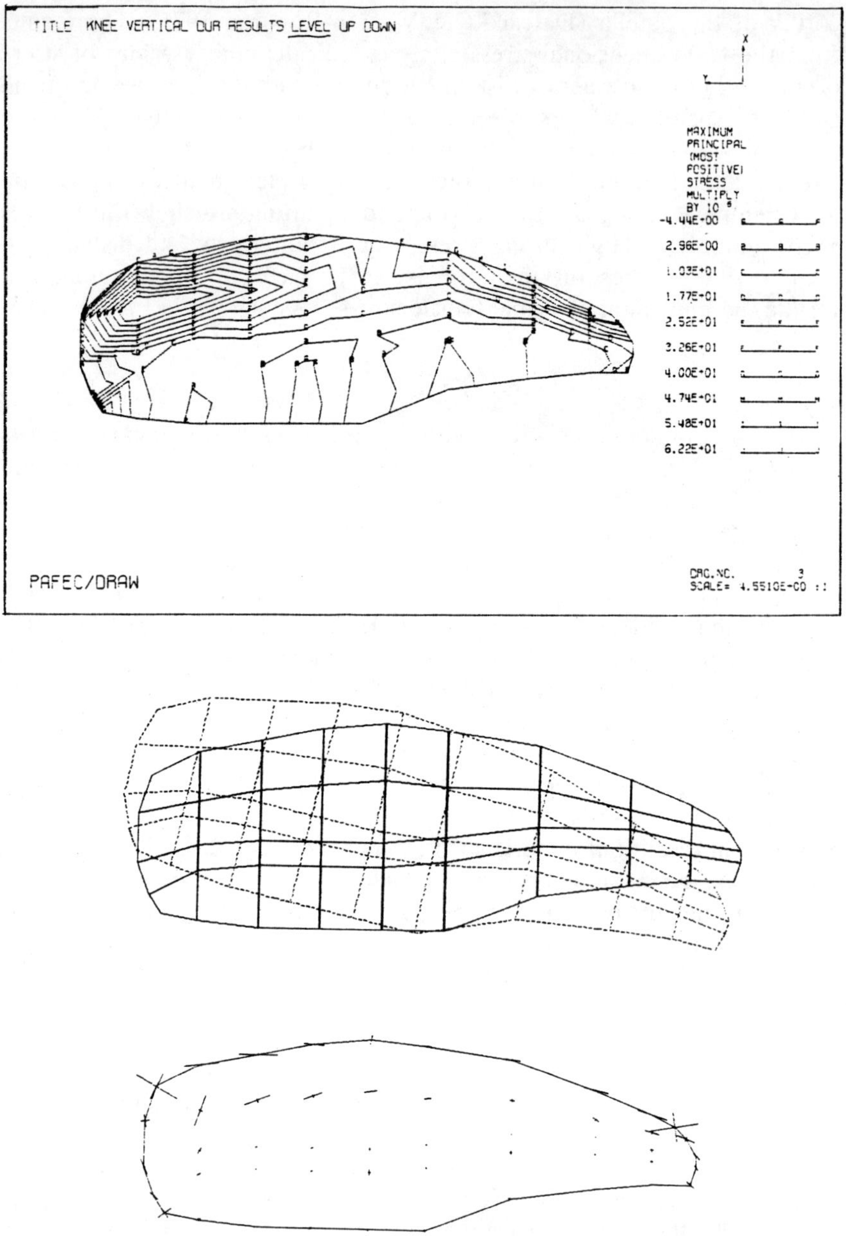

Fig. 16.10 Output of the PAFEC stress analysis with the maximum principal stress contours (top), the gross deflection (middle) and the Von Mises equivalent stress vectors for each node (bottom) for a position of maximum joint force (P_J) during level walking.

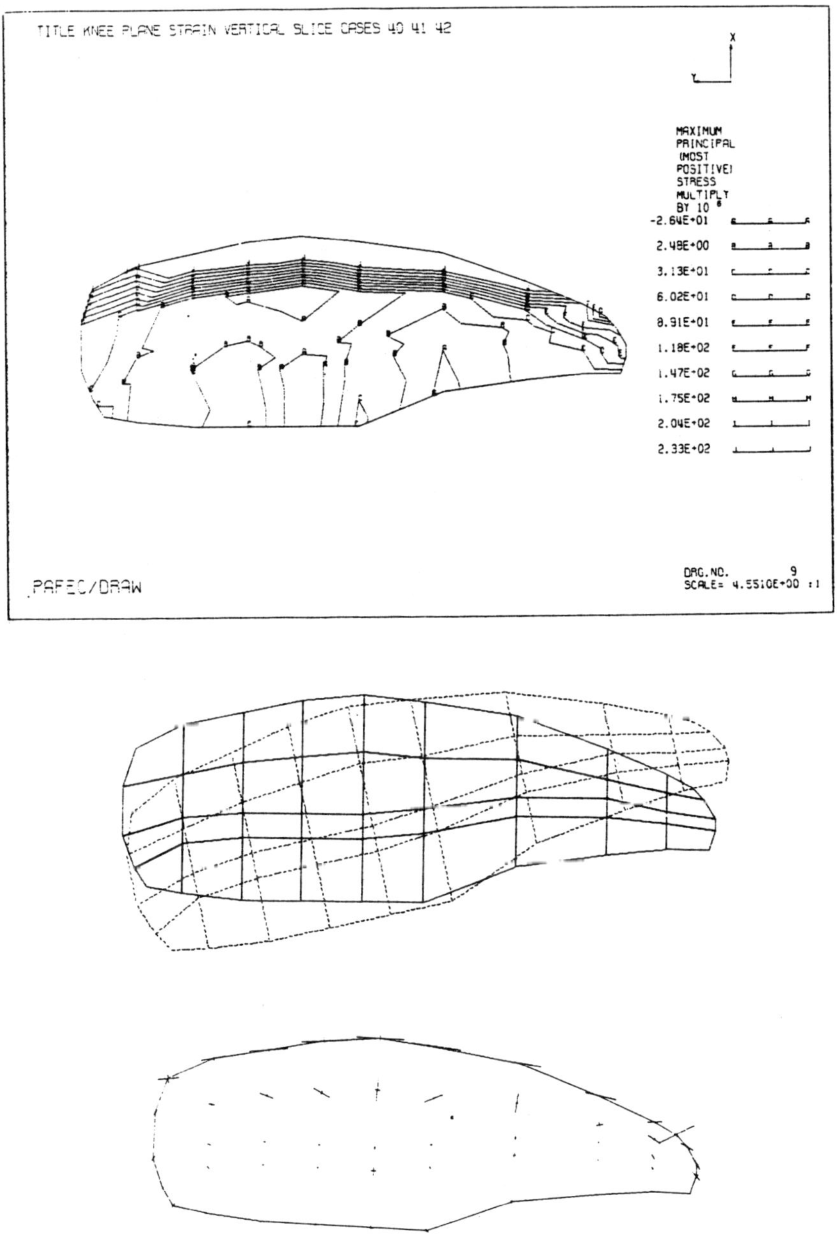

Fig. 16.11 Output of the PAFEC stress analysis at 80° of knee flexion during a knee bend squat.

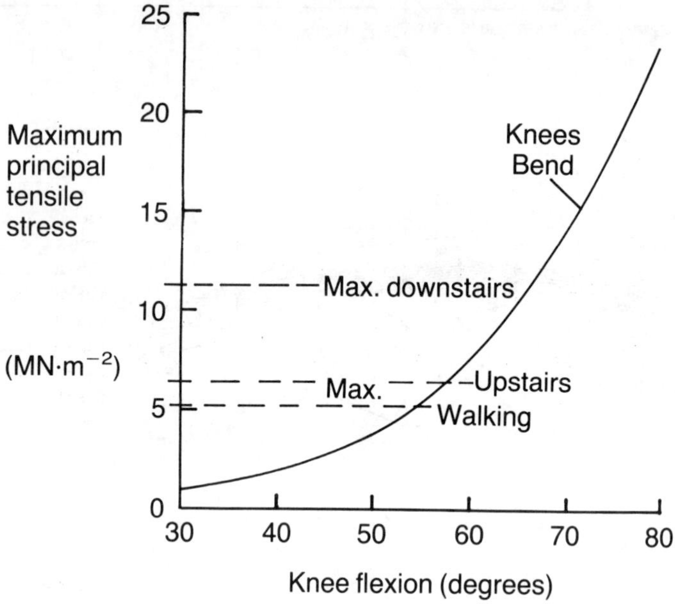

Fig. 16.12 The maximum principal stresses in the patella during the four activities analysed.

alignment must be reduced accurately since a small lateral component of load would still produce harmfully high tensile stresses medially. Mechanically, it appears that one should not remove the articular cartilage as in chondrectomy; this would decrease the patellar tendon lever arm and consequently increase the patella loads. but also subject exposed subchondral bone to higher tensile and shear stresses.

There are many procedures for realignment, the most commonly used being tibial tubercule elevation as well as displacement medially of the insertion of the patellar tendon. It may be contraindicated to use one of these methods for realignment as many results from these realignment procedures have mixed success (Bentley, 1970). Although it is quite clear that there are many syndromes associated with overuse or excessive exercise, one must clearly distinguish that it may be primary idiopathic osteoarthritis or degenerative joint disease, which has a different course of treatment than for a localised isolated compartmental syndrome such as the excessive lateral pressure syndrome described by Ficat & Hungerford (1975). The surgery for excessive lateral pressure syndrome may simply be lateral release of the retinacular structures (Merchant & Mercer, 1974), but for osteoarthritis one may have to extend the surgery to the articular cartilage and bony structures of the patella. Direct intervention surgically, on the articular surface and substructures of the articular surface of the

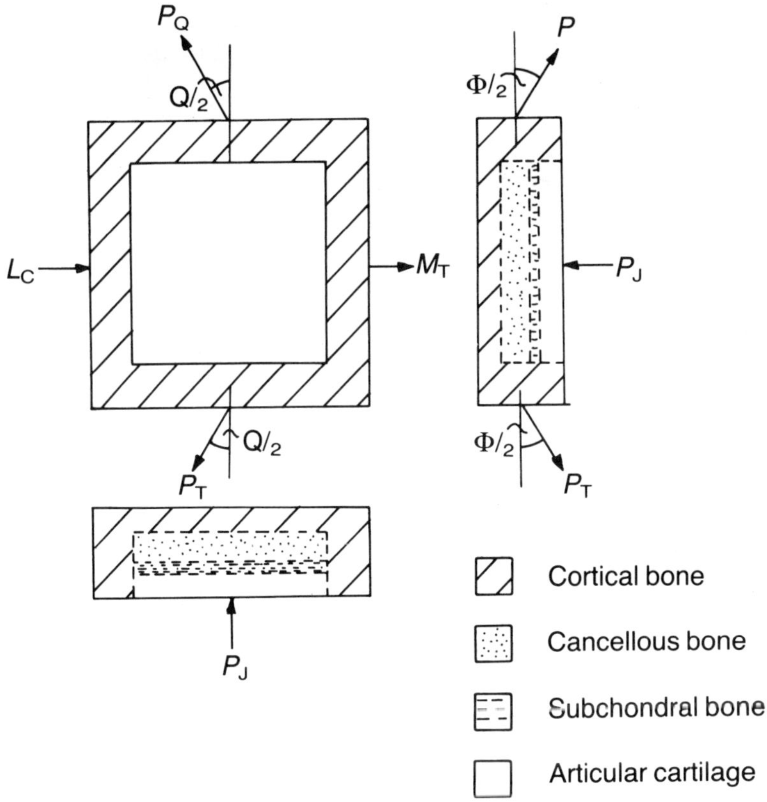

Fig. 16.13 Rectangular block model of the patella and the distribution of the four materials within this block.

patella, may be a much more direct approach to the relief of pain, if one is to believe the cause of retropatellar pain may be basal degeneration of the articular cartilage and excessive loading of the subchrondral bone due to mechanical factors causing microfractures of the trabeculae. Surgical techniques for realignment or lateral release may have a stress relief effect on articular cartilage and consequent pain relief.

A summary of the mechanical conditions that may prevail with the different features of patellar structure is outlined in Figure 16.16, cartilage damage occurring with excessive shear or tensile stresses of the matrix components. Articular cartilage has the ability to repair, especially if the defects or fibrillation seen on the surface are small (Bentley, 1985) and if by arthroscopic examination the degeneration of the articular cartilage is not extensive; conservative treatment includes inactivity and decreasing the load by less repetitive flexion of the knee, which may produce spon-

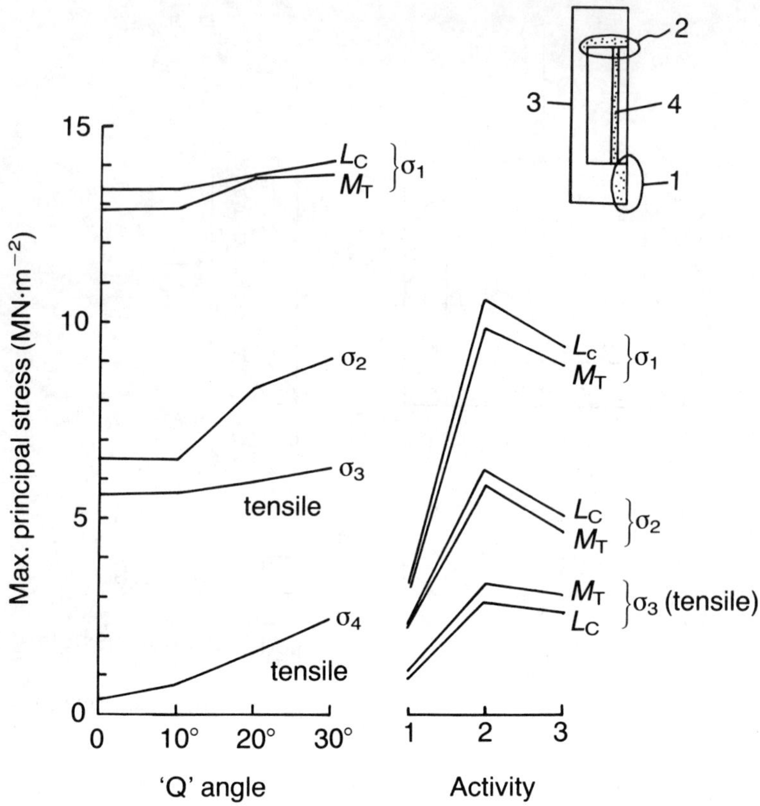

Fig. 16.14 Maximum principal compressive stresses within cortical bone of zone 1 (σ_1) and zone 2 (σ_2) and tensile stresses within zone 3 (σ_3) and subchondral bone (σ_4).

taneous healing and repair of the articular surface. If, by arthroscopic and radiological means the articular cartilage and the subchondral bone are seen to be severely affected, then conservative treatment may not produce any success and one may have to resort to more extensive treatment including surgery to create a new articular surface. Shaving does not appear to produce new articular cartilage, but drilling through to subchondral bone may be one solution, being mechanically strong enough to be sufficient to resist the loads encountered during knee flexion. Prosthetic replacement of the whole of the patella surface has been attempted but problems are usually encountered with the articular cartilage on the femoral patellar groove. One design does incorporate a metallic surface on the femur to articulate with a polyethylene patellar button and this may

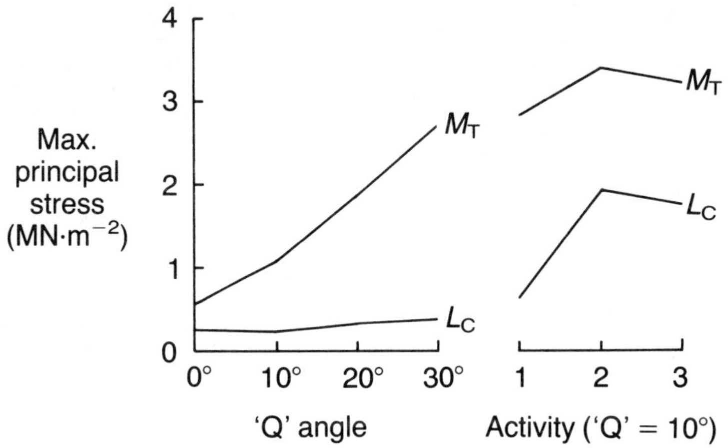

Fig. 16.15 Maximum principal tensile stresses calculated from the stress analysis that occur within the articular cartilage when the patella is restrained by tension of the medial border (M_T), or compression laterally (L_C).

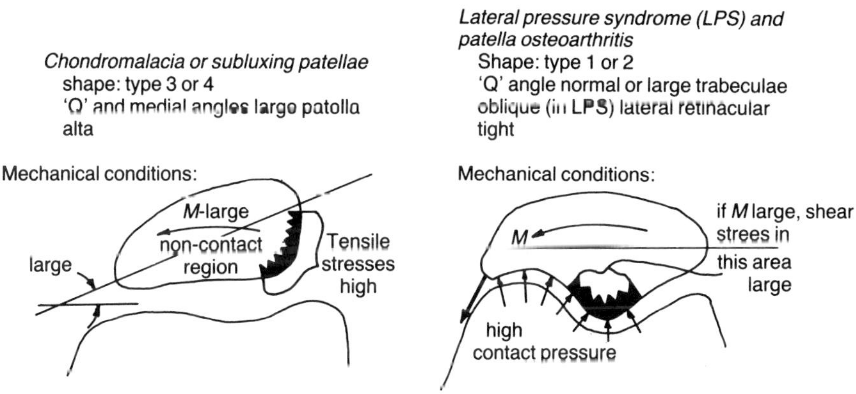

Fig. 16.16 The mechanical and anatomical relationships seen in this study for chondromalacic or subluxing patellae, lateral pressure syndrome and patellar osteoarthritis.

be a solution for advanced osteoarthritis of the patella (Hanslik, 1974). However, care has to be taken to adjust the relationship between the two articulating surfaces such that the pressure on the patello-femoral joint is lowered to prevent loosening and excessive wear of the polyethylene component during articulation.

References

Abernethy, P. J., Townsend, P. R., Rose, R. M. & Radin, E. L. (1978). Is chondromalacia patellae a separate clinical entity? *Journal of Bone and Joint Surgery*, **60B**, 205–10.

Aglietti, P., Insall, J. N. & Cerulli, G. (1983). Patellar pain and incongruence. 1, Measurements of incongruence. *Clinical Orthopaedics and Aclated Research*, **176**, 217–24.

Bandi, W. & Brenwald, J. (1974). The significance of femoropatellar pressure in the pathogenesis and treatment of chondromalacia patellae and femoropatellar arthrosis. In: *The Knee Joint*, ed. O. S. Ingwerson *et al.* Excerpta Medica, Amsterdam.

Baumgartl, F. (1964). *Das Kniegelenk; Erkrankumgun Verletzungun und ihre Behandeing mit Hinweisen fur die Begatachtung,* p. 452. Springer-Verlag, Berlin.

Bentley, G. (1985). Articular cartilage changes in chondromalacia patellae. *Journal of Bone and Joint Surgery*, **67B**, 769–74.

Blackburne, J. S. & Peel, T. E. (1977). A new method of measuring patellar height. *Journal of Bone and Joint Surgery*, **59B**, 241–2.

Crosby, E. B. & Insall, J. (1979). Recurrent dislocation of the patella. Relation of treatment to osteoarthritis. *Journal of Bone and Joint Surgery*, **61A**, 714–9.

Darracott, J. & Vernon-Roberts, B. (1971). The bony changes in 'chondromalacia patellae'. *Rheumatism and Physical Medicine*, **11**, 175–9.

Delgardo-Martins, H. (1979). A study of the position of the patella using computerised tomography. *Journal of Bone and Joint Surgery*, **61B**, 443–4.

Denham, R. A. & Bishop, R. E. D. (1978). Mechanics of the knee and problems in reconstructive surgery. *Journal Bone and Joint Surgery*, **60B**, 345–8.

Emery, I. H. & Meachim, G. (1973). Surface morphology and topography of patello-femoral cartilage fibrillation in Liverpool necropsies. *Journal of Anatomy*, **116**, 103–20.

Ferguson, A. B., Brown, T. D., Fu, F. H. & Rutkowski, R. (1979). Relief of patellofemoral contact stress by anterior displacement of the tibial tubercle. *Journal of Bone and Joint Surgery*, **61A**, 159–65.

Ficat, C. & Maroudas, A. (1975). Cartilage of the patella. Topographical variation of glycosaminoglycan content in normal and fibrillated tissue. *Annuals of Rheumatic Disease*, **34**, 515–9.

Ficat, R. P. & Hungerford, D. S. (1977). *Disorders of the Patello-Femoral Joint*, 1st Edition. Williams and Wilkins, Philadelphia.

Goodfellow, J., Hungerford, D. S. & Zindel, M. (1976a). Patello-femoral joint mechanics and pathology. 1, Functional anatomy of the patello-femoral joint. *Journal of Bone and Joint Surgery*, **58B**, 287–90.

Goodfellow, J., Hungerford, D. S. & Woods, G. (1976b). Patello-femoral joint mechanics and pathology. 2, Chondromalacia patellae. *Journal of Bone and Joint Surgery*, **58B**, 291–6.

Haasters, J. (1974). Functional analysis of the spongiosa structure of the human patella. In: *The Knee Joint*, ed. O. S. Ingwerson *et al.* Excerpta Medica, Amsterdam.

Hanslik, L. (1974). A preliminary report on the use of the McKeever patellar prosthesis combined with high tibial osteotomy. In: *The Knee Joint*, ed. O. S. Ingwerson *et al.*, Excerpta Medica, Amsterdam.

Hori, R. Y. & Mockros, L. F. (1976). Indentation tests of human articular cartilage. *Journal of Biomechanics*, **9**, 259–68.

Hvid, I., Anderson, L. I. & Schmidt, H. (1983). Patellar height and femoral trochlear development. *Acta Orthopaedica Scandinavica*, **54**, 91–3.

Insall, J. (1982). Current concepts review. Patellar pain. *Journal of Bone and Joint Surgery*, **64A**, 147–52.
Insall, J. & Salvati, T. (1971). Patellar position in the normal knee joint. *Diagnostic Radiology*, **101**, 101–4.
Insall, J., Falvo, K. A. & Wise, D. W. (1976). Chondromalacia patellae. A prospective study. *Journal of Bone and Joint Surgery*, **58A**, 1–11.
Marks, K. E. & Bentley, G. (1978). Patellar alta and chondromalacia. *Journal of Bone and Joint Surgery*, **60B**, 71–3.
Matthews, L. S., Sonstegard, D. A. & Henke, J. A. (1977). Load bearing characteristics of the patello-femoral joint *Acta Orthopaedica Scandinavica*, **48**, 511–7.
McElhaney, J. H., Alem, N. & Roberts, V. (1970). A porous block model for cancellous bone. ASME Publ 70-WA/BHF-2.
Meachim, G., Bentley, G. & Baker, R. (1977). Effect of age on thickness of adult patellar articular cartilage *Annals of Rheumatic Disease*, **36**, 563–8.
Merchant, A. C. & Mercer, R. L. (1974). Lateral release of the patella. A preliminary report. *Clinical Orthopaedics and Related Research*, **103**, 40–6.
Minns, R. J. & Braiden, P. M. (1980) A loading and stress analysis of the patella. In: *Mechanical Factors and the Skeleton*, ed. I. A. F. Stokes. John Libbey & Co., London.
Minns, R. J., Birnie, A. J. M. & Abernethy, P. J. (1979). A stress analysis of the patella and how it relates to patellar articular cartilage lesions. *Journal of Biomechanics*, **12**, 699–708.
Minns, R. J., Steven, F. S. & Hardinge, K. (1977). Osteoarthrotic articular cartilage lesions of the femoral head observed in the scanning electron microscope. *Journal of Pathology*, **122**, 63–70.
Minns, R. J., Walsh, W. K. and Clark, J. A. (1989). Techniques for measuring the static and dynamic properties of the patella. *J. Biomedical Engineering*. (In press.)
Outerbridge, R. E. (1961). The etiology of chondromalacia patella. *Journal of Bone and Joint Surgery*, **43B**, 752–6.
Reider, B., Marshall, J. L. & Ring, B. (1981). Patellar tracking. *Clin. Orthop.* **157**, 143–8.
Reilly, D. T. and Burstein, A. H. (1974). The mechanical properties of cortical bone. *Journal of Bone and Joint Surgery*, **56A**, 1001–18.
Stougard, J. (1975). Chondromalacia of the patella – incidence, macroscopical and radiological findings at autopsy. *Acta orthopaedica Scandinavica*, **46**, 809–22.
Townsend, P. R., Raux, P., Rose, R. M., Miegel, R. E. & Radin, E. L. (1975). The distribution and anistropy of the stiffness of cancellous bone in the human patella. *Journal of Biomechamics*, **8**, 363–7.
Townsend, P. R., Rose, R. M., Radin, E. L. & Raux, P. (1977). The biomechanics of the human patella and its implications for chondromalacia. *Journal of Biomechanics*, **10**, 403–11.
Veress, S. A., Lippert III, F. G., Hou, M. C. Y. & Takamoto, T. (1979). Patellar tracking patterns measurement by analytical X-ray photogrammetry. *Journal of Biomechanics*, **12**, 639–47.
Wiberg, G. (1941). Roentgenographic and anatomic studies on the femoropatellar joint with special reference to chondromalacia patellae. *Acta Orthopaedica Scandinavice*, **12**, 319–44.
Wiles, P., Andrews, P. S. & Devas, M. B. (1956). Chondromalacia of the patella. *Journal of Bone and Joint Surgery*, **38B**, 95–113.

17

The distribution of stresses in the human lumbar spine

H. S. Ranu

17.1 **Introduction**

Stress analysis of the human spine is useful for a number of clinically relevant situations for example: in designing and improvising prostheses for spinal fixation, development of new materials for replacement of the disc or the vertebral bone, and to refine the finite element modelling of the spinal segments. There is also a need to develop new techniques to predict and understand fracture of the vertebra. Therefore, quantifying the stresses in different regions of the human spine is important in order to prevent spinal injuries. In this chapter the non-linear and time-dependent response of the disc and the vertebra to loading is reported.

17.2 **Description of an intradiscal pressure transducer (IDPT) and an intervertebral pressure transducer (IVPT)**

A needle tipped with an ultra miniature pressure transducer (diameter 2.03 mm) was developed in order to measure the intradiscal pressure (IDPT). The calibration of the needle was carried out using a dead-weight tester. It was found to be linear for both loading and unloading up to a pressure of 2760 kN·m^{-2}. It was capable of measuring intradiscal pressures up to 3450 kN·m^{-2}.

In order to map the pressure distribution under the annulus fibrosus of a spinal disc and its adjacent endplates, an intervertebral pressure transducer (IVPT) was developed and fabricated. It was a diaphragm-type pressure transducer with a slender box section. The intervertebral pressure transducer was calibrated as described by Ranu *et al.* (1979) and its response was linear up to a pressure of 4830 kN·m^{-2}. The IVPTs and their holder (a partial ring) have unique features controlling depth of penetration under the annulus to permit attachment of multiple units to the spinal

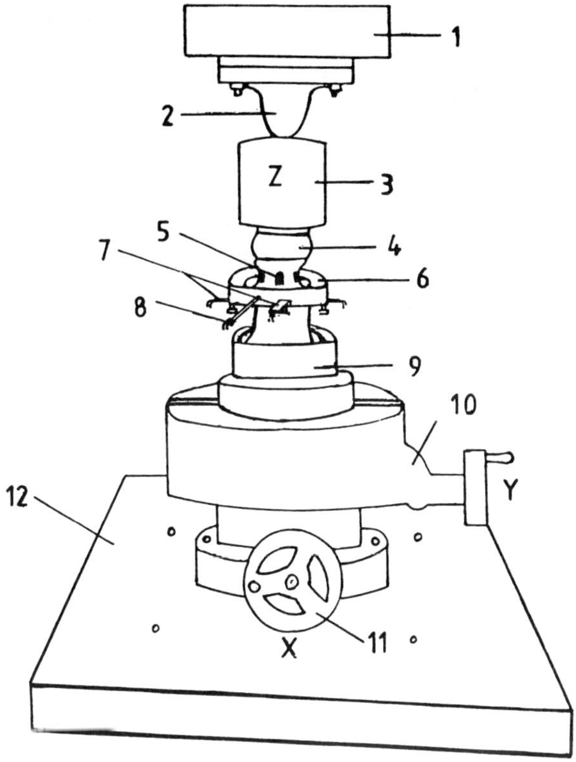

Fig. 17.1 Schematic arrangement of the experimental test set up: (1), cross head of universal testing machine; (2), special loading head; (3), upper vertebral body holder; (4), specimen (spinal segments); (5), strain gauges on vertebral body; (6), partial ring holder for IVPTs; (7), intervertebral pressure transducers (IVPTs); (8), intradiscal pressure transducer (IDPT); (9), bottom vertebral body holder; (10), moment applicator in lateral direction (y); (11), moment applicator in anterio-posterior direction (x); (12), six axis load cell attached to the base of the universal testing machine.

segment. The holder was also equipped with positioning screws which held it firmly to a vertebral body during the loading cycle. Figure 17.1 shows the schematic test set up in which three IVPTs were used. The radial movement of the IVPTs was prevented by a set of knurled screws on the bottom of the partial ring as shown in Figure 17.1.

17.3 **Specimen selection and preparation**

Sections of spine were obtained from unembalmed human bodies within a short period after death. Their suitability for experimental use was determined by taking radiographs. After this, the lumbar sections of the spine were stripped of all muscles making sure that none of the ligaments, bone and discs were damaged.

Preparation of the segments began by moulding the top and bottom ends of the segment in an alloy (Ostalloy; m.p.t. = 47.3°C), thus providing an adequate load-bearing surface for the specimen. However, during the moulding process two conditions were satisfied:

(1) the natural curvature of the segment was kept the same as *in situ;*

(2) The top and bottom surfaces of the fixture were kept parallel so that a normal load could be applied to the upper surface.

The disc was carefully dissected and the IVPTs were inserted to a controlled depth of penetration. Inserting the IVPT caused minimal amount of change in the disc properties because of the small size of the transducers. The whole assembly was kept moist with normal saline solution and was covered by a plastic sheet before testing.

In order to implant the instrumented needle into a disc a specially developed probe was used. This guided the initial insertion of the needle and also helped to control the depth of penetration of the needle into the nucleus pulposus. Strain gauges were mounted on the anterior, lateral right, and lateral left sites of the vertebra. They were prepared as described by Ranu & King (1980).

17.4 ***In vitro* testing procedure**

A quasi-static compressive load was applied on the specimen using a Riehle testing machine via a specially designed loading head which could control the eccentricity of the applied load. In order to establish the line of action and the eccentricity of the applied load, a three-dimensional loading platform equipped with a six-axis load cell was incorporated into the experimental test set up (Figure 17.1). The eccentric loading of the segment was carried out by moving the platform in controlled increments along the two orthogonal axes (x and y). The rate of loading used was approximately $0.13\,mm{\cdot}s^{-1}$. The entire segment was loaded to just over 2.0 kN. During each test, the following data were recorded:

(1) intradiscal pressure as a function of time;

(2) intervertebral pressure as a function of time for anterior and lateral edges of the vertebra at a depth of 3.175 mm;

(3) strain as a function of time for the anterior, lateral right and lateral left sites of the vertebra;

(4) applied compressive load as a function of time;

(5) stress–relaxation of the entire segment as a function of time.

The tests were also carried out on another set of specimens to the point of bony fracture of the vertebra. The following data were recorded for these tests:

(6) intradiscal pressure as a function of load, intervertebral pressure as a function of load for the anterior and lateral edges of the vertebra and at two depths of approximately 3.175 and 6.35 mm, these depths were insufficient to penetrate into the nucleus;

(7) strain as a function of load was also obtained for the anterior, lateral right and lateral left sites of the vertebra.

After testing, the discs were sectioned transversely through the plane of the disc and their degree of degeneration was assessed. The scheme used by Galante (1967) was adopted for classification purposes: Group I (no degeneration) to Group IV (severely degenerated). During this procedure it was observed that this classification scheme could be further improved. Therefore, if the disc degeneration did not exactly match with Galante's classification, a further positive or negative sign was assigned to the graded discs.

17.5 **Results**

Typical results from several cadavera are presented to illustrate the loading behaviour of the lumbar spinal segment. Figures 17.2 to 17.5 show the variation of load, intradiscal pressure, strains, intervertebral pressures and stress-relaxation with time for a 32-year-old male cadaver having no disc degeneration (Group I) when loaded to a compressive load of over 2.0 kN. Figures 17.2 to 17.4 all show the same response to loading and to stress–relaxation. However, annulus pressure (Figure 17.5) does not respond to stress–relaxation. This is due to the functional aspect of the annulus, i.e. it distributes the pressures evenly in all directions when the non-degenerated nucleus of the intervertebral disc is loaded in compression. It is further observed that the annulus only responds to stress–relaxation in the direction in which bulging of the disc takes place. Annulus fibres also respond to stress–relaxation.

Stress–relaxation results (Figures 17.2 to 17.4) show that in 96s, load relaxed by 736 N (36%), nucleus pressure by 0.42 $MN \cdot m^{-2}$ (44%), and average strain by 0.23 $mm \cdot M^{-1}$ (28%). These specimens were allowed to relax for more than 30 min. In figures 17.2 to 17.4, stress–relaxation of

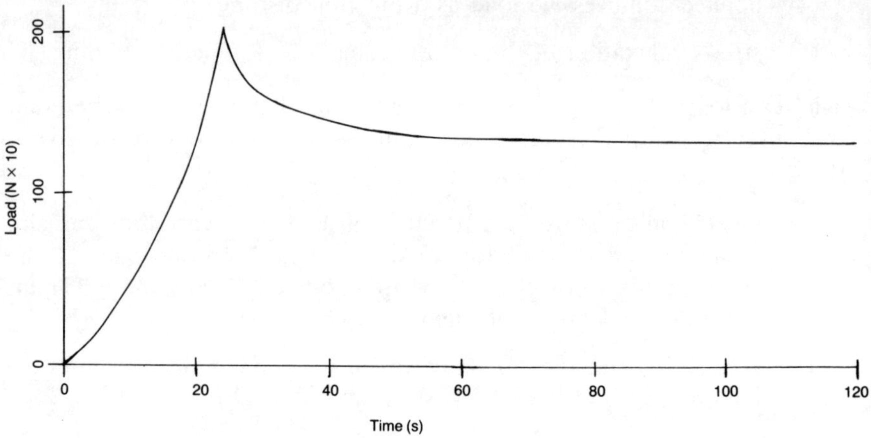

Fig. 17.2 Load *versus* time curve for a normal disc.

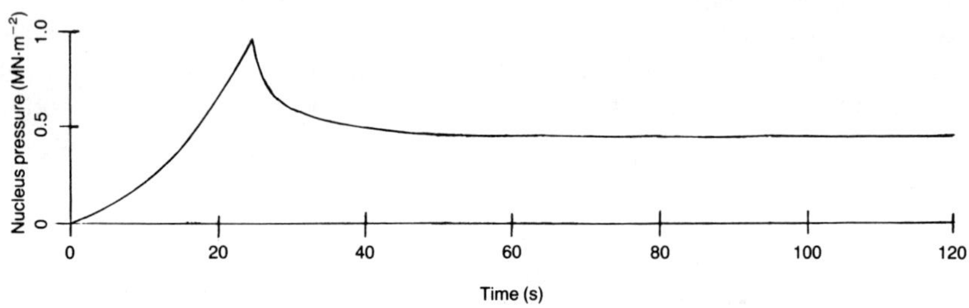

Fig. 17.3 Nucleus pressure *versus* time curve for a normal disc.

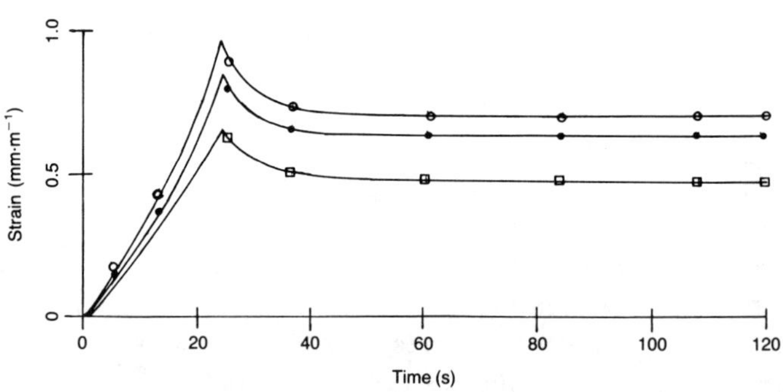

Fig. 17.4 Strain *versus* time curves for (○) lateral left, (●) lateral right and (□) anterior sites of the vertebra.

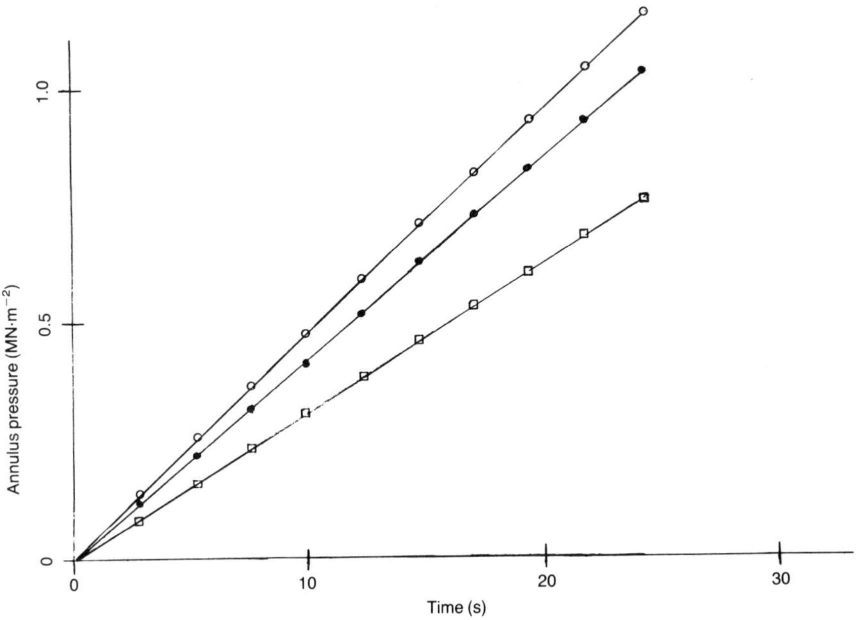

Fig. 17.5 Annulus fibrosus pressure *versus* time curve: (○) IV3 lateral left; (●) IV5 lateral right; (□) IV7 anterior.

only up to 120 s is recorded as there was no change in this phenomenon after this time period. It is further suggested that this stress–relaxation phenomenon can be used to quantify the degree of degeneration of the spinal disc.

Figure 17.6 illustrates the pressures in the nucleus and different regions of the annulus up to a fracture load of 3.3 kN for a 67-year-old male cadaver having disc degeneration of +II. The pressures are linear but non-linearity at extreme loads can also be seen. The pressure in the nucleus is somewhat less than the average annulus pressure. Figure 17.7 shows that the vertebral strains are linear up to a fracture load of 3.3 kN. Figure 17.8 shows the change in nucleus pressure against the annulus pressure to the point of fracture. Again a linear relationship exists, but non-linearity near and up to point of fracture can be seen.

It was noted that at the point of fracture bone marrow oozed out of the vertebral bodies and there was a cracking noise at the moment of yielding. This also resulted in blood discharging from the vertebral body surface pores. However, for normal discs there was no failure of the annulus and the majority of the fractures occurred in the endplate. After fracture

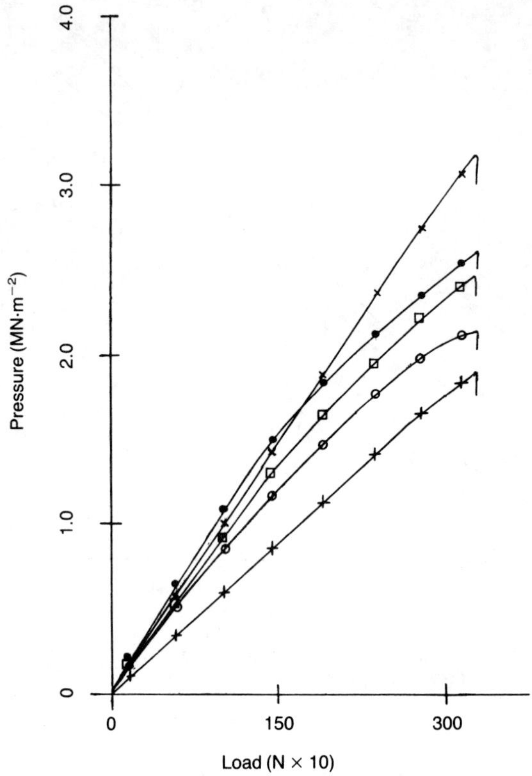

Fig. 17.6 Nucleus and annulus pressures *versus* load to the point of fracture: (○) IDPT; (×) IV1 lateral left; (+) IV2 anterior; (●) IV3 lateral right; (□) average IVPT.

of the vertebral bodies, there was no evidence of stress–relaxation of the spinal segments.

Figure 17.9 illustrates the variation of nucleus pressure against the average annulus pressure at two depths of penetration (3.175 and 6.33 mm) and for compressive loads of 1.3 kN and 3.95 kN from a 25-year-old male cadaver having no disc degeneration (Group I). Again it can be seen that the pressures are linear. However, at 6.33 mm depth of penetration and at a load of 3.95 kN, the tendency is for the relationship to be non-linear.

17.6 Conclusions

(1) The response of the normal and young lumbar spine to loading is viscoelastic.

(2) The normal disc and the vertebral body show a significant amount of stress–relaxation.

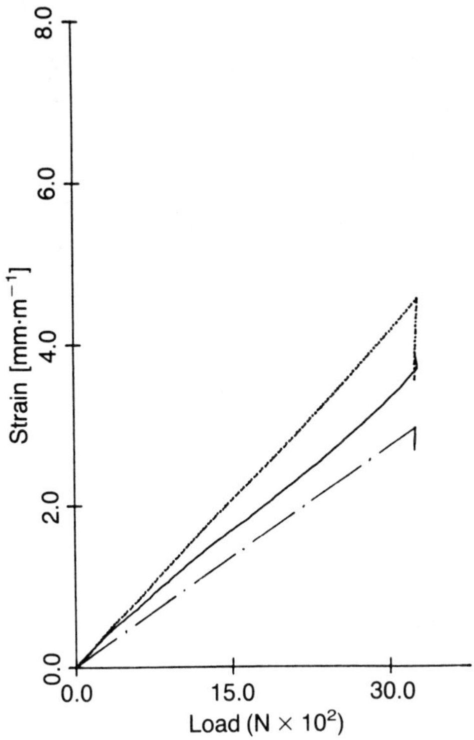

Fig. 17.7 Strain *versus* load to the point of fracture: (——) strain gange lateral right; (···) lateral left; (—·—) anterior.

(3) The nucleus and the annulus pressures have a linear relationship with the applied compressive load. Somewhat higher pressures are developed under the annulus at higher depth of penetration and at the point of bony fracture the nucleus and the annulus pressures are not linearly related to the load. Nachemson (1960) postulated that pressure in the nucleus is three times the annulus pressure. However, the findings here suggest that there is almost a one-to-one relationship between the annulus and the nucleus pressures.

(4) Normal discs failed through the vertebral body endplate and no rupture of the annulus was observed.

(5) After fracture of the vertebral bodies, no stress–relaxation of the spinal segment was observed.

Acknowledgements

The assistance of Dr A. I. King and Dr Harpal Kaur Ranu is fully acknowledged.

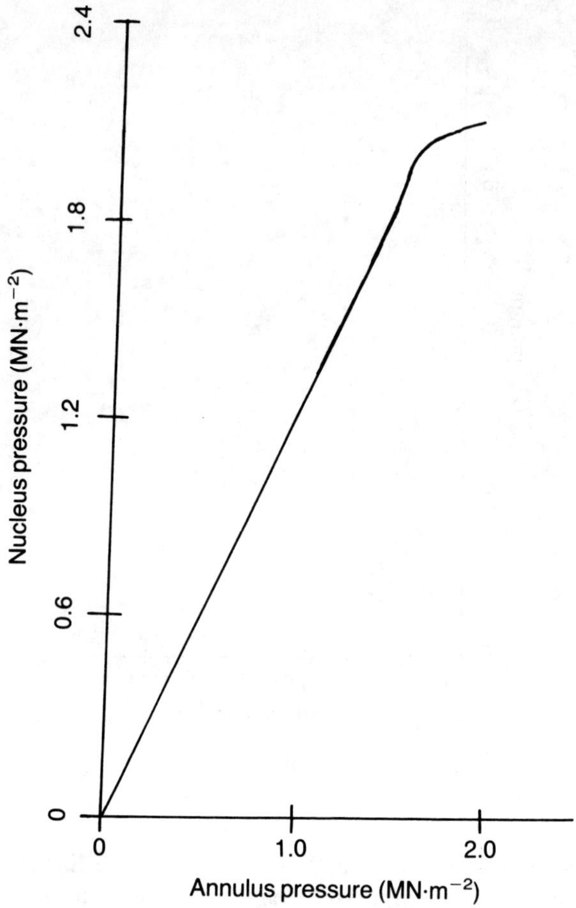

Fig. 17.8 Nucleus pressure *versus* annulus pressure in a spinal disc to the point of fracture.

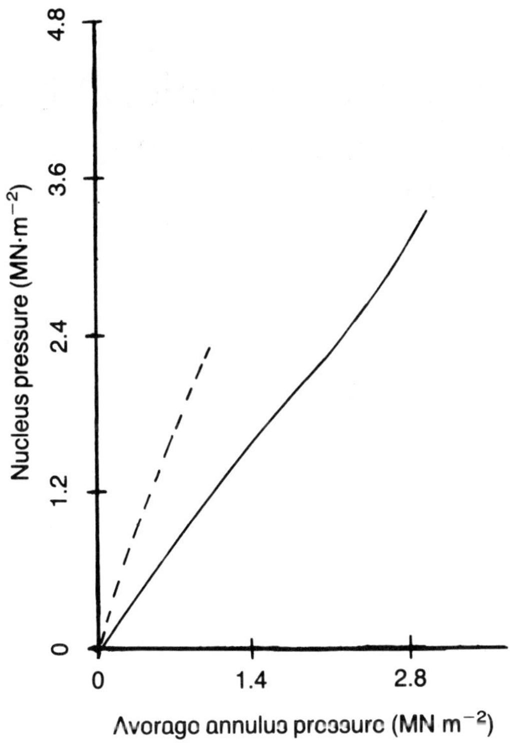

Fig. 17.9 Nucleus pressure *versus* annulus pressure at depths of (——) 6.330 mm and (---) 3.175 mm.

References

Galante, J. O. (1967). Tensile properties of the human lumbar annulus fibrosus. *Acta Orthopaedica Scandinavica*, Suppl. 100.

Nachemson, A. (1960). Lumbar intradiscal pressure. *Acta Orthopaedica Scandinavica*, Suppl. 43.

Ranu, H. S., Denton, R. A. & King, A. I. (1979). Pressure distribution under an intervertebral disc – an experimental study. *Journal of Biomechanics*, **12**, 807–12.

Ranu, H. S. & King, A. I. (1980). Correlaton of intradiscal pressure with vertebral endplate pressure. In: *Engineering Aspects of the Spine*, pp. 37–42. Institution of Mechanical Engineers, London.

18

Stress–strain characteristics of flat bones in relation to non-penetrating injuries of the chest and head

D. E. M. Taylor, J. S. Whamond, G. J. Cooper, V. A. Evans and R. L. Maynard

18.1 **Introduction**

Non-penetrating (blunt) injuries of the chest and the head are major causes of death, prolonged hospital treatment and permanent disability, particularly in road traffic accidents despite the gratifiable reductions following legislation on motor-cyclist helmets and car seat-belts. The latest figures showed that in England and Wales in 1985 there were about 3000 and 9000 deaths due primarily to chest and to head injuries, respectively (HMSO, 1986). In addition to fractures of the ribs and the skull, secondary injuries can present major clinical problems. Common lesions and complications of chest injury include lung contusion, cardiac contusion and rupture of the aortic arch, while those of head injury include extradural and subdural haemorrhages, prolonged unconsciousness, permanent paralysis, speech defect and personality change. Despite considerable discussion on the biomechanical basis of these injuries, experimental observations have been mainly qualitative; there is little in the way of quantitative data on the distribution, propagation and time course of the stresses and resultant strains involved. The chest and head have anatomical features in common: both cavities contain vital organs, enclosed and partly protected by bony walls. Two principal mechanisms might cause injury remote from the site of impact: shear stress due to tissue deformation and a propagated stress wave. Although factors determining the propagation and attenuation of deformation and of stress waves differ, both depend primarily on the magnitude and rate of deformation of the bony wall on blunt impact.

Work carried out by major car manufacturers, particularly in the United States, has used principally manikins or cadavers, with extrapolation to the living subject: only limited studies have been conducted using anaesthetised animals. Whether manikins or animals are used, the relevance of extrapolation to human subjects depends on firm evidence of mechanical equivalence. Anatomically the thorax of man and pig differ from each other in general geometry and in the orientation of the ribs (Figure 18.1), while the mechanical properties of the flat bones in both species are ill defined. Viano (1978) and Viano & Warner (1976) studied impact force and acceleration in anaesthetised pigs, but did not fully quantify either strain distribution or its time course. Similarly, the studies of chest distortion on mid-sternal impact in the dog (Roberts *et al.*, 1967), and in Macaque monkeys (Shatsky *et al.*, 1974), were purely qualitative. It has been shown in earlier studies on anaesthetised pigs (Cooper *et al.*, 1982, 1984) that a blunt injury to the chest can result in the antero-posterior diameter being reduced by more than 30% in about 3 ms and in a rapidly propagated stress wave of over 150 $kN{\cdot}m^{-2}$ occuring within 3 ms in the lungs, and there are indications that a similar stress wave occurs in heart chambers and large blood vessels. Shatsky *et al.* (1974) also studied head movements in monkeys; but again in a predominantly qualitative manner. Head injuries are the most difficult for which to derive experimental *in vivo* data, for there is no suitable animal with skull geometry and dimensions analogous to man. The closest geometry is that of much smaller laboratory primate skulls, but available mechanical data are insufficient to establish a valid dynamic equivalence.

To translate these qualitative and preliminary quantitative observations into an understanding of the mechanism of injury it was necessary first to determine the stress – strain characteristics of the various structures involved. The biomechanics of long limb bones has been studied extensively in many species (Yamada, 1970); similar data were not available for flat bones, such as the ribs, sternum and cranium. It was necessary, therefore, to study both the stresses and strains produced by experimental trauma using deeply anaesthetised animals, and to determine the biomechanical properties of the flat bones concerned, before stress analysis could be attempted. For head injury, only preliminary studies have been completed. Because of the absence of an adequate animal model, conclusions needed to be based on isolated bone biomechanical studies and a combination of extrapolation from the results with experimental chest trauma plus an *ex vivo* model using dried human skulls.

The development and validation of mathematical models of the biomechanics of blunt chest and head injuries has involved three parallel investigations:

(1) quantification of impact kinetics and the consequent strains, including their time course, and injuries to the chest wall, lungs and cardiovascular system following blunt trauma to the chest in the anaesthetised pig;

(2) determination of the dynamic stress–strain characteristics of the

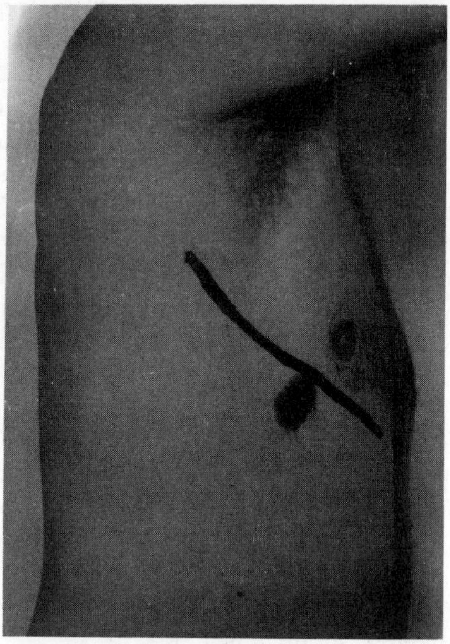

a (man)

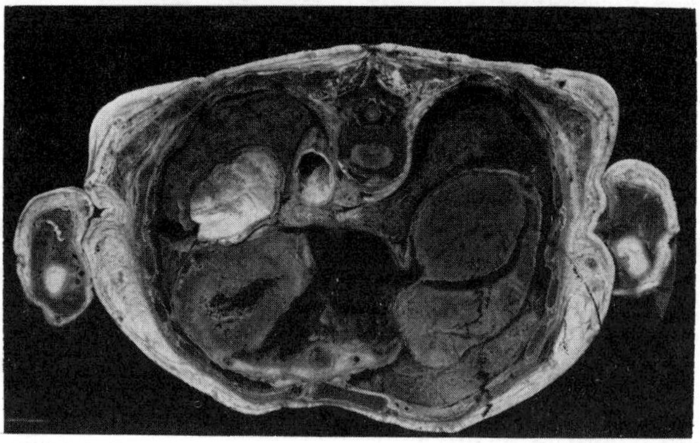

b (man)

Fig. 18.1 Thoracic geometry in the pig and in man. (a) Lateral view with position of 5th rib marked. Note the angulation to the transverse plane in man (left) as compared to the pig (right). (b) Thoracic cross-section at the level of the 6th thoracic vertebrum. Note the difference in eccentricity and that the major axis is lateral in man (left) as compared to antero-posterior in the pig (right).

ribs and sternum in the pig and in man, and of the bones of the human skull;

(3) derivation of a mathematical model suitable for detailed stress analysis studies and capable of being used in a predictive manner.

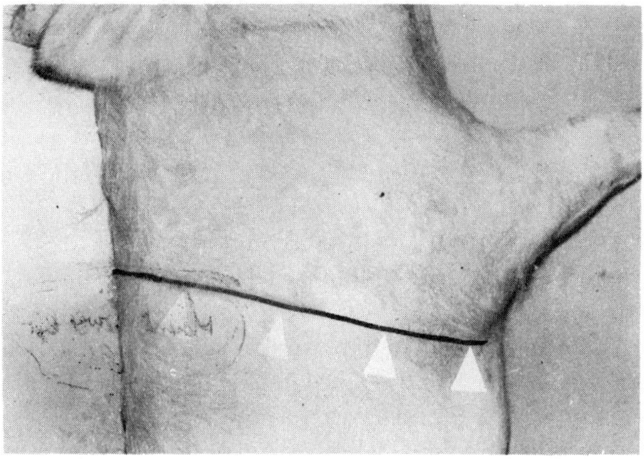

a

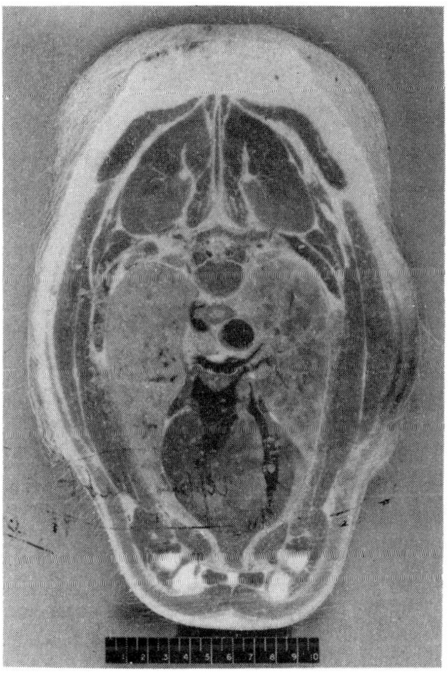

b

18.2 **Strain and impact kinetics**

The response of the thoracic wall to non-penetrating impact loading was determined following the impact of free-flying projectiles upon anaesthetised pigs (body weight (W) 22–74 kg). Five projectiles, each of impact diameter of 37 mm, were used; their mass (M) being 3.0, 1.0, 0.38, 0.14 or 0.069 kg. The projectile was fired by compressed air to give an impact velocity of 5.8–81.2 m·s^{-1}. Details of the experimental methods have been reported previously (Cooper *et al.*, 1982), together with the response to anterior impact and some preliminary lateral impacts. This has been extended in this study by a series of right thoracic impacts at the level of the xiphisternum. The previous studies (Cooper *et al.*, 1982, 1984) were concerned principally with cardiovascular injuries, while those reported in this chapter concentrated on the biomechanics of pulmonary contusion. The distortion of the thoracic wall at the point of impact (Y') was acquired as a function of time by a polynomial fit of displacement/time data digitised from high-speed cinephotography (up to 12 000 prints·s^{-1} (Cooper *et al.*, 1982)). The duration of the dynamic compressive loading of the thorax was taken as the time after contact that the velocity of the impactor was reduced by 95% of its pre-impact value ($t_{0.95}$).

Distortion of the thoracic wall was expressed as the decimal fractional reduction of the diameter in the axis of the applied force ($Y'' = Y'/a$). Y''_{max} and $t_{0.95}$ were the principal determinants of the severity of shear-determined internal injury, while the severity of stress propagation injuries was determined principally by impact velocity (V). The term $t_{0.95}$ was related to the mass of the impactor and to the weight of the animal, but was independent of the impact velocity. Increasing the mass of the impactor extended the duration of dynamic compressive strain, while in larger animals $t_{0.95}$ was slightly reduced. The relationship was:

$$t_{0.95} = 141.3 \times (M/W)^{0.711} \tag{18.1}$$

The lightest impactor (0.069 kg) had a $t_{0.95}$ of 1.0–1.6 ms, while with the heaviest impactor (3.0 kg) it was 16–22 ms. There was negligible rebound velocity of the impactor and the body wall exhibited a highly damped response; subsequent analysis of the deflection/time data showed that viscoelastic elements would be required to model the data adequately. A similar conclusion was reached from the tensiometric studies.

Y''_{max} could be related to pre-impact momentum ($M \times V$) and body weight (W) (Figure 18.2). Over the observed range of impact momentum (3.01 to 36.0 kg·m·s^{-1}) the relationship was:

$$Y''_{max} = 0.686 \times ((M \times V)/W)^{0.593} \tag{18.2}$$

Fracture of the rib under the point of contact occurred with most impacts; at increasing Y''_{max} adjacent ribs became involved and displaced fractures were more frequent. With severe impact, fractures occured occasionally at the posterior angle of the rib. The predominant injury to the underlying lung was contusion, without gross parenchymal laceration; the latter only being common with a large Y''_{max} produced by a high mass, high momentum impact.

The severity of pulmonary contusion was dependent on dY''/dt rather

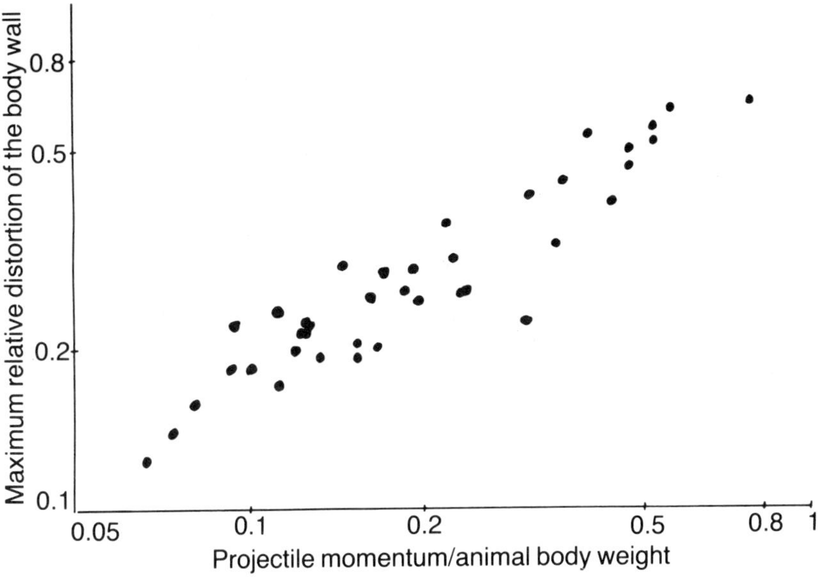

Fig. 18.2 The relationship between the maximum relative thoracic wall distortion (Y''_{max}) and the impact severity (pre-impact momentum/body weight). The impact diameter was 37 mm, the projectile mass varied from 0.069 to 3.000 kg and the impact velocity from 5.8 to 81.2 $m \cdot s^{-1}$.

than Y''_{max}. Severe compression ($Y''_{max} = 0.5$) produced by the 3 kg impactor at low velocity ($t_{0.95} = 20$ ms) resulted in only minor pulmonary contusion. In contrast, using the 0.0695 kg impactor at high velocity ($t_{0.95} = 1–2$ ms), only half the displacement occurred ($Y''_{max} = 0.25$) but severe pulmonary contusions were produced. Degrees of severity of pulmonary contusion formed well-defined clusters when plotted against Y''_{max} and $t_{0.95}$ (Figure 18.3).

Low velocity impact produced contusion immediately under the impact area with little, if any, involvement of distant sites. On the other hand, high velocity impact, with small values of $t_{0.95}$, produced contusions extending some distance beyond the area of impact – even spreading to involve the entire right lung. Although indirect lung collections of blood may have been due to simple reflux from the area under the impact, it was considered that these represented true indirect lesions: probably the result of propagated stress waves. It is likely that with increasing impact velocity and reducing $t_{0.95}$, the aetiology of the internal injuries changes from being a shear-dominated, direct mechanism to include an indirect, stress wave-mediated phenomenon. This interpretation would agree with previous observations on cardiovascular lesions with anterior impact (Cooper *et al.*,

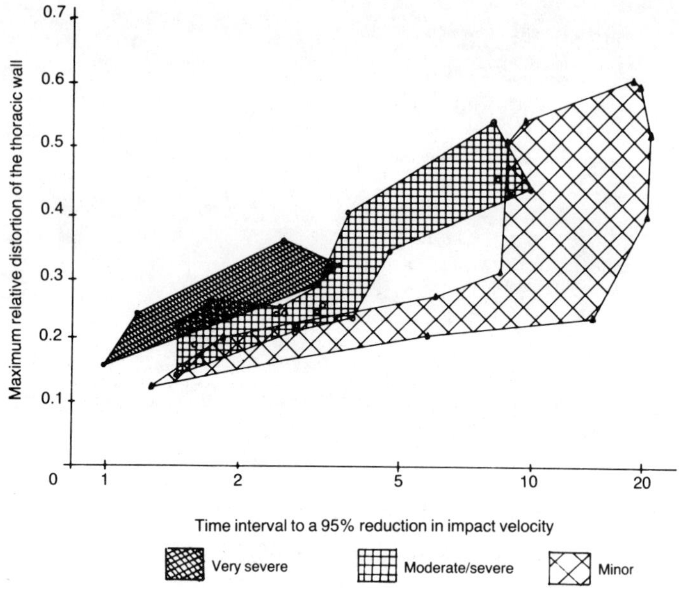

Fig. 18.3 The relation of the severity of pulmonary contusion to the time interval to a 95% reduction in impact velocity ($t_{0.95}$) and to the maximum relative displacement of the thoracic wall (Y''_{max}). Impact kinetics as in text: the severity of pulmonary contusion was assessed by comparison between the weight of the injured lung and the predicted normal lung weight in a pig of identical body mass (Cooper & Maynard, 1986).

1984). Further support for the role of internal stress propagation in high velocity impact has come from experiments in which the injury to the lungs of rats exposed to blast loading was assessed from serial transverse sections. Lung parenchyma immediately underlying the thoracic wall could be uninjured, but indirect injuries were identified on the distal side of the lobes adjacent to the mediastinum – where stress concentration resulting from reflection and focusing from interfaces would be anticipated.

18.3 **Biomechanical database**

Before modelling studies could be carried out it was necessary to obtain a biomechanical database. A major text, *Strength of Biological Materials* (Yamada, 1970), contains extensive information on the limb bones of more than 20 species, including man and pig, but no data on ribs, sternum or cranium. The precise variables to be determined depended on the types of stress likely to be applied and on the anatomical differences in the thoracic cages (Figures 18.1 and 18.4). Blunt trauma to either the thorax or the

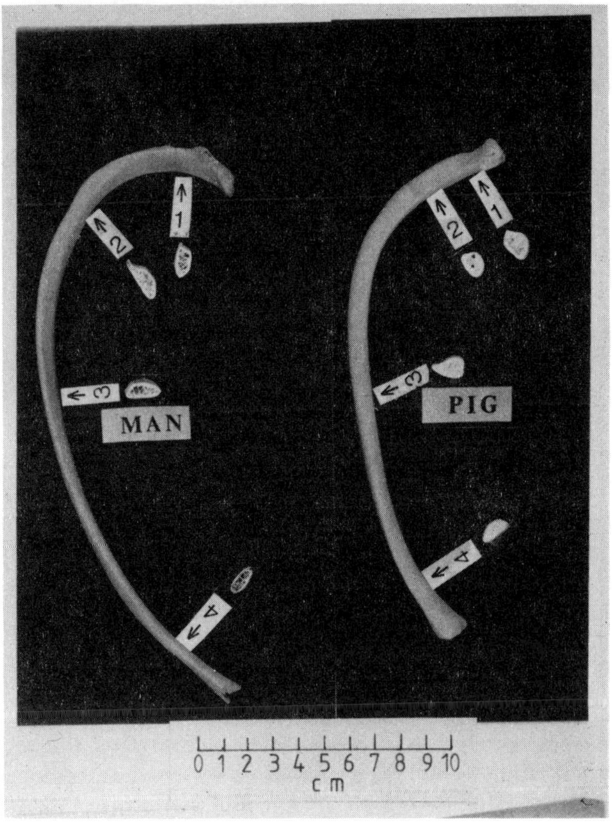

Fig. 18.4 The 5th rib in man and the pig, with cross-sections at four levels. Note the greater posterior curvature of the human rib and the different eccentricity between the species at all levels of cross section. The differences in cross-sectional geometry were the major cause of the difference in geometric moment of inertia between man and the pig at any level.

skull will apply a bending stress to the bones, rather than direct longitudinal compressive and extensive stresses. The greater caudal angulation of human ribs (Figure 18.1) also meant that torsional stresses needed to be incorporated into the models.

Custom jigs were designed and constructed to enable a length of rib or sternum, or a strip of skull, to be studied as a freely suspended beam under concentrated load (beam mode); the jigs being mounted in an Instron type 1026 tensiometer. Two test beam lengths were used, 50 and 70 mm; the load being applied to the mid-point at a rate of traverse of either 50 or 500 mm·min^{-1}. This enabled mechanical equivalence to be established over thoracic or skull impact velocities of 1–70 $m \cdot s^{-1}$. Bone samples were stressed continuously to break; the stress – strain curve was used to determine apparent elastic modulus (E), break load (F_{max}), break strength (P_{max}) and displacement at break (Y''_{max}), with measurement at the site of fracture to determine geometric moment of inertia (I). In further studies, test specimens were stressed rhythmically in the beam mode to 50% of Y''_{max} using a symmetrical saw tooth displacement profile; stress – strain loops at the two rates of traverse were analysed to determine viscous damping factors.

Studies were carried out on ribs 2–10 in man and 2–11 in the pig. Estimates were made at nine approximately equidistant points along the typical ribs (4–7 man; 4–9 pig), with the same or an appropriate lesser number of equivalent points on other ribs. Sternae had measurements carried out at the manubrio-sternal angle and between the insertions of the costal cartilages. Fresh (wet) and dry bones from both pigs and humans were studied, to produce hemithoracic grids of 97 and 86 points for pig and man, respectively. From these it was possible to derive estimates of variables at any position on the thoracic cage (Taylor *et al.*, 1987).

Combined bending and rotational stresses on ribs required another custom jig, which held and constrained the posterior end of the test sample as a cantilever, while the concentrated load was applied normal to the free end. The same rates of traverse were employed as for beam mode testing. From the initial posture of the specimen, the cantilever length and the force/displacement curve it was possible to determine the torque stress–strain characteristics.

Skull measurements were carried out mainly on dry human anatomical material, with a smaller number of wet samples; no other species were studied. Beam mode measurements were made on both sagittal and coronal strips of approximately 20 mm width, to yield a hemicranial grid of 37 points.

For both thorax and skull only results from wet bones will be presented in this chapter, together with their use in subsequent stress simulation.

18.4 Elastic moment

Elastic moment ($E \times I$), the product of apparent elastance (E) and geometric moment of inertia (I), is a recurring function in stress analysis. It would seem, therefore, appropriate and simpler to determine ($E \times I$) as a single variable. However, on analysis of the tensiometric data for ribs, it proved more advantageous for subsequent interactive simulation to obtain separate equations for surfaces to describe E and I in terms of the relative distance along the bone (x') and the number of the rib (N). In view of the curvature of the bones, the direction of the applied load will be classified as convex or concave, depending on the face to which it was applied.

Table 18.1 *Values for Young's elastic modulus on compression and extension, apparent elastic modulus on bending, and ultimate yield stress on bending, of ribs and sternum in man and pig and skull in man. Equivalent data for femur on bending, derived from the results of Yamada (1970), are given for comparison.*

A. Elastic modulus (E)

	Rib		Sternum		Skull		Femur	
	Comp.	Ext.	Comp.	Ext.	Comp.	Ext.	Comp.	Ext.
Man	7.57 (0.45)	8.99 (0.52)	—		9.97 (0.68)	8.22 (0.98)	10.40	17.26
Pig	5.54 (0.33)	6.07 (0.40)	—		—		14.61	4.81

B. Apparent elastic modulus on bending (E_{app})

	Rib		Sternum	Skull	Femur
	Convex	Concave	Anterior		Anterior
Man	2.21 (0.49)	2.25 (0.52)	2.26 (0.29)	2.08 (0.97–7.74)	18.34
Pig	2.30 (0.49)	2.21 (0.52)	0.270 (0.067)	—	3.53

C. Ultimate yield stress on bending (UTS)

	Rib		Sternum	Skull	Femur
	Convex	Concave	Anterior	Median (Range)	Anterior
Man	82.7 (11.6)	54.2 (8.4)	79.3 (6.1)	75.4 (38.2–150.7)	207.9 (7.6)
Pig	89.3 (9.7)		18.2 (4.32)	—	119.6 (3.6)

Values are expressed in $GN{\cdot}m^{-2}$ (E, E_{app}) and $MN{\cdot}m^{-2}$ (UTS): mean (SD) or, for skull, median (range). For ribs, n = 293 (pig) and 83 (man); for sternum, n = 18 (pig) and 4 (man); for both, n is the number of bones tested. For skull, n = 37 (man); n is the number of samples tested.

Elastic moduli of compact bone from the ribs and skull, together with figures from the literature for the shaft of the femur (Yamada, 1970), are given in Table 18.1. The cortical bone of the ribs was tested along the axis of the bone in both compression and extension. Skull cortical bone was tested in both coronal and the sagittal axes, but showed no evidence of anisotropy: pooled data are presented.

18.4.1 *Apparent elastance (E)*

For ribs in both species, E decreased exponentially from posterior to anterior, and decreased linearly from cephalad to caudal. The apparent elastance of the rib cage could be described by a surface:

$$E_{x'}(N) = (E' - A \times N) \times \exp(B \times (1 - C \times N) \times x') \quad (18.3)$$

Parameter values for the two species are given in Table 18.2. It had been found previously (Cooper & Maynard, 1986) that E for dry and wet bones was related by the simple linear identity:

$$E_{x'}(N)_{\text{wet}} = 0.0056 + 0.9110 \times E_{x'}(N)_{\text{dry}} \quad (18.4)$$

The same relation was also found to apply to sternae and skull. It also proved possible for ribs to transpose between species using the relation:

$$E_{x'}(N)_{\text{man}} = E_{x'}(N)_{\text{pig}} \times \left(\frac{(3.679 - 0.130 \times N)}{(3.734 - 0.114 \times N)} \times \exp(-0.0371 + 0.0047 \times x'\right) \quad (18.5)$$

Table 18.2 *Parameters for regression equations of surfaces to describe: the apparent elastance, and the geometric moment of inertia of porcine and of human ribs as functions of rib number* (N) *and relative distance along the rib from the posterior end* (x').

A. Apparent elastance ($GN{\cdot}m^{-2}$)

$$E_{x'}(N) = (E' - A \times N) \times \exp(B \times (1 - C \times N) \times x')$$

	E'	A	B	C
Pig (n = 293)	3.734 (0.009)	0.137 (0.004)	−0.715 (0.004)	0.0308 (0.0015)
Man (n = 83)	3.679 (0.016)	0.130 (0.006)	−0.752 (0.007)	0.0355 (0.0029)

B. Geometric moment of inertia (mm^4)

$$I_x'(N) = I' \times \exp[A \times (\ln(N) - B]^2) \times [1 + C \times \exp(D \times \exp(E \times N) \times x')]$$

	I'	A	B	C	D	E
Pig (n = 293)	36.00 (0.89)	−0.870 (0.039)	1.539 (0.025)	5.209 (0.469)	−6.354 (0.021)	−0.0699 (0.0023)
Man (n = 83)	48.09 (3.70)	−0.896 (0.123)	1.435 (0.150)	5.004 (0.059)	−5.717 (0.039)	−0.0789 (0.0046)

Values are means (SD).

In some simulations, particularly when attempting to extrapolate from a stress analysis of the porcine thoracic cage based on experimental observation, to the possible equivalent situation in man, it proved more effective to use eqn (18.5), and similar transforms for other variables, as embedded optional subroutines rather than to reset parameter values.

It has not been possible to derive similar relationships for the sternae and skull. These variables have been stored and accessed as arrays. The human sternum showed a slight decrease in E from cephalad to caudal, but in the pig the presence of separate sternubrae with cartilaginous junctions resulted in E being 20–55% of that in man, with fracture always occurring between sternubrae. The skull showed areas of high E in the frontal and occipital regions, with two areas where it was low, in the temporal region and adjacent to the midline anteriorly. The differences correlated directly with the ratio of compact to spongy bone.

18.4.2 *Geometric moment of inertia (I)*

Despite the wide variability in cross sectional geometry between sites on ribs and between species (Figure 18.4), it was found that I showed a consistent pattern of behaviour, enabling equations to be derived in terms of x' and N. From posterior to anterior the geometric moment of inertia again decreased exponentially, so that:

$$I_{x'} = A' + B' \times \exp(C' \times f(x')) \tag{18.6}$$

the function $f(x')$ also being an exponential, $\exp(D')$. From upper to lower ribs the relationship was log normal; but with the peak more cephalad in man. The general equation was:

$$I_{x'}(N) = I' \times \exp[A \times (\ln(N) - B)^2] \times [1 + C \times \exp(D \times \exp(E \times N) \times x')] \tag{18.7}$$

Parameter values for pig and man are given in Table 18.2. In a similar manner to eqn (18.5), a transfer function between species was derived, which could be executed as an optional subroutine in simulations. It has not yet been possible to derive similar relationships for sternum and for skull; these variables are stored and accessed as arrays.

18.4.3 *Elastic moment (E × I)*

Values for $(E \times I)_{x'}(N)$ derived from the individual measurements by tensiometry (Figure 18.5) were within the 95% confidence limits for the mean of that computed from the product of the surface equations for $E_{x'}(N)$ and $I_{x'}(N)$. The use of the mathematical model for the thoracic cage was therefore validated for subsequent stress analysis. For computational purposes it proved both simpler and faster to use the product of the two surface equations for the ribs, together with the transposition equations

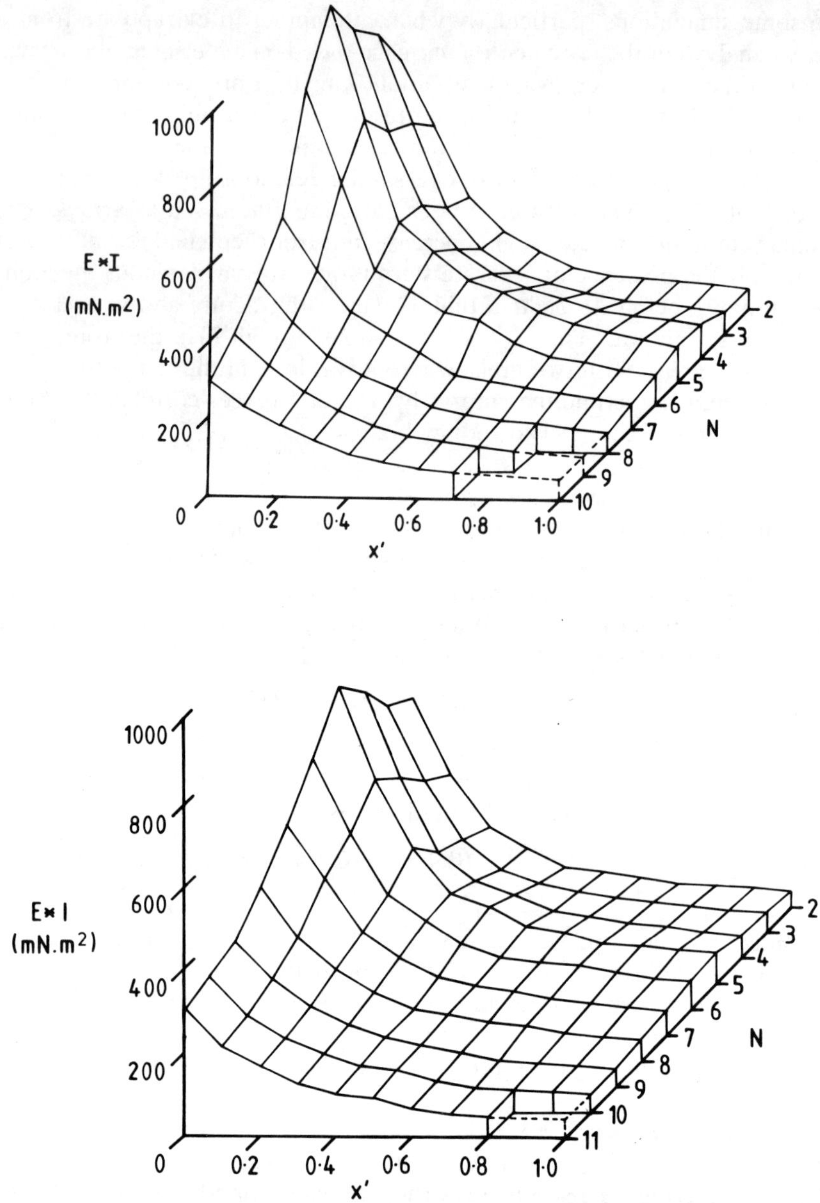

Fig. 18.5 The elastic moment ($E \times I$) for different rib numbers (N) and relative distance along the rib (x') determined experimentally in (a) man and (b) pig. The individual points and their variability showed good agreement with those predicted from the surface equations (see text).

between species. The product of cells in the arrays for $E_{x'}(N)$ and $I_{x'}(N)$ were employed for sternum and skull.

18.5 Ultimate tensile strength/yield strength

Virtually all skull specimens behaved as a brittle material, with no evidence of yield prior to break, whereas ribs and sternae in both species often behaved as plastic materials. For the purposes of this chapter, ultimate tensile strength (UTS) will be taken as the tensile strength at yield or break, whichever occurred first; this is the value of more relevance in injury and also showing a smaller coefficient of variation than the UTS at failure. The values of UTS for ribs and sternae in the two species and the skull in man are given in Table 18.1.

18.5.1 *Ribs*

For porcine ribs, where there was relatively little curvature apart from the region of the neck and angle (Figure 18.4), the UTS did not vary significantly with N, x' or the mode of stressing (concave/convex). A mean value of 89.31 MN·m^{-2} was used in the stress analysis of the procine chest. No single value for UTS could be assumed for human ribs; possibly this is a consequence of their greater curvature and angulation (Figure 18.2). when subjected to a convex stress, human ribs had a similar UTS to porcine, with an average value of 82.67 MN m^{-2}, although with a progressive cephalo-caudal decrease, described by:

$$P_{max}(N) = 99.66 - 2.832 \times N \tag{18.8}$$

Within-group variability was larger than between-group, so the average value could be assumed in many simulations. On concave stressing UTS was less at 54.17 MN·m^{-2}; again with a gentle cephalo-caudal reduction. The compound movement of a rib on impact loading necessitated the incorporation of both convex and concave values in simulations. The difference in UTS between pig and man again underlines the potential danger of direct extrapolation between species without proof of mechanical equivalence.

18.5.2 *Sternum*

There were marked differences in UTS between the two species, due to the major anatomical differences already described. In man the UTS for the sternum did not differ significantly from that for the ribs subjected to convex stressing, and the latter value was assumed for modelling, while in the pig the UTS was only 18.17 MN·m^{-2}, with dislocation occurring between sternubrae.

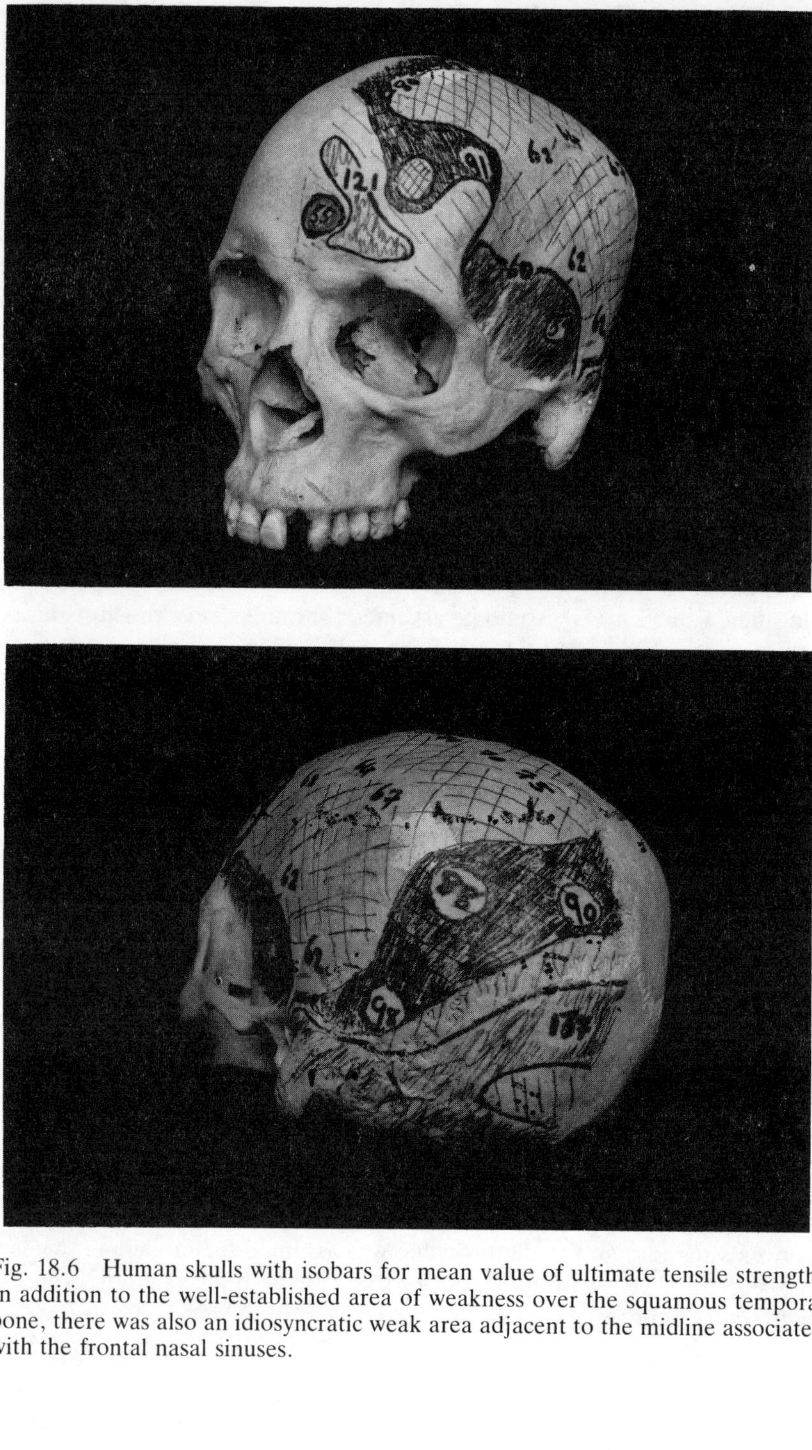

Fig. 18.6 Human skulls with isobars for mean value of ultimate tensile strength. In addition to the well-established area of weakness over the squamous temporal bone, there was also an idiosyncratic weak area adjacent to the midline associated with the frontal nasal sinuses.

18.5.3 *Skull*

The UTS of the skull showed considerable regional variation (Figure 18.6), with maxima in the parietal and supraorbital regions on convex stressing (137.4 $MN \cdot m^{-2}$), and a minimum at the squamous temporal (55.1 $MN \cdot m^{-2}$). There was also a variable zone of reduced UTS, in the same frontal region as was associated with a low apparent elastic moment. UTS was lower with concave than with convex stress in the frontal and occipital regions; these are the areas of both greatest thickness and greatest curvature. In the parietal and temporal regions there were no significant differences between modes of stressing. As for the other cranial parameters, a unitary value could not be assumed for UTS; simulations again employed values stored in an array.

18.6 **Viscoelasticity**

Analysis of the the stress-strain loops on cyclic stressing at two rates of strain showed similar viscoelastic properties for ribs and sternae in both species, and for the human skull. All could be modelled adequately as a simple system with first order damping; having an undamped natural frequency dependent on elastic modulus and length, and a median damping factor of $\beta = 0.92$ (range 0.74–1.22).

18.7 **Simulation studies**

Formal stress analysis employing the database parameters could be carried out using both finite element digital and multi-time iterative full hybrid computer simulations. The type of effective applied load (convex or concave) was of most importance in the human thorax and skull, due to the differences in UTS described earlier.

18.7.1 *Single rib*

Whereas a simple cantilever model under concentrated load at the distal end produced the anticipated stress concentration at the neck of the rib in both species (Taylor *et al.*, 1987), analysis of a rib under rapid loading (impact energy > 150 J: $t_{0.95} < 1.2$ ms) was less simple. Under such loading, the site of contact tended to move in the line of impact due to inertia. The result was that the rib deformed effectively as a series of beams and cantilevers, with a mixture of convex and concave loads. The resulting stress distributions in the porcine rib on anterior impact exhibited two secondary zones of maximum stress concentration: laterally and postero-laterally near the angle of the rib (Figure 18.7). Human ribs also had two zones of secondary maximum stress concentration, but because of the greater posterior curvature and caudal angulation of the human ribs

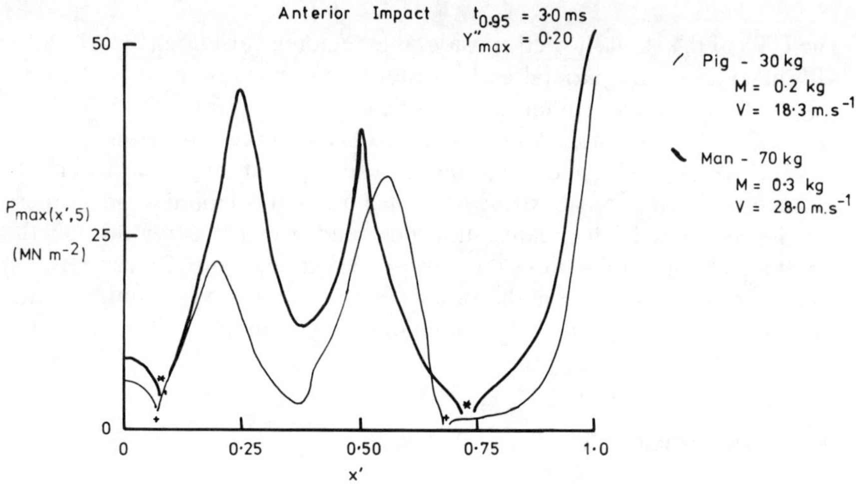

Fig. 18.7 Stress distribution along the model of the 5th rib in man and pig subjected to anterior impact. The two secondary peaks corresponded to the common sites of indirect fracture encountered clinically and experimentally. The apparent discontinuities (* man; + pig) corresponded to transition points where the bending stress changed between concave and convex.

there were important differences from the pig. At both sites the peaks were higher in man, and the postero-lateral peak in man occured at a higher value of x' (Figure 18.7). These interspecies differences were more pronounced when related to UTS, as the effective stress at both the lateral and postero-lateral sites on anterior impact was effectively concave. Therefore, the risk of indirect rib fracture on anterior impact would be greater in man than in the pig. Further, when indirect fracture did occur it would most probably be lateral in pig but at the angle of the rib in man. With lateral impact the primary convex stress concentration was greater in the pig, with a consequently greater risk of primary fracture in that species. In both species there would be a secondary stress concentration in the postero-lateral region, but of much smaller amplitude than that under the impact. The consequent risk of indirect fracture following lateral impact was much less than with anterior impact in both species, being likely only with high energy, high velocity impact, and with a greater probability of occurrence again in man.

18.7.2 *Thorax*

The thorax was modelled over the range of ribs for which biomechanical data were available, the thoracic contents being approximated as three compartments: two of compressible viscoelastic gas laterally, for the

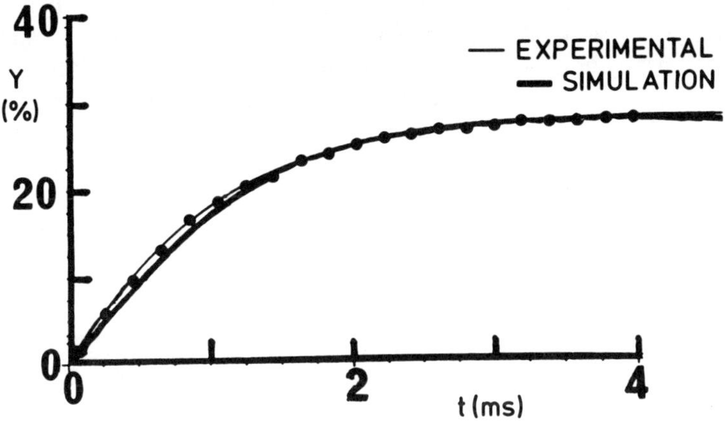

Fig. 18.8 Time-displacement curve observed for the site of contact in a pig subjected to anterior impact, and the best fit optimisation obtained from a simulation with similar impact kinetics. (Reproduced from Taylor *et al.* (1987) by kind permission of the UKSC).

lungs, with an incompressible viscous fluid section between for the mediastinum, including the heart and major blood vessels. The use of lumped parameters was considered justified as a first approximation in the study of chest-wall stress and displacement. Anterior and mid lateral impact has been modelled.

The predicted time-displacement curves of the chest wall under the point of impact, using mean values for parameters, were similar to those observed experimentally. For a given experiment, it proved possible to obtain an almost perfect fit for the time-displacement under the site of impact (Figure 18.8) using the measured thoracic dimensions and optimising the biomechanical variables within their 95% confidence distributions; a modified 'drunkard's walk' technique was employed. A predicted probability of the time-displacement distribution within a population to a given impact could also be computed together with a probability of fracture both direct and indirect. At present displacement prediction is only attempted when fracture does not occur up to $t_{0.95}$ or to the time of fracture. This has restricted these studies, particularly of anterior impact in the pig, where fracture between sternubria was predicted to occur at comparatively low impact kinetics (<100 J).

Lateral impact simulation in both species with a small impact area produced a maximum displacement and stress in the rib under the site of impact. The attenuation relative to the same load on a single rib was related not to the impact kinetic energy, but rather to the impact momentum; a low velocity impact produced a greater dissemination of effect to adjacent structures than the same energy exchange, but at a higher velocity. Because of the viscoelastic factors in the thoracic wall and contents, it was further predicted that a smaller Y''_{max} would result from a high velocity than with a low velocity impact of the same kinetic energy: the energy exchange in the former probably being partly converted into stress wave energy. The properties of the chest wall appeared the more important factor, little change occurring in either $t_{0.95}$ or Y''_{max} when the thoracic contents were removed, either experimentally or in the simulation (Taylor *et al.*, 1987)

18.7.3 *Skull*

As yet, simulations of the skull are still under development and at too early a stage to be reported.

18.8 **Discussion and summary**

Mathematical models are well established for the time/energy profile of impact, shock and blast; these are utilised in the study of the strains produced and the stress distributions in inanimate structures. Such simulation techniques have only rarely been used in the study of biological systems, particularly in the elucidation of the biomechanics of blunt trauma in man. Major factors inhibiting such studies have been the complicated geometry and anisotropy of the organs and tissues, and biological variability. Because of this, prediction of the effects of a known non-penetrating impact on man has relied predominantly on retrospective analysis of clinical records, supplemented by measurements on manikins and a few animal experiments; for both the latter the dynamic equivalence to man has been largely speculative.

In the studies here it has been found that time/displacement curves following non-penetrating impact to the porcine thorax could be described mathematically in terms of the impact kinetics. Further, the probability and severity of associated direct and indirect injuries to the thoracic cage and contents could be predicted. Once these relationships had been established satisfactorily by a minumum number of animal experiments, and in a form which took account of biological variability, it became feasible to consider a computer model for such injuries. The only remaining deficit to the development of such a simulation was the lack of a satisfactory database of all the basic biomechanical variables required; this again incorporates some measure of the biological variability. The multiformity of structure and form necessitated the use of lumped variables for composite structures, such as the bone of the ribs, or the lung and contained air. For the bones of the thoracic cage and the skull, at any point an apparent elastance, elastic moment and UTS of a full thickness finite element was assumed, rather than taking account of the relative dispositions of cortical and medullary bone. From this premise, it was possible to describe the ribs, in both man and pig, by surface equations with N and x' as the independent variables: biological variability is included as a confidence envelope and viscoelastic damping as a dynamic component. With the addition of similar data for the sternum, but as an array rather than an equation, it has been possible to produce simulations of the thorax in both man and pig, together with functions to transform between species. Preliminary models are also being developed for the human skull.

Simulations of the stress distributions in the bones of the thorax in response to a range of non-penetrating impacts have been described briefly. These were able to be verified against experimental observations in the pig. Using mean values of variables in the simulation, a general agreement of observed to predicted time/displacement curves was obtained, which could be of a high order in a particular case with parameter optimisation. The correctness of the latter to the particular values in an individual animal was not verifiable, for obvious reasons. The availability of estimates with a measure of biological variability has also permitted modelling to predict the spread of responses likely in a population, rather than a notional 'average subject'.

The predicted stress distribution for anterior thoracic impact showed a compound rib displacement, resulting in two zones of peak stress in addition to that under the impact site. Although in general these secondary stress peaks were lateral and postero-lateral, the location in terms of x' and amplitude differed between pig and man (Figure 18.7). The reason appeared twofold: the greater curvature of the human rib in its posterior third; and the greater angulation of the human rib to the transverse plane of the body (Figures 18.1 and 18.4). There were also differences in UTS between the species so that, for a given impact, indirect rib fracture appeared most likely in the region of the angle of the human rib, with a slightly less probability of fracture of the pig rib laterally. The prediction of indirect rib fracture following anterior impact in the pig was further compounded by the high probability of fracture between sternubrae. Lateral chest impact showed a single secondary stress peak in the region of the angle of the rib in both species – but of much lower amplitude than with anterior impact. Under the site of lateral impact the bending stress on the rib was of greater amplitude in pig than in man with a consequent greater probability of fracture. The predicted sites of secondary maximum stress concentration agreed with the common sites of indirect fractures observed clinically and experimentally in man and pig, respectively. As both direct and indirect rib fractures were not only directly related to Y''_{max} but also inversely to $t_{0.95}$ the probability and site of fracture could not be predicted from a single parameter. A high velocity impact would produce less displacement than one of low velocity with the same kinetic energy, but the former would usually carry the higher probability of fracture. From equations (18.1) and (18.3), it can be deduced that any function of $Y''_{max}/t_{0.95}$ would be related non-linearly directly to impact momentum and indirectly to subject weight. The precise function to separate clusters of severity of injury has still to be determined.

In both the experimental and the modelling studies the risk of visceral injury, either direct or indirect, has so far proved to be even less readily predictable. Impact velocity, dY''/dt and $t_{0.95}$ all appear to be relevant, with impactor momentum being of more importance than kinetic energy within the range of impactor mass and velocity studied (Figure 18.3). Some cluster separation was obtained for cardiac injury following anterior impact with respect to Y''_{max} (Cooper *et al.*, 1984), but with considerable overlap. Even this finding was not simple to interpret, for there was a parallel correlation with sternal fracture. The much lower UTS of the pig sternum meant that Y''_{max} was less dependent on chest-wall stiffness and viscoelasticity than on lateral impact. This may have accounted for the direct correlation of Y''_{max} to impact kinetic energy and inverse correlation to effective impact diameter in these studies. More work is needed in this area on the interaction of the various factors to provide more specific indicators of the probability and severity of visceral injury. It is likely that these will vary between direct and indirect injuries, and between shear- and stress wave-determind aetiologies.

The present studies have demonstrated that it is possible to apply classical engineering methods to the study of acute injuries based on accurate knowledge of chest-wall biomechanical parameters. It has also been shown that it is feasible to translate observations from experimental animals to the human clinical situation by computer simulation, so as to begin to predict the probability and severity of injuries. The latter simulations have afforded an insight into the mechanics of a variety of indirect injuries. It has been found that single predictive indices are unlikely to be sustainable over the range of kinetics and site of impact which may occur. Further, the risk and severity of direct shear-determined and of indirect shear-determined and stress wave-determined injuries are likely to depend principally on different dynamic factors. The improved understanding of the biomechanics of non-penetrating injuries resulting from these studies will aid clinical practice, both by enabling a more accurate prediction of the probability, severity and nature of secondary injuries, and by aiding the development of improved protective measures.

Acknowledgements

Thanks are due to the technical staff at both sites for their skilled assistance, and to the engineering department at CDE for the construction of the various jigs. The investigations at the Royal College of Surgeons of England were supported by generous grants from the Ministry of Defence. © Controller, Her Majesty's Stationery Office, 1989.

References

Cooper, G. J., Pearce, B. P., Stainer, M. C. & Maynard, R. L. (1982). The biomechanical response of the chest wall to impact with particular reference to cardiac injuries. *Journal of Trauma*, **22**, 994–1008.

Cooper, G. J., Maynard, R. L., Pearce, B. P., Stainer, M. C. & Taylor, D. E. M. (1984). Cardiovascular distortion in experimental nonpenetrating chest impacts. *Journal of Trauma*, **24**, 188–200.

Cooper, G. J. & Maynard, R. L. (1986). An experimental investigation of the biokinetic principles governing non-penetrating impact to the chest and the influence of the rate of body wall distortion upon the severity of lung injury. *Proceedings International Research Committee on the Biokinetics of Impacts*, pp. 331–42. Zurich.

HMSO (1988). Office of Population Census and Surveys, *Annual Reports and Mortality Tables*. London.

Roberts, V. L., Jackson, F. R. & Berkas, E. M. (1967). Heart motion due to blunt trauma to the thorax. *Proceedings of the 10th Stapp Car Crash Conference*, paper 660800. Society of Automotive Engineers Inc.

Shatsky, S. A., Alter, W. A. III, Evans, D. E. *et al.* (1974). Traumatic distortion of the primate head and chest: correlation of biomechanical, radiological and pathological data. *Proceedings of the 18th Stapp Car Crash Conference*, paper 741186. Society of Automotive Engineers Inc.

Taylor, D. E. M., Cooper, G. J., Evans, V. A. & Whamond, J. S. (1987). Modelling of the ribs and thoracic cage in studies of the biomechanics of non-penetrating chest injury. In *UKCS Conference on Computer Simulation – 1987 Proceedings*, ed. R. N. Zobel, pp. 220–6. Manchester University Press Manchester.

Viano, D. C. & Warner, C. Y. (1976). Thoracic impact response of live porcine subjects. *Proceedings of the 20th Stapp Car Crash Conference*, paper 760823. Society of Automotive Engineers Inc.

Viano, D. C. (1978). Traumatic aortic ruptures. *General Motors Research Publication GMR-2717*. GM Research Laboratories, Warren, Michigan.

Yamada H. (1970). Mechanical properties of locomotor organs and tissues. In: *Strength of Biological Materials*, ed. F. G. Evans, pp. 49–73. Williams and Wilkins, Baltimore.

Index

alumina 123
annulus fibrosus 78, 164, 246
apatite crystals 98
articular cartilage 44, 139, 217

biocompatibility 2
Björk-Shiley heart valve 1

cancellous bone 16, 100
carotid artery 160
chest 250
collagen 14, 25, 44, 211
constitutive relationships 28
contact stress 134
cortical bone 100

Dacron 3, 151
Delrin 6
dental material properties 179
dentine 60
dynamic compliance 157
dynamic modulus 159

elastin 14

femoral artery 160
fracture 61
friction 123

glutaraldehyde 14

head 250
heart valve 1, 14

hip joint forces 190

impact kinetics 253
indentation testing 141
interface stress 203
intravertebral disc 78

knee joint forces 226

left ventricle 25

mandible 103
meniscus 109
metal fatigue 9
middle ear 174
myocardium 25

nucleus pulposus 78, 164, 240

octahedral shear stress 192
osteoarthritis 44

patella 49, 214
pericardium 14
plasticity 20

ribs 263

scanning electron microscopy 4, 27, 94
scapula 99
spinal motion segment 77, 164, 240
sternum 263
stress trajectories 98, 118

Teflon 3, 151
tensile strength 18
thrombosis 2
tooth 174
total hip replacement 187, 203

vertebra 76, 163, 241
viscoelasticity 19, 143, 244, 265

X-ray diffraction 49